MATERIALS SCIENCE AND TECHNOLOGIES

ANALYTICAL MODELS OF THERMAL STRESSES IN COMPOSITE MATERIALS IV

Materials Science and Technologies

Additional books in this series can be found on Nova's website
under the Series tab.

Additional e-books in this series can be found on Nova's website
under the e-book tab.

MATERIALS SCIENCE AND TECHNOLOGIES

ANALYTICAL MODELS OF THERMAL STRESSES IN COMPOSITE MATERIALS IV

LADISLAV CENIGA

nova publishers
New York

Library of Congress Cataloging-in-Publication Data

ISBN: 978-1-63463-750-3

Published by Nova Science Publishers, Inc. † New York

Contents

Táto kniha je venovaná s láskou mojim najdrahším rodičom.
This book is dedicated with love to my dearest parents.

Preface

This book presents analytical models of thermal stresses and thermal-stress induced phenomena in components of two- and three-component materials (see Section 2.1.). With regard to the analytical modelling, these real multi-component materials with finite dimensions are replaced by two- and three-component model systems with infinite dimensions (see Section 2.2.), i.e., multi-particle-matrix and multi-particle-envelope-matrix systems, respectively (see Figure 2.1). The analytical determination of a thermal stress-strain state is based on a cell model which considers a cubic cell (see Figure 2.1). The cell model is usually used in case of the analytical and computational modelling of phenomena in periodic model systems [1]–[24].

Additionally, as presented in [25], the case when an infinite matrix is considered within analytical modelling of phenomena in real multi-component materials with finite dimensions is of particular interest for the mathematical simplicity of analytical solutions. As presented in [25], such analytical solutions are assumed to exhibit sufficient accuracy due to the size of material components (e.g., precipitates, envelopes) which is relatively small in comparison with the size of macroscopic material samples, macroscopic structural elements, etc.

The thermal stresses which originate below relaxation temperature (see Section 3.5.1.1) during a cooling process are a consequence of the difference in dimensions of the components. This difference is a consequence of different thermal expansion coefficients and/or a consequence of the phase-transformation induced strain which is determined in Section 3.5.2. for anisotropic and isotropic crystal lattices. This coefficient and this strain are included in the coefficient β_q for the spherical particle ($q = p$), the spherical envelope ($q = e$) and the cell matrix ($q = m$) (see Sections 3.5.1., 3.5.2.).

The analytical modelling results from fundamental equations of solid continuum mechanics which are represented by the Hooke's law for an anisotropic and isotropic elastic solid continuum (see Section 3.3.), and by the Cauchy's, compatibility and equilibrium equations which are determined by the spherical coordinates (r, φ, ν) (see Figure 3.1).

The analytical models of the thermal stresses are determined for the model systems which consists of either anisotropic, or isotropic, or anisotropic and isotropic components (see Chapters 4–6). The thermal stress-strain state in each component of the model systems is determined by several mutually different solutions which fulfil the boundary conditions which are determined in Section 4.6.. In case of the cell matrix, mandatory and additional boundary conditions are determined (see Section 4.6.3.). Due to these different solutions, a principle of minimal total potential energy of an elastic solid body [26] is then required to be considered (see Section 3.4.). Strictly speaking, such solutions or a combination of solutions are considered to exhibit minimal total potential energy of the model systems (see Figure 2.1). As analysed in Section 3.4., the total potential energy W_t of the model systems is represented by the deformation energy W_d. The deformation energy of the model systems is represented by the thermal-stress induced elastic energy W_c (see Eqs. (3.52), (3.53)) which is accumulated in the cubic cell, and then $W_t = W_d = W_c$

Additionally, analytical models of thermal-stress induced phenomena are also determined. These phenomena include the crack formation (see Section 7.1.), the energy barrier (see Section 7.2.), the micro- and macro-strengthening (see Section 7.3.), and the analytical-computational and analytical-computational-experimental methods of the lifetime prediction (see Section 7.4.).

In addition to experimental methods [27]–[39], crack formation is also investigated analytically and/or computationally [40]–[70]. This analytical and/or computational investigation is usually applied to an existing crack. A shape of such crack which is present in a model system is required to be mathematically defined, e.g., a penny-shaped crack. Strictly speaking, this mathematically defined crack exists in a model system before the loading of the model system. Additionally, such analytical and/or computational determination is based on e.g., the finite and boundary element methods, the Green's functions, J-integrals, semi-smooth Newton methods, singular integral equations (techniques) which result from Lekhnitskii's complex variable formulation, weight functions and Bueckner's work conjugate integral, the mean-field theory, the Mori-Tanaka's homogenization scheme, the Stroh formalism [40]–[70].

As presented in Section 7.1., the crack formation in the multi-particle-matrix system includes crack initiation which is followed by crack propagation. With regard to the crack initiation, *no* mathematically defined crack is present in this model system *before* the thermal-stress loading.

The crack formation analysis in Section 7.1. is based on the comparison of energy which is accumulated in the cubic cell with energy for the creation of a new surface (i.e., a surface of a crack). This comparison is used e.g., in [71]–[74]. The crack formation analysis then considers a curve integral of the thermal-stress induced elastic energy density w (see Eq. (3.50)) along a curve in the cubic cell (see Eq. (7.3)). The condition which defines a limit state with respect to the crack initiation in the cracking plane x_ix_j $(i, j = 1,2,3)$ in the spherical particle or cell matrix is determined (see Eq. (7.10)). The limit state is thus defined by the critical particle radius $R_{1cq}^{(ij)}$. With regard to the crack propagation at $R_1 > R_{1cq}^{(ij)}$, the condition for the determination of a position of the crack tip in x_ix_j (see Figure 7.1a) in components of the multi-particle-matrix system is determined (see Eq. (7.11)). Formulae which define the crack shape in a plane which is perpendicular to the cracking plane x_ix_j in components of the multi-particle-matrix system are also determined (see Eqs. (7.12)–(7.14)). These results concerning the crack propagation are valid for ceramic components which are characterized by a high-speed crack propagation. Additionally, the crack formation analysis in Section 7.1.3..1 (see Method 2, p. 116; see Section 8.1.) explains paradoxical behaviour of the cracking which is experimentally observed in a real two-component material [71]–[74].

In contrast to the crack propagation results, the determination of the limit state is applicable disregarding a 'character' of components of the multi-particle-matrix system (ceramic=brittle, elastic, elastoplastic components).

The energy barrier represents a surface integral of thermal-stress induced elastic energy density over a surface in the cubic cell (see Eqs. (7.37)–(7.39)). This energy barrier is determined for both model systems (see Sections 7.2.2., 7.2.3.).

In addition to experimental methods [75]–[78], strengthening of multi-component materials is also investigated analytically and/or computationally. Such analytical and/or computational determination is based on e.g., the finite element methods, modified Oldroyd model, a simulation of dislocation dynamics [79]–[90].

As presented in Section 7.3.1., the micro- and macro-strengthening (see Eqs. (7.50), (7.51)) is also based on a surface integral of the thermal-stress induced elastic energy density w_i ($i = 1,2,3$) over a surface in the cubic cell (see Eq. (7.48)). In this case, w_i represents such elastic energy density which is induced by the thermal stress σ_i which act along the axis x_i (see Eqs. (4.76)–(4.80)). The macro-strengthening represents a mean value of the micro-strengthening in the cubic cell. The micro- and macro-strengthening is also determined for both model systems (see Sections 7.3.2.1, 7.3.2.2).

With regard to the analytical modelling, the lifetime prediction methods are based on a transformation of the resistive effect of the thermal stresses to the contributory effect with respect to mechanical loading (see Section 7.4.1.). This transformation results in the analytical determination of critical microstructural parameters (a radius of grains and thickness of an envelope which is segregated on a surface of the grains). The lifetime prediction methods also consider experimental results as well as results which are obtained by a computational simulation of the microstructural parameters during a time-temperature-dependent development of microstructure.

The analytical-computational lifetime prediction method considers the analytical and computational results (see Section 7.4.2.). The analytical-computational-experimental lifetime prediction method considers the analytical, computational and experimental results (see Section 7.4.3.).

Additionally, the analytical models of the crack formation, of the energy barrier and of the micro- and macro-strengthening along with the methods of lifetime prediction exhibit a general validity. These analytical models and these lifetime prediction methods are then valid for the thermal-stress induced elastic energy density as well as for energy density which is induced by any stresses acting in the multi-particle-matrix and multi-particle-envelope-matrix systems. In case of the lifetime prediction methods, conditions which are required with respect to this general validity are presented (see page 103).

Illustrative examples of applications of the analytical models of these phenomena to real engineering materials (superconductive and structural ceramics, a creep-resistant steel) are also presented (see Chapter 8).

Results of this book are applicable within *basic research* (solid continuum mechanics, theoretical physics, materials science and engineering) as well as *engineering practice*.

With regard to the basic research, the analytical models of the thermal stresses in the model systems with either anisotropic components, or isotropic components, or anisotropic and isotropic components can be incorporated to the Eshelby's model [91]. The Eshelby's model and its development [1, 2, 25, 91] which are based on the Green's function, ordinary

Newtonian potential and biharmonic potential define the disturbance of an applied stress-field in a solid continuum. The applied stress-field (e.g., mechanical loading) is disturbed due to the presence of inclusions in a solid continuum. This incorporation thus defines a stress-strain state in a multi-component material which is loaded by thermal and mechanical stresses [25].

The analytical models of the thermal stresses can be incorporated into experimental and computational methods for the determination of residual stresses [92]–[97].

The analytical model of the thermal-stress induced energy barrier can be also incorporated into analytical and/or computational models which are created by e.g. theoretical physicists, and which describe an interaction of an energy barrier with dislocations or magnetic domain walls [98, 99].

Similarly, the analytical model of the thermal-stress induced micro- and macro-strengthening can be also incorporated into analytical, computational or experimental models of strengthening.

The thermal stresses can be incorporated into these models in a form of a thermal-stress induced field of stresses and/or strains, or in a form of thermal-stress induced elastic energy density.

The analytical-computational and analytical-computational-experimental lifetime prediction methods can be also connected with other techniques for the determination of lifetime [100]–[103].

As presented in [71, 104, 105], material scientists often numerically determine a thermal stress-strain state in a real two-component material by an analytical model which is determined for a one-particle-matrix system [106, 107]. As presented in [106, 107], this one-particle-matrix system with the particle volume fraction $v = 0$ consists of an isotropic spherical particle and an isotropic infinite matrix. However, real two- and three components materials are characterized by $v > 0$. The analytical models in Chapters 4–6 for the multi-particle model systems in Figure 2.1, both with $v > 0$, thus enable to numerically determine the thermal stress-strain state with respect to different values of the particle volume fraction. Additionally, such numerical determination for the multi-particle model systems with anisotropic and/or isotropic components is assumed to be more considerable in comparison with numerical results for the one-particle-matrix system with isotropic components.

Concerning engineering practice, materials scientists and engineers are able to numerically determine the limit state (the critical particle radius) with respect to the crack initiation. Consequently, this numerical determination can be helpful at the determination of 'suitable' heat treatment parameters. These 'suitable' heat treatment parameters are expected to lead to non-critical radii of precipitates (particles) in microstructure of real two-component material of the precipitate-matrix type (see Item 1, Section 2.1.).

Materials scientists and engineers are also able to numerically determine the lifetime of creep-resistant steels by the analytical-computational and analytical-computational-experimental prediction methods. This lifetime is related to the transformation of the resistance effect to the contributory effect of the thermal stresses with respect to mechanical loading. This numerical determination requires experimental and/or computational determination of all material components and/or phases which create the envelope on a surface of grains. Additionally, the volume fraction of each of these material components and/or phases are also required to be experimentally and/or computationally determined. These volume fractions are then considered within a numerical determination of elastic moduli and a thermal expansion coefficient of this multi-component envelope. Finally, this 'thermal-stress induced' lifetime represents a part of total lifetime [1].

[1]This work was supported by the Slovak Grant Agency under the contracts VEGA 2/0083/13, 2/0075/13 and by the Slovak Research and Development Agency under the contract APVV-0147-11.

Chapter 1

Outline of Principles

This book deals with the analytical modelling of thermal stresses and of related phenomena in multi-component materials, i.e., in two- and three-component materials. This analytical modelling is considered from a viewpoint of microstructure of the multi-component material. Strictly speaking, this analytical modelling thus represents the determination of the thermal stresses in each component of the multi-component material. These thermal stresses which originate during a cooling process are then a consequence of mutually different changes of dimensions of components of the multi-component material.

The analytical modelling of the thermal stresses includes:

- the definition of types of two- and three-component materials, and the determination of such model systems which correspond to these types (see Chapter 2).

- the determination of such coordinate system which corresponds to the geometry of the model systems (see Section 3.1.).

- an analysis of displacements in an arbitrary point of the model systems (see Section 3.2.).

- the determination of the Cauchy's, compatibility and equilibrium equations for this coordinate system. These equations along with the Hooke's law represent fundamental equations of solid continuum mechanics (see Section 3.3.).

- the determination of a suitable mathematical procedure (see Sections 4.2., 4.3., 5.2.–5.6.) which is applied to these fundamental equations to obtain a solution for an anisotropic and isotropic elastic solid continuum (see Chapter 4, 5). If several different mathematical procedures which result in several different solutions are determined, then a principle of minimal total potential energy [26] of the model systems (see Section 3.4.) is required to be considered. This principle [26] means that an elastic solid body (i.e., the model system) tends to exhibit minimal total potential energy.

- the determination of boundary conditions for each anisotropic or isotropic component of the model systems (see Section 4.6.).

- the determination of final formulae for the thermal stresses in the model systems (see Sections 4.5., 4.7., 4.8.).

- the determination of the thermal stresses in the model system which consists of anisotropic and isotropic components (see Chapter 6).

The analytical modelling of the related phenomena (see Chapter 7) includes

- the determination of conditions for crack formation which also define the limit state of the model systems (see Section 7.1.),

- the determination of an analytical model of energy barrier in the model systems (see Section 7.2.),

- the determination of micro- and macro-strengthening in the model systems (see Section 7.3.),

- the determination of analytical-computational and analytical-experimental-computational methods of lifetime prediction (see Section 7.4.).

CHAPTER 2 presents types of multi-component materials (i.e., two- and three-component materials) (see Section 2.1.) along with model systems which are considered within the analytical modelling (see Section 2.2.).

This analytical modelling is then applied to **two-** and **three-component model systems** which correspond to these two- and three-component materials (see Section 2.2.), respectively. The two-component model system is represented by a **multi-particle-matrix system** (see Figure 2.1a). This system consists of spherical particles with the radius R_1 which are periodically distributed in an infinite matrix, where d is an inter-particle distance along the axes x_1, x_2, x_3 of the Cartesian system $(Ox_1x_2x_3)$. The beginning O of the Cartesian system is identical with the particle centre. In case of the **multi-particle-envelope-matrix system** (see Figure 2.1b), each spherical particle is covered by a spherical envelope, where $R_1 < R_2$ are radii and $t = R_2 - R_1 > 0$ is thickness of the spherical envelope.

As presented in Section 2.3., the radii R_1, R_2, the inter-particle distance d along with the particle volume fraction $v \in (0, v_{max}\rangle$ are characteristics of these model systems. These characteristics represent **microstructural parameters** of the two- and three-component materials defined in Section 2.1. Consequently, relationships between components of the multi-particle-matrix and multi-particle-envelope-matrix systems and components of the two- and three-component materials are required to be determined, respectively. These relationships are presented in Sections 2.4.1., 2.4.2. As an example, let a real two-component material of the precipitate-matrix type be considered (see Item 1, Section 2.1.). In this case, the precipitate and the matrix (grains) of this real two-component material correspond to the spherical particles and the infinite matrix of the multi-particle-matrix system (see Figure 2.1a).

The infinite matrix of the model systems (see Figure 2.1) is **imaginarily** divided into identical cubic cells with the dimension d. Each cubic cells contains a central spherical particle without or with the spherical envelope on the particle surface. The beginning O of the Cartesian system is identical with the particle centre as well as with the centre of the cubic cell. The cubic cell represents a part of the infinite matrix, and this part is related to one spherical particle. Due to the infinity of the matrix, formulae for the thermal stresses which are determined for a certain cubic cell are also valid for any cubic cell of the infinite matrix. The thermal stresses are thus investigated within the cubic cell, i.e., within the spherical particle, spherical envelope and cell matrix. The thermal stresses are thus functions of the microstructural parameters.

This 'cell approach' (i.e., **cell model**) is used within mathematical procedures which are applied to the analytical and computational modelling of phenomena in periodic model systems [1]–[24]. The periodic model systems which consider spherical cells [1]–[3], cubic cells [4]–[21], as well as prismatic cells (e.g., in case of a fibre-matrix system) [22]–[24] are thus used for the description of real multi-component materials. However, two spherical cells are mutually connected in one point [1]–[3]. In contrast to [4]–[21], the matrix between two spherical cells is thus replaced by vacuum what is physically unacceptable.

CHAPTER 3 presents selected topics on solid continuum mechanics. As usual in solid continuum mechanics, the determination of a stress-strain state of a elastic solid continuum is represented by the determination of a stress-strain state in each point of this continuum, i.e., in an arbitrary point of this continuum. Consequently, an infinitesimal part of the continuum is considered in this arbitrary point. An influence of the continuum on this infinitesimal part is represented by displacements of the infinitesimal part along individual directions as well as by stresses acting on surfaces of this infinitesimal part [106, 108]–[111]. These stresses act along these individual directions. A position of the arbitrary point is determined by such suitable coordinates which consider the geometry of the continuum. A shape of the infinitesimal part is also determined with respect to the geometry of the continuum as well as with respect to this suitable coordinate system. Due to the spherical shape of the particles and envelopes in the model systems (see Figure 2.1), the **spherical coordinates** (r, φ, ν) and an **infinitesimal spherical cap** are considered. The infinitesimal spherical cap is described in Section 3.1.1. (see Figure 3.1b).

As shown in Figure 3.1a, the thermal stresses are then investigated at the arbitrary point P of a solid continuum along the axes x_1', x_2', x_3' of the Cartesian system $(Px_1'x_2'x_3')$. A position of the arbitrary point P regarding the Cartesian system $(Ox_1x_2x_3)$ is determined by the spherical coordinates (r, φ, ν), where O is a centre of the spherical particle (see Figure 2.1). The axes $x_1' = x_r$ and $x_2' = x_\varphi$, $x_3' = x_\nu$ at the point P represent radial and tangential directions related to the spherical surface with the radius $r = |OP|$, respectively, where $x_\varphi \parallel x_1x_2$. With regard to $x_1' = x_r$, $x_2' = x_\varphi$, $x_3' = x_\nu$ (see Figure 3.1), the conventional subscripts r, φ, ν in a connection with the spherical coordinates (r, φ, ν) are replaced by the subscripts 1, 2, 3, respectively, due to the mathematical procedures in Section 4.2. (see e.g., Eqs. (4.1)–(4.12)).

The model systems in Figure 2.1 are symmetric. As presented in Section 3.1.2., this symmetry results from the matrix infinity as well as from the periodical distribution of the spherical particles and spherical envelopes. Due to this symmetry, the thermal stresses are thus sufficient to be investigated within one eighth of the cubic cell (see Figure 3.2), i.e., for $\varphi \in \langle 0, \pi/2 \rangle$ and $\nu \in \langle 0, \pi/2 \rangle$. Intervals of the variable r for components of the model systems are determined in Section 3.1.2.

Section 3.2. presents an analysis of **displacements** of the infinitesimal spherical cap. These displacements are considered within the Cartesian system $(Px'_1x'_2x'_3)$ (see Figure 3.1a).

The thermal stresses in multi-component materials are a consequence of mutually different changes of dimensions of material components. These dimension changes originate during a cooling process at the temperature $T < T_r$. As defined in [104], the **relaxation temperature** T_r is such temperature below that the stress relaxation as a consequence of thermal-activated processes does not occur in components of the model systems (see Section 2.2.). The determination of T_r of the model systems is presented in Section 3.5.1.1.

These mutually different changes mentioned above thus result in an interaction on the surface of a boundary of two material components. In case of the multi-particle-matrix system, this interaction is represented by the pulling or pushing of the spherical particle by the cell matrix. Let P represent an arbitrary point at the particle-matrix boundary. Due to the symmetry of the multi-particle-matrix system, the pulling or pushing of the arbitrary point P is realized along a normal to the boundary surface (i.e., to the particle-matrix boundary) at this arbitrary point. Due to the spherical shapes of the particles and envelopes, this normal at any point on the boundary surface is identical to the axis x'_1 which define a radial direction regarding $(Px'_1x'_2x'_3)$. The arbitrary point P at the particle-matrix boundary is thus shifted (i.e., P is pulled or pushed) along the radial direction. As analysed in Section 3.2., the infinitesimal spherical cap at any point on the axis x'_1 exhibits the **radial displacement** $u'_1 = u_r$ within $(Px'_1x'_2x'_3)$ (see Figure 3.1), where $q = p,e,m$ [1].

Additionally, Section 3.2. also presents a concept of the **imaginary separation of components** of the model system. This concept which is also considered for the determination of the Eshelby's model [25, 91]. Let the multi-particle-matrix system be considered. Let the components of this system be imaginarily separated at the relaxation temperature T_r. In this case, the spherical particles in the infinite matrix are replaced by spherical hollows. Let $R_1 = R_1(T)$ be a temperature dependence of a radius of these separated particles. Let $R_{1m} = R_{1m}(T)$ be a temperature dependence of a radius of the spherical hollows. Let these separated components be cooled down from the relaxation temperature T_r to the temperature T, where $T < T_r$. A radius of the separated particle is thus changed from $R_1(T_r)$ to $R_1(T)$, where $R_1(T) < R_1(T_r)$. A radius of the hollow is thus changed from $R_{1m}(T_r)$ to $R_{1m}(T)$, where $R_{1m}(T) < R_{1m}(T_r)$. Additionally, the condition $R_1(T_r) = R_{1m}(T_r)$ is valid. A dimension of the cubic cell is thus changed from $d(T_r)$ to $d(T)$, where $d(T) < d(T_r)$. Let the condition $\beta_p \lesseqgtr \beta_m$ be considered, where the coefficient β_q $(q = p,e,m)$, mentioned below, is determined in Section 3.5.1..2. Consequently, if $\beta_p \lesseqgtr \beta_m$, then we get $R_1(T) \gtreqless R_{1m}(T)$ for $T < T_r$.

Let the separated particle with the radius $R_1(T)$ be put into the spherical hollow with the radius $R_1(T)$ of the cubic cell with with the dimension $d(T)$. After the embedding of the separated particle in the hollow at the temperature $T < T_r$, a surface of the particle is pushed or pulled by a surface of the matrix due to $R_1(T) > R_{1m}(T)$ or $R_1(T) < R_{1m}(T)$, respectively. Similarly, a surface of the matrix is pushed or pulled by a surface of the particle due to $R_1(T) > R_{1m}(T)$ or $R_1(T) < R_{1m}(T)$, respectively. The pushing or pulling at $T < T_r$ within the spherical particle and cell matrix is expressed by the **thermal-stress**

[1]The **subscripts** $q = p$, $q = e$ and $q = m$ are related to the spherical particle, the spherical envelope and the cell matrix, respectively.

induced radial displacements u'_{1p} and u'_{1m} along the axis x'_1, i.e., along a radial direction, respectively. The same is also valid for the multi-particle-envelope-matrix system as well as for the thermal-stress induced radial displacement u'_{1e} in the spherical envelope.

As mentioned above, the pushing or pulling, which is induced in the cubic cell *after* the embedding at $T < T_r$, is related to the cell dimension $d(T)$ at the temperature $T < T_r$. This fact is required to be considered.

Section 3.3. deals with the determination of the **Cauchy's equations, compatibility and equilibrium equations** as well as with the **Hooke's law** for an anisotropic and isotropic elastic solid continuum (see Section 3.3.2.–3.3.5.). These equations along with the Hooke's law represent fundamental equations of solid continuum mechanics. Due to the analysis in Section 3.2., the Cauchy's equations are determined for the infinitesimal spherical cap which exhibits the radial displacement u'_1 along the axis x'_1.

As presented in Sections 4.3., 4.4., a stress-strain state in an anisotropic component of the model systems is determined by different solutions which fulfil boundary conditions. The same is also valid for an isotropic component of the model systems (see Sections 5.3.–5.6.).

In general, if a system of these fundamental equations exhibits different solutions, then a principle of minimal total potential energy (i.e., the tendency of an elastic solid body to exhibit minimal total potential energy) [26] is required to be considered (see Section 3.4.). Strictly speaking, such solution is considered which exhibits minimal total potential energy of an elastic solid body, i.e., such stress-strain state which results in minimal total potential energy of an elastic solid body is only realized.

In case of the model systems, this energy principle results in the determination of thermal-stress induced elastic energy which is accumulated in each component of the cubic cell (see Eq. (3.51)). As presented in Section 3.4., the total potential energy $W_t = W_d - (W_v + W_s)$ [26] of the model systems is then represented by the thermal-stress induced elastic energy W_c which is accumulated in the cubic cell (see Eqs. (3.52), (3.53)), i.e., $W_t = W_c$, where the energy W_v and W_s is induced by the volume and surface forces, $\vec{F_v}$ and $\vec{F_s}$, respectively, and W_d is deformation energy. In case of the model systems (see Figure 2.1), we get $\vec{F_v} = \vec{F_s} = 0$, and then $W_v = W_s = 0$, $W_t = W_d$. Additionally, the deformation energy W_d of the model systems is thus identical with W_c (see Eqs. (3.51)–(3.53)), and then $W_t = W_d = W_c$.

As presented in Section 3.5.1.2, the coefficient β_q $(q = p,e,m)$ mentioned above is a function of the temperature T, the relaxation temperature T_r, the **thermal expansion coefficient** α_q and the **phase-transformation induced radial strain** ε'_{11tq} along the axis x'_1 (see Figure 3.1a), where $T \leq T_r$. This radial strain for different types of an **anisotropic crystal lattice** (i.e., orthorhombic, tetragonal and hexagonal crystal lattices) is determined in Section 3.5.2..1. In case of an **isotropic crystal lattice** (i.e., cubic crystal lattice), ε_{11tq} is determined in Section 3.5.2..2. The radial strain ε_{11tq} originates at the phase-transformation temperature $T_{tq} \in \langle T_f, T_r \rangle$, where T_f is final temperature of the cooling process.

As presented in Section 3.6., the mutually different changes of dimensions of material components mentioned above induce **radial stresses** p_1, and p_2 acting along the axis x'_1 at the particle-matrix or particle-envelope boundary, and at the matrix-envelope boundary, respectively. These radial stresses are determined by formulae (3.81)–(3.83). Finally, the temperature dependence of the thermal stresses is analysed in Section 3.7..

CHAPTER 4 deals with the determination of the thermal stresses in an **anisotropic component** of the model systems (see Figure 2.1). A part of the fundamental equations is thus considered for the determination of the tangential stresses $\sigma'_{22} = \sigma'_{22}(\sigma'_{11})$, $\sigma'_{33} = \sigma'_{33}(\sigma'_{11})$ and the shear stresses $\sigma'_{12} = \sigma'_{12}(\sigma'_{11})$, $\sigma'_{13} = \sigma'_{13}(\sigma'_{11})$ (see Section 4.2.) which represent functions of the radial stress σ'_{11}. These stresses are determined in the Cartesian system $(Px'_1x'_2x'_3)$ (see Figure 3.1) [2].

Consequently, the radial stress σ'_{11} is assumed in the form $\sigma'_{11} = \sum_{j=1}^{n} C_j r^{\lambda_j}$, where n is a number of boundary conditions, and C_i ($j = 1, \ldots, n$) is an integration constant determined by the boundary conditions. As presented in Section 4.3., the application of a suitable mathematical procedure to the other part of the fundamental equations along with the assumption for σ'_{11} results in the determination of the characteristic equation (4.36) with the roots $\lambda_1, \ldots, \lambda_n$. Additionally, Equation (4.36) represents six different characteristic equations due to the subscript i which is included in this 'main' characteristic equation (4.36), where $i = 0, \ldots, 5$. Accordingly, $\left\{\lambda_1^{(4.36)(0)}, \lambda_2^{(4.36)(0)}\right\}, \ldots, \left\{\lambda_1^{(4.36)(5)}, \lambda_2^{(4.36)(5)}\right\}$ represent six sets of solutions (roots) of these six characteristic equations represented by Eq. (4.36).

Let both exponents of each of the sets be imaginary, or none of these six sets consists of such real exponents which fulfil boundary conditions (see Section 4.6.) for a component of the model systems (see Section 2.2.). In this case, other mathematical procedure is considered for the determination of the characteristic equation (4.41) which include the subscript $i = 0, \ldots, 29$. If the same is also valid for roots of Eq. (4.41), other two mathematical procedures are considered for the determination of the characteristic equations (4.45), (4.49).

Consequently, the analysis of a number of solutions for each of Eqs. (4.36), (4.41), (4.45), (4.49) is presented in Section 4.4.. As presented in Sections 4.1., 4.4., this analysis can be performed numerically for a real two- or three-component material defined in Section 2.1..

Mathematical solutions for the thermal stresses in an anisotropic component of the model systems is presented in Section 4.5.. These solutions along with boundary conditions result in **final formulae** for the thermal stresses, thermal strains, thermal-stress induced elastic energy density in a component, and thermal-stress induced elastic energy of a component which are presented in Sections 4.7., 4.8.. In case of the multi-particle-envelope-matrix system (see Figure 2.1), the conditions $\beta_p \neq \beta_e = \beta_m$, $\beta_p \neq \beta_e \neq \beta_m$, $\beta_p = \beta_e \neq \beta_m$ are also considered (see Sections 4.8.1.–4.8.3.).

The **boundary conditions** for each component of the model systems are determined in Section 4.6.. Due to $\sigma'_{11} = \sum_{j=1}^{n} C_j r^{\lambda_j}$, the analysis of a course of the function $f(r) = r^{\lambda_j}$ ($j = 1, \ldots, n$) with respect to the boundary conditions is also presented. In case of the cell matrix, **mandatory** as well as **additional boundary conditions** are determined (see Sections 4.6.3..1, 4.6.3..2). A detailed analysis of one of the mandatory condition is also presented.

The additional boundary condition is based on the fact that the division of the matrix of

[2]With regard to the notation of subscripts presented in Section 3.3.1., we get $\sigma'_{11} = \sigma'_r$, $\sigma'_{22} = \sigma'_\varphi$, $\sigma'_{33} = \sigma'_\nu$, $\sigma'_{12} = \sigma'_{r\varphi}$, $\sigma'_{13} = \sigma'_{r\nu}$ due to $x'_1 = x_r$ and $x'_2 = x_\varphi$, $x'_3 = x_\nu$ (see Figure 3.1).

the model systems into identical cubic cells is imaginary (see Figure 2.1). The surface of the cubic cell is *not* a physical boundary. Let u'_{1mA} and u'_{1mB} represent thermal-stress induced radial displacements in the cell matrix which are related to neighbouring cubic cells A and B, respectively. Let C be an arbitrary point on a common surface of these neighbouring cells (see Figure 4.1). A connection of these identical functions $u'_{1mA} = u'_{1mA}(r, \varphi, \nu)$ and $u'_{1mB} = u'_{1mB}(r, \varphi, \nu)$ in C is assumed to be 'smooth'. Due to this assumption of the 'smooth' connection, the dependences $u'_{1mA} - r$ and $u'_{1mB} - r$ are thus required *not* to mutually create a singular point at the point C, i.e., for $r = r_c$ (see Section 3.1.2..1). The same analysis concerning the smooth course of the dependence $u'_{1m} - r$ for $r = r_c$ is also valid in case of the dependence $w_m - r$ for $r = r_c$, where $w_m = w_m(r, \varphi, \nu)$ is thermal-stress induced elastic energy density (see Eq. (4.28)).

To fulfil this **non-singularity** assumption, the functions $u'_{1m} = u'_{1m}(r, \varphi, \nu)$, $w_m = w_m(r, \varphi, \nu)$ are required to be extremal on the surface of the cubic cell, i.e $r = r_c$. In case of $|u'_{1m}|$ and w_m, this extreme for $r = r_c$ represents the minimum, where the dependences $|u'_{1m}| - r$ and $w_m - r$ are decreasing. If $u'_{1m} < 0$, then this extreme for $r = r_c$ represents the maximum. The additional boundary conditions $(\partial u'_{1m}/\partial r)_{r=r_c} = 0$ and $(\partial w_m/\partial r)_{r=r_c} = 0$ (see Eqs. (4.99), (4.100)) **might** be considered to represent reasonable assumptions.

CHAPTER 5 deals with the determination of the thermal stresses in an **isotropic** components of the model systems (see Figure 2.1). The determination results from the Hooke's law for an isotropic elastic solid continuum as well as from the Cauchy's and equilibrium equations (see Sections 3.3.2., 3.3.4., 3.3.5.). In contrast to the determination for an anisotropic component (see Chapter 4), the compatibility equations (see Section 3.3.3.) are not considered in case of an isotropic component.

As presented in Section 5.2., the radial, tangential and shear stresses are determined as functions of the thermal-stress induced radial displacement u'_1 along the axis x'_1 (see Figure (3.1)), i.e., $\sigma'_{ij} = \sigma'_{ij}(u'_1)$ $(i, j = 1,2,3)$ (see Eqs. (5.1)–(5.4)). Due to $\sigma'_{ij} = \sigma'_{ij}(u'_1)$, the equilibrium equations are then transformed to a system of differential equations in terms of u'_1 (see Eqs. (5.6), (5.7)). This system is solved by four different mathematical procedures (see Sections 5.3.–5.6.) which result in four different solutions for u'_1. The determination of stresses and strains then results from the functions $\sigma'_{ij} = \sigma'_{ij}(u'_1)$ and $\varepsilon'_{ij} = \varepsilon'_{ij}(u'_1)$ $(i, j = 1,2,3)$ (see Eqs. (3.19)–(3.23), respectively.

The radial displacement $u'_{1q} = u'_{1q}(r, \varphi, \nu)$ which is related to the i-th mathematical procedure $(i = 1, \ldots, 4)$ is derived as $u'_{1q} = \sum_{j=1}^{n_q} C^{(i)}_{jq} \xi^{(i)}_{jq} u^{(i)}_{jrq}$, where n_q is a number of boundary conditions, $\xi^{(i)}_{jq}$ is a coefficient and $u^{(i)}_{jrq} = u^{(i)}_{jrq}(r)$ $(j = 1, \ldots, n_q)$ is a function of r. The integration constants $C^{(i)}_{1q}, \ldots, C^{(i)}_{n_q q}$ are then determined by the boundary conditions for the spherical particle $(q = p)$ (see Section 4.6.1.), the spherical envelope $(q = e)$ (see Section 4.6.2.) and the cell matrix $(q = m)$ (see Section 4.6.3.). In contrast to an anisotropic component, a number of solutions for an isotropic component is based on an analysis of a course of the dependence $u^{(i)}_{jrq} - r$ $(i = 1, \ldots, 4;\ j = 1, \ldots, n_q)$. As presented in Section 5.7., this analysis for each isotropic component of the model systems determines such functions of the set $\left\{u^{(i)}_{jrq}, \ldots, u^{(i)}_{nrq}\right\}$ which fulfil the boundary conditions, and which are thus considered for the determination of corresponding integration constants. Due to a range of this book, final formulae for the thermal stresses in isotropic components of the model systems (see Figure 2.1) are not presented. These final formulae can be then determined with respect to the analysis in Section 5.7.

CHAPTER 6 explains the determination of formulae for the thermal stresses in such model system (see Figure 2.1) which consists of anisotropic components as well as isotropic components.

CHAPTER 7 deals with the analytical modelling of the thermal-stress induced phenomena mentioned above.

As analysed in the beginning of Chapter 7 (see page 103), the analytical models of the crack formation, of the energy barrier and of the micro- and macro-strengthening along with the methods of lifetime prediction exhibit a general validity. Strictly speaking, these analytical models and these lifetime prediction methods are valid for the thermal-stress induced elastic energy density as well as for energy density which is induced by any stresses acting in the multi-particle-matrix and multi-particle-envelope-matrix systems. In case of the methods of lifetime prediction, the radial stress σ'_{11q}, the tangential stresses σ'_{22q}, σ'_{33q} and the shear stresses σ'_{12q}, σ'_{13q} $(q=p,m)$ are required to be derived in the forms $\sigma'_{11p} = -p_1 f_{11p}$, $\sigma'_{11m} = -p_2 f_{11m}$; $\sigma'_{22p} = -p_1 f_{22p}$, $\sigma'_{33p} = -p_1 f_{33p}$, $\sigma'_{22m} = -p_2 f_{22m}$, $\sigma'_{33m} = -p_2 f_{33m}$; $\sigma'_{12p} = -p_1 f_{12p}$, $\sigma'_{13p} = -p_1 f_{13p}$, $\sigma'_{12m} = -p_2 f_{12m}$, $\sigma'_{13m} = -p_2 f_{13m}$, where f_{ijp} and f_{ijm} $(i,j=1,2,3)$ are functions of a position in the spherical particle and cell matrix, respectively.

The analytical modelling of crack formation in Section 7.1. is based on the comparison of thermal-stress induced elastic energy in the cubic cell with energy for the creation of the surface of a crack. This comparison is used e.g., in [71]–[74]. This comparison results from a general analysis (see Section 7.1.1.) which is then applied to the multi-particle-matrix system (see Figure 2.1a). This analytical modelling includes the determination of a condition for the crack initiation (see Eq. (7.10)) in the plane $x_i x_j$ of the Cartesian system $(x_i x_j x_k)$ $(i,j,k=1,2,3;\ i \neq j \neq k)$ (see Figure 7.1a). The condition (7.10) is thus used for the determination of the critical particle radius $R_{1cq}^{(ij)}$ which is a reason of the crack initiation in the plane $x_i x_j$ in the spherical particle ($q=p$; for $\beta_p > \beta_m$) or cell matrix ($q=m$; for $\beta_p < \beta_m$).

The critical particle radius $R_{1cq}^{(ij)}$ thus defines a **limit state** with respect to the crack initiation in the plane $x_i x_j$ in the spherical particle $(q=p)$ and cell matrix $(q=m)$. The crack initiation is then followed by the crack propagation for $R_1 > R_{1cq}^{(ij)}$. Along with the crack initiation condition (see Eq. (7.10)), the condition (7.11) for $R_1 > R_{1cq}^{(ij)}$ is used for the determination of the crack shape in the plane $x_i x_j$. Additionally, the function $f_q^{(ij)}$ describing the crack shape in a plane which is perpendicular to $x_i x_j$ is also determined.

As presented in Section 7.1.3., the determination of these crack parameters considers the curve integral $W_{cq}^{(ij)}$ $(q=p,m)$ of thermal-stress induced elastic energy density (see Eq. (3.50)) along a curve within the cubic cell. The determination of $W_{cp}^{(ij)}$, $W_{cm}^{(ij)}$ considers the conditions $\beta_p(\varphi,\nu) \neq \beta_m(\varphi,\nu)$ and $\beta_p(\varphi,\nu) = \beta_m(\varphi,\nu)$ (see Sections 7.1.3.1, 7.1.3.2). The condition $\beta_p(\varphi,\nu) = \beta_m(\varphi,\nu)$ results in the determination of the angle $\nu_{pm} = \nu_{pm}(\varphi)$ along with definition intervals for the variable φ (see Eqs. (7.18), (7.19)). The function

$\nu_{pm} = \nu_{pm}(\varphi)$ thus determines two volumes in the cubic cells. These two volumes which are defined by $\nu \in \langle 0, \nu_{pm})$ and $\nu \in (\nu_{pm}, \pi/2\rangle$ are characterized by the conditions $\alpha_p \lessgtr \alpha_m$ and $\alpha_p \gtrless \alpha_m$, respectively.

The analytical results in Sections 7.1.1., 7.1.2. are valid disregarding the determination of $W_{cq}^{(ij)}$ $(q = p,m)$. On the one hand, the thermal stresses are modified (changed) during the crack propagation for $R_1 > R_{1cq}^{(ij)}$, and then $W_{cq}^{(ij)}$ is also modified (changed) during the crack propagation. If such modification is determined by e.g., analytical and/or computational methods, then this modification can be applied to the analysis in Sections 7.1.1., 7.1.2.. On the other hand, formulae for $W_{cq}^{(ij)}$ in Section 7.1.3. do not consider the modification of $W_{cq}^{(ij)}$ during the crack propagation. These formulae for $W_{cq}^{(ij)}$ are thus determined for a stress-strain state which is induced by the thermal stresses before the crack initiation. Consequently, these formulae are then required to be considered only in case of a two-component material with ceramic components. In general, a ceramic component exhibits a high-speed crack propagation [71]–[74] during which the modification (change) of $W_{cq}^{(ij)}$ can be assumed to be neglected.

The determination of $R_{1cq}^{(ij)}$ which considers the thermal stresses *before* the crack initiation is valid for any two-component material.

Additionally, an analysis of the crack formation in the spherical particle or cell matrix in spite of the condition $\beta_p < \beta_m$ or $\beta_p > \beta_m$, respectively, is also presented (see Section 7.1.3.2). This analysis which considers the angle $\nu_{pm} = \nu_{pm}(\varphi)$ can explain the paradoxical behaviour concerning the crack formation which is experimentally observed in real two-component materials [71]–[74]. As presented in Section 7.1.3.1 (see Method 2, p. 116), this analysis which is determined for the spherical particle and cell matrix is based on an assumption that the release of energy of a system considers 'minimal resistance' of the system, i.e., the energy of a system is released through 'minimal resistance' of the system.

The transformations concerning the cracking in the planes $x_i x_j$, $x_j x_k$, $x_k x_i$ $(i, j, k = 1,2,3;$ $i \neq j \neq k)$ are also determined (see Section 7.1.4.).

Due to a range of this book, the thermal-stress induced crack formation is analysed for the multi-particle-matrix system with anisotropic and/or isotropic components (see Fig 2.1a). Results for the multi-particle-envelope-matrix system (see Fig 2.1b) can be determined analogically.

Section 7.2. presents the determination of an **energy barrier**. As presented in Section 7.2.1., this determination results from a solid continuum of a general shape (see Figure 7.8). The energy barrier along the axis x_i $(i = 1,2,3)$ of the Cartesian system $(Ox_1x_2x_3)$ (see Figure 7.8) represents a surface integral of the energy density $w = w(x_1, x_2, x_3)$ (see Eqs. (7.37)–(7.39)). The integration surface $S_i = S_i(x_i)$ which is perpendicular to the axis x_i represents a cross-section of the solid continuum. This definition of an energy barrier given by Eqs. (7.37)–(7.39) is also presented in [112].

The analysis in Section 7.2.1. exhibits a general character with respect to a reason of $w = w(x_1, x_2, x_3)$. Accordingly, if the energy density $w = w(x_1, x_2, x_3)$ is induced by the thermal stresses, then such barrier is thermal-stress induced energy barrier.

In general, energy barriers influence motion of dislocations, magnetic domain walls (see Figure 7.9) [112].

This general analysis is then applied to a cubic cell of both model systems (see Figure 2.1). The surface integrals within the cubic cell of the multi-particle-matrix and multi-particle-envelope-matrix systems along with integration boundaries are presented in Sections 7.2.2. and 7.2.3. (see Eqs. (7.43)–(7.46)), respectively.

Section 7.3. presents an analytical model of **strengthening** (i.e., **micro- and macro-strengthening**) along the axis x_i $(i=1,2,3)$ of the Cartesian system $(Ox_1x_2x_3)$ (see Figure 7.8) is based on the surface integral $W_i = W_i(x_1,x_2,x_3)$ (see Eqs. (7.48)) of the energy density $w_i = w_i(x_1,x_2,x_3)$ over the surface $S_i = S_i(x_i)$. In contrast to the energy barrier, the energy density $w_i = w_i(x_1,x_2,x_3)$ is induced by the stress $\sigma_i = \sigma_i(x_1,x_2,x_3)$ which acts along the axis x_i. Accordingly, $W_i = W_i(x_1,x_2,x_3)$ thus represents energy which is accumulated on the surface area $S_i = S_i(x_i)$, and which is induced by the stress $\sigma_i = \sigma_i(x_1,x_2,x_3)$.

The micro-strengthening $\sigma_{si} = \sigma_{si}(x_i)$ (see Eq. (7.50)) along the axis x_i represents such stress which acts along the axis x_i, and which is constant at each point of the surface $S_i = S_i(x_i)$. Accordingly, let $W_{\sigma_{si}} = W_{\sigma_{si}}(x_1,x_2,x_3)$ given by Eq. (7.47) represents energy which is also accumulated on the surface area $S_i = S_i(x_i)$, and which is induced by the stress $\sigma_{si} = \sigma_{si}(x_i)$.

The determination of $\sigma_{si} = \sigma_{si}(x_i)$ is thus based on the comparison of $W_i = W_i(x_1,x_2,x_3)$ with $W_{\sigma_{si}} = W_{\sigma_{si}}(x_1,x_2,x_3)$. The where the macro-strengthening $\overline{\sigma_{si}}$ represents a mean value of $\sigma_{si} = \sigma_{si}(x_i)$ (see Eq. (7.51)) regarding the interval $x_i \in \langle x_{1i}, x_{2i} \rangle$.

Analyses of signs of the micro-/macro-strengthening regarding compressive or tensile mechanical loading are presented in Items 1–6, Section 7.3.1.

The analysis in Section 7.3.1. also exhibits a general character with respect to a reason of $w_i = w_i(x_1,x_2,x_3)$. Accordingly, if the energy density $w_i = w_i(x_1,x_2,x_3)$ is induced by the thermal stresses, then micro- and macro-strengthening is thermal-stress induced micro- and macro-strengthening.

This general analysis is then applied to a cubic cell of both model systems (see Figure 2.1). The micro-/macro-strengthening within the cubic cells of the multi-particle-matrix and multi-particle-envelope-matrix systems along with integration boundaries is presented in Sections 7.3.2.1 and 7.3.2.2 (see Eqs. (7.52)–(7.70)), respectively.

Section 7.3. deals with the determination of methods of the **lifetime prediction**. These **analytical-computational** and **analytical-computational-experimental** lifetime prediction methods (see Section 7.4.2., 7.4.3.) are applicable to the three-component material defined in Item 3, Section 2.1., which consists of grains with and without a continuous component on a grain surface. The grains with the continuous component and the grains without the continuous component are identical or different from microstructural point of view, and thus exhibit identical or different thermal expansion coefficients.

As analysed in Items 1–3, Section 7.4. (see p. 132) this three-component material is replaced by the multi-particle-envelope-matrix system (see Figure 2.1b). Strictly speaking, the grain with the continuous component on the grain surface corresponds to a spherical particle (with the radius R_1) of the multi-particle-envelope-matrix system. The grain without the continuous component on the grain surface corresponds to the cell matrix. The continuous component corresponds to the spherical envelope with the radii $R_1 < R_2$, i.e., with the thickness $t = R_2 - R_1 > 0$.

As presented in Section 7.4.1., these lifetime prediction methods are based on the conditions $\overline{p_i}(R_1,t,v) = 0$ and $|\overline{p_i}(R_1,t,v)| = |\overline{\sigma_{ri}}(R_1,t,v)|$ $(i=1,2)$, where $\overline{p_1} = \overline{p_1}(R_1,t,v)$ and $\overline{p_2} = \overline{p_2}(R_1,t,v)$ (see Eq. (7.71)) are mean values of the φ, ν-dependent radial stresses

$p_1 = p_1(\varphi, \nu, R_1, t, v)$ and $p_2 = p_2(\varphi, \nu, R_1, t, v)$ which act at the particle-envelope and matrix-envelope boundaries, respectively.

Let $\overline{\sigma_{r1}} = \overline{\sigma_{r1}}(R_1, t, v)$ and $\overline{\sigma_{r2}} = \overline{\sigma_{r2}}(R_1, t, v)$ represent mean values of the φ, ν-dependent radial stresses $\sigma_{r1} = \sigma_{r1}(\varphi, \nu, R_1, t, v)$ and $\sigma_{r2} = \sigma_{r2}(\varphi, \nu, R_1, t, v)$ which act at the particle-envelope and matrix-envelope boundaries, respectively. Let the radial stress $\sigma_{ri} = \sigma_{ri}(\varphi, \nu, R_1, t, v)$ $(i = 1{,}2)$ represent a 'response' of the mechanical loading $\sigma_{1\,mech}$, $\sigma_{2\,mech}$, $\sigma_{3\,mech}$ acting along the axes x_1, x_2, x_3 (see Figure 3.1), respectively [3].

Consequently, the conditions $\overline{p_i}(R_1, t, v) = 0$ and $|\overline{p_i}(R_1, t, v)| = |\overline{\sigma_{ri}}(R_1, t, v)|$ result in the functions $R_{1Bc} = f_i(t_{Bc})$ and $R_{1Ac} = f_i(t_{Ac}, \overline{\sigma_{ri}})$ $(i = 1{,}2)$, respectively. The functions $R_{1Bc} = f_i(t_{Bc})$ and $R_{1Ac} = f_i(t_{Ac}, \overline{\sigma_{ri}})$ define dependences between the critical microstructural parameters R_{1Bc}, t_{Bc} and R_{1Ac}, t_{Ac}, respectively. The microstructural parameters R_{1Bc}, t_{Bc} and R_{1Ac}, t_{Ac} represent critical values of the variables R_1, t of the functions $\overline{p_i} = \overline{p_i}(R_1, t, v)$, $\overline{\sigma_{ri}} = \overline{\sigma_{ri}}(R_1, t, v)$, respectively.

The dependences $R_{1Ac} = f_i(t_{Ac}, \overline{\sigma_{ri}})$, $R_{1Bc} = f_i(t_{Bc})$ $(i = 1{,}2)$ define the areas A, B, C shown in Figure 7.10. The areas A, B and C are characterized by such coordinates (R_1, t) for which the thermal stress exhibit resistive, quasi-resistive and contributory effects regarding the mechanical loading.

As presented in Section 7.4.2., the *analytical-computational* lifetime prediction method results from the function $F(R_1, t, \tau, T) = 0$,. This function is determined by a computational simulation of the time-temperature-dependent development of microstructure. The grain radius R_1, the thickness t and the time τ represent variables of $F(R_1, t, \tau, T) = 0$ at the exploitation temperature T $(\equiv T_e)$ which is a parameter of $F(R_1, t, \tau, T) = 0$, where $T < T_r$, and T_r is relaxation temperature (see Section 3.5.1..1). With regard to Figure 7.11, intersections of $R_{1Ac} = f_i(t_{Ac}, \overline{\sigma_{ri}})$ and $R_{1Bc} = f_i(t_{Bc})$ with the function $F_p(R_1, t, T) = 0$ define the time $\tau_{Ac} = \tau_{Ac}(\overline{\sigma_r})$ $(\equiv \tau_{Aci} = \tau_{Aci}(\overline{\sigma_{ri}}))$ and τ_{Bc} $(\equiv \tau_{Bci})$ $(i = 1{,}2)$. The function $F_p(R_1, t, T) = 0$ represents a projection of $F(R_1, t, \tau, T) = 0$ into the plane $R_1 - t$.

Consequently, the intervals $\tau < \tau_{Ac}(\overline{\sigma_r})$, $\tau \in \langle \tau_{Ac}(\overline{\sigma_r}), \tau_{Bc} \rangle$, $\tau > \tau_{Bc}$ defines the time τ which is related to the resistive, quasi-resistive, contributory effects of the thermal stresses against the mechanical loading, respectively. The intervals $\tau < \tau_{Ac}(\overline{\sigma_r})$, $\tau \in \langle \tau_{Ac}(\overline{\sigma_r}), \tau_{Bc} \rangle$, $\tau > \tau_{Bc}$ determine non-critical, quasi-critical, critical time periods of the exploitation, respectively.

Finally, $\tau_{Ac\,min}$ and $\tau_{Bc\,min}$ which represent minimal values of the sets $\{\tau_{Ac1}(\overline{\sigma_{r1}}), \tau_{Ac2}(\overline{\sigma_{r2}})\}$ and $\{\tau_{Bc1}, \tau_{Bc2}\}$ are considered to represent the lifetime regarding the 'resistance–quasi-resistance' and 'quasi-resistance–contribution' transformations (see Figure 7.10), respectively.

As presented in Section 7.4.3., the *analytical-computational-experimental* lifetime prediction method (see Eqs. (7.72)–(7.77)) represents a connection of the analytical dependences $R_{1Ac} = f_i(t_{Ac}, \overline{\sigma_{ri}})$, $R_{1Bc} = f_i(t_{Bc})$ $(i = 1{,}2)$

- with the function $\tau = f(t)$ which is determined by a computational simulation, and with the function $R_1 = f(\tau)$ which is determined by an experiment.

[3] Such radial stresses (i.e., σ_{r1}, σ_{r2}, σ_{r3}) can be determined by e.g., the Eshelby's model. The Eshelby's model and its development [2, 3, 25, 91] which are based on the Green's function, ordinary Newtonian potential and biharmonic potential define the disturbance of an applied stress-field in a solid continuum, where the applied stress-field is disturbed due to the presence of inclusions in a solid continuum.

- or with the function $\tau = f(t)$ which is determined by an experiment, and with the function $R_1 = f(\tau)$ which is determined by a computational simulation.

CHAPTER 8 deals with illustrative examples of applications of the analytical models of the thermal-stress induced phenomena which are determined in Chapter 7. These illustrative examples of applications are as follows.

Section 8.1. presents an application of the analytical model of the crack formation by the method 2, in Section 7.1.3.2 (see p. 116) to the YBaCuO superconductor with ceramic components [71]–[74]. YBaCuO represents a two-component material of the precipitate-matrix type which is defined in Item 1, Section 2.1. (see Figure 2.1a). YBaCuO consists of the Y_2BaCuO_5 precipitates and the $YBa_2Cu_3O_7$ matrix, where Y_2BaCuO_5 and $YBa_2Cu_3O_7$ are anisotropic.

The papers [71]–[74] present experimental results of the crack formation in YBaCuO. In spite of the condition $\alpha_p > \alpha_m$, the cracks which are observed in the YBaCuO superconductor are formed in the $YBa_2Cu_3O_7$ matrix during a cooling process.

Due to the condition $\alpha_p > \alpha_m$, the crack formation in YBaCuO is expected to be observed in the Y_2BaCuO_5 particle, where α_p and α_m are thermal expansion coefficients of the Y_2BaCuO_5 precipitate and the $YBa_2Cu_3O_7$ matrix, respectively. However, as presented in [71]–[74], the crack formation in YBaCuO is observed in the $YBa_2Cu_3O_7$ matrix in spite of $\alpha_p > \alpha_m$. This paradoxical behaviour is supported by the numerical results in Figure 8.1. A shape of the crack in the $YBa_2Cu_3O_7$ matrix is described by Eqs. (8.2)–(8.3) (see Figures 8.2, 8.3).

Section 8.2. presents an application of the analytical model of the micro- and macro-strengthening (see Section 7.3.2.1) to the SiC-Si_3N_4 two-component material. SiC-Si_3N_4 also represents a two-component material of the precipitate-matrix type [27]–[29]. SiC-Si_3N_4 thus consists of SiC precipitates and Si_3N_4 grains (matrix), where both components of SiC-Si_3N_4 are also ceramic. The micro-strengthening along with the thermal stresses in Figures 8.4–8.7 are determined for the analytical model by the mathematical procedure 1 in Section 5.3.

In case of YBaCuO, the numerical determination in Section 8.1. is performed for only one of combinations of solutions of the thermal stresses in the multi-particle-matrix system (see Figure 2.1a) with anisotropic components (see Section 4.7.). As analysed in Section 4.4., the thermal stress-strain state in this model system with anisotropic components is determined by several combinations of solutions.

On the one hand, these different combinations result in different thermal-stress induced elastic energy of the multi-particle-matrix system (see Eqs. (3.51), (3.52)). Accordingly, such combination is considered to result in minimal thermal-stress induced elastic energy of this model system. This combination can be determined numerically for a real two-component material. On the other hand, such numerical determination is time-consuming. Accordingly, a comprehensive software product would be required.

Finally, due to the analysis in Section 5.7.4., the numerical determination in Section 8.2. is also performed for only one of combinations related to the multi-particle-matrix and multi-particle-envelope-matrix systems both with isotropic components, respectively.

Chapter 2

Types of Multi-component Materials. Model Systems and Cell Model

2.1. Types of Multi-component Materials

The analytical models of the thermal stresses and the related phenomena (see Chapter 7) are applicable to real multi-component materials which consist of two or three components.

The following two types of the two-component materials with finite dimensions consist of

1. precipitates and grains, where the precipitates are distributed in the grains,
2. two types of grains. Crystal lattices of grains of these two types are mutually different.

The following two types of the three-component materials with finite dimensions consist of

3. precipitates, grains and a continuous component on a surface of each of the precipitates, where the precipitates are distributed in the grains,
4. grains with a continuous component on their surface and grains without the continuous component on their surface. Crystal lattices of grains of these two types (i.e., with and without the continuous component) are mutually identical or mutually different.

2.2. Model Systems and Cell Model

With regard to analytical modelling of the thermal stresses, the two- and three-component materials with finite dimensions, defined in Items 1–4, Section 2.1., are replaced by multi-particle-matrix and multi-particle-envelope-matrix systems with infinite dimensions (see Figure 2.1), respectively.

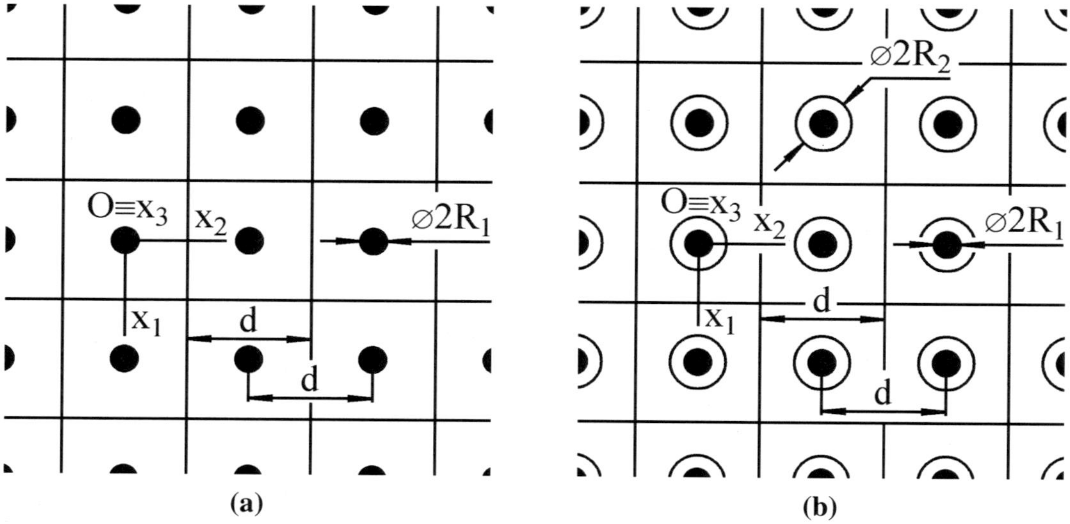

Figure 2.1. (a) The multi-particle-matrix system and (b) multi-particle-envelope-matrix system as model systems which are applicable to the two- and three-component materials defined in Items 1, 2 and 3, 4, Section 2.1., respectively. The infinite matrix of these systems is imaginarily divided into identical cubic cells with a central spherical particle without or with a spherical envelope on the particle surface. The beginning O of the Cartesian system $(Ox_1x_2x_3)$ is identical with the particle centre. The cubic cell dimension d along the axes x_1, x_2, x_3 which is identical to inter-particle distance, the particle radius R_1, the envelope radii R_1, R_2, the envelope thickness $t > 0$ and the particle volume fraction $v \in (0, v_{max}\rangle$ (see Eqs. (2.1)–(2.4)) represent characteristics of these model systems, where $R_1 < R_2$.

The multi-particle-matrix and multi-particle-envelope-matrix systems which represent **model systems** regarding the analytical modelling consist of periodically distributed spherical particles without and with a spherical envelope on a surface of each of the spherical particle surfaces, respectively. The spherical particles with or without the spherical envelopes are embedded in an infinite matrix. These systems are characterized by the inter-particle distance d, the particle radius R_1, the envelope radii R_1, R_2, where $R_1 < R_2$.

To derive the thermal stresses acting in the multi-particle-matrix and multi-particle-envelope-matrix systems, the infinite matrix is **imaginarily** divided into identical cubic cells with the dimension d along each of the axes x_1, x_2, x_3 of the Cartesian system $(Ox_1x_2x_3)$. Each cubic cells contains a central spherical particle without or with the spherical envelope on the particle surface. The beginning O of the Cartesian system is identical with the particle centre as well as with the centre of the cubic cell.

The thermal stresses are thus investigated within the cubic cell. Strictly speaking, the thermal stresses are determined within the spherical particle, spherical envelope and cell matrix. The cubic cell represents a part of the infinite matrix, and this part is related to one spherical particle. Due to infinity of the matrix, formulae for the thermal stresses which are determined for a certain cubic cell are also valid for any cubic cell of the infinite matrix (see Figure 2.1a,b).

The multi-particle-matrix and multi-particle-envelope-matrix systems shown in Figure 2.1 are depicted in the plane x_1x_2. Due to the matrix infinity regarding each of the axes x_1, x_2, x_3, the same figure is also considered for the planes x_1x_3 and x_2x_3.

As analysed in Section 4.6.3., the surface of the cubic cell represents a set of points for which one of the mandatory boundary conditions (see Eq. (4.96)) as well as the additional boundary conditions (see Eqs. (4.99), (4.100)) for the cell matrix are determined.

Let the multi-particle-matrix and multi-particle-envelope-matrix systems have a finite matrix. The analytical modelling of the thermal stresses in the model systems with a finite matrix is required to consider a shape and dimensions of the finite matrix as well as a position of each cell in the finite matrix. Boundary conditions related to the surface of the cubic cell are then required to be separately determined for each cell.

On the one hand, let the boundary conditions for the surface of the cubic cell in the model systems with a finite matrix be defined. An application of such analytical model to real two- and three-component materials results in numerical dependences of the thermal stresses and the related phenomena analysed in Chapter 7 on characteristics of the model systems (see Section 2.3.), i.e., on microstructural parameters of the two- and three-component materials. On the other hand, such application would be probably time-consuming.

Additionally, as presented in [111], the case when an infinite matrix is considered within analytical modelling of phenomena in real multi-component materials with finite dimensions is of particular interest for the mathematical simplicity of analytical solutions. As presented in [111], such analytical solutions are assumed to exhibit sufficient accuracy due to the size of material components (e.g., precipitates, envelopes) which is relatively small in comparison with the size of macroscopic material samples, macroscopic structural elements, etc.

2.3. Characteristics of Model Systems

Let $V_p = 4\pi R_1^3/3$ and $V_c = d^3$ represent volume of the spherical particle and cubic cell, respectively. The particle volume fraction v of the multi-particle-matrix and multi-particle-envelope-matrix systems is derived as

$$v = \frac{V_p}{V_c} = \frac{4\pi}{3}\left(\frac{R_1}{d}\right)^3 \in \left(0, v_{max}\right\rangle. \tag{2.1}$$

In case of the multi-particle-envelope-matrix system, the maximal value v_{max} which is determined on the condition $d = 2R_2$ has the form

$$v_{max} = \frac{\pi}{6}\left(1 - \frac{2t}{d}\right)^3 = \frac{\pi}{6}\left(1 - \frac{t}{R_1 + t}\right)^3, \tag{2.2}$$

where the envelope thickness $t > 0$ is derived as

$$t = R_2 - R_1 > 0. \tag{2.3}$$

In case of the multi-particle-matrix system, i.e., for $t = 0$, we get

$$v_{max} = \frac{\pi}{6}. \tag{2.4}$$

The characteristics R_1, t, d, v of the model systems also represent microstructural parameters of the two- and three-component materials defined in Section 2.1.

In case of the analytical modelling of the thermal stresses, these characteristics are related to the temperature $T \in \langle T_f, T_r \rangle$, where T_f is final temperature of a cooling process, and the relaxation temperature T_r is analysed in Section 3.5.1.1.

In case of the numerical determination of the thermal stresses in a real two- or three-component material for the temperature $T \in \langle T_f, T_r \rangle$, the temperature dependences $R_1 = R_1(T)$, $t = t(T)$ and $d = d(T)$ are required to be determined for the interval $T \in \langle T_f, T_r \rangle$ by a suitable experimental-computational method.

2.4. Multi-Component Materials versus Model Systems

Section 2.4. deals with relationship between components of the multi-component materials defined in Section 2.1. and components of the model systems defined in Section 2.2. The following relationships are required to be determined for the analytical modelling of the thermal stresses.

2.4.1. Two-Component Materials

The precipitates and grains of the two-component material defined in Item 1, Section 2.1. correspond to the spherical particles and infinite matrix of the multi-particle-matrix system, respectively.

Let the grains A and B of the two-component material defined in Item 2, Section 2.1. be characterized by the volume fractions v_A and $v_B = 1 - v_A$, respectively. If $v_A > v_{max} = \pi/6$ (see Eq. (2.4)), then the grains A are considered to represent the infinite matrix of the multi-particle-matrix system. If $v_B > v_{max} = \pi/6$, then the grains B are considered to represent the infinite matrix of the multi-particle-matrix system.

Let $v_A < v_{max}$ and $v_B < v_{max}$ be valid. Let $W_{cAB} = W_{pA} + W_{mB}$ (see Eq. (3.52)) represent thermal-stress induced elastic energy of the cubic cell of such multi-particle-matrix system when the grains A and B are considered to represent the spherical particle and cell matrix within the analytical modelling of the thermal stresses, respectively.

Let $W_{cBA} = W_{pB} + W_{mA}$ (see Eq. (3.52)) represent thermal-stress induced elastic energy of the cubic cell of such multi-particle-matrix system when the grains B and A are considered to represent the spherical particle and cell matrix, respectively.

Let W_{cAB} and W_{BA} be numerically determined for a real two-component material defined in Item 2, Section 2.1..

If $W_{cAB} < W_{cBA}$, then the grains A and B of this real two-component material are considered to represent the spherical particle and cell matrix, respectively. If $W_{cAB} > W_{cBA}$, then the grains B and A of this real two-component material are considered to represent the spherical particle and cell matrix, respectively.

2.4.2. Three-Component Materials

The precipitates, continuous component and grains of the three-component material defined in Item 3, Section 2.1. correspond to the spherical particles, spherical envelope and infinite matrix of the multi-particle-envelope-matrix system, respectively.

The continuous component, the grains with the continuous component and the grains without the continuous component of the three-component material defined in Item 4, Section 2.1. correspond to the spherical envelope, spherical particle and infinite matrix of the multi-particle-envelope-matrix system, respectively.

Chapter 3

Selected Topics on Solid Continuum Mechanics

As usual in solid continuum mechanics, a stress-strain state of a solid continuum is investigated within an infinitesimal part of the solid continuum at an arbitrary point. A position of this arbitrary point is determined by a suitable system of coordinates. The infinitesimal part has a suitable shape regarding the geometry of the solid continuum.

An influence of the solid continuum on the infinitesimal part is represented by displacements of the infinitesimal part along individual directions, and by stresses acting on surfaces of the infinitesimal part [106, 108]. The stresses act along these individual directions.

An analysis of the displacements of the infinitesimal part results in the determination of the Cauchy's equations which define relationships between the displacements and strains of the infinitesimal part (see Section 3.2.). The Cauchy's equations then result in equations of compatibility of the strains (see Section 3.3.2.). The equilibrium principle results in equilibrium equations (see Section 3.3.4.) which define relationships between stresses acting on the surface of the infinitesimal part. The Cauchy's, compatibility and equilibrium equations along with the Hooke's law (see Section 3.3.5.) which defines relationships between the strains and stresses in anisotropic and isotropic elastic solid continuum represent fundamental equations of solid continuum mechanics.

If a system of these fundamental equations exhibits different solutions, then a principle of minimal total potential energy of an elastic solid body [26] is required to be considered (see Section 3.4.). Strictly speaking, such solution is considered which exhibits minimal total potential energy of an elastic solid body, i.e., such stress-strain state which results in minimal total potential energy of an elastic solid body (i.e., the model system (see Figure 2.1)) is only realized.

Chapter 3 deals with

- the determination of such coordinate system and such infinitesimal part which consider the geometry of the model systems (see Section 3.1.1.), i.e., the geometry of the multi-particle-matrix and multi-particle-envelope-matrix systems, defined in Section 2.2. (see Figure 2.1),

- the determination of intervals of coordinates within the cubic cell of the model systems (see Section 3.1.2.),

- an analysis of displacements of the infinitesimal part in the model systems (see Section 3.2.),
- the determination of the fundamental equations for the infinitesimal part (see Section 3.3.),
- the determination of elastic energy of the model systems (see Section 3.4.) regarding the minimal total potential energy principle [26],
- an analysis of a reason of the thermal stresses (see Section 3.5.) which includes
  - an analysis of the relaxation temperature of the model systems (see Section 3.5.1.),
  - the determination of the phase-transformation induced radial strain for anisotropic and isotropic crystal lattices (see Section 3.5.2.),
- the determination of mathematical conditions which are considered for the determination of radial stresses acting at the particle-matrix, particle-envelope and matrix-envelope boundaries in the model systems (see Section 3.6.),
- an analysis of a temperature dependence of the thermal stresses (see Section 3.7.).

3.1. Coordinate System and Intervals of Coordinates

3.1.1. Coordinate System

Thermal stresses are investigated at the arbitrary point P of a solid continuum along the axes x_1', x_2', x_3' of the Cartesian system $(Px_1'x_2'x_3')$ (see Figure 3.1). A position of the arbitrary point P regarding the Cartesian system $(Ox_1x_2x_3)$ is determined by the spherical coordinates (r, φ, ν), where O is a centre of the spherical particle (see Figure 2.1).

The axes $x_1' = x_r$ and $x_2' = x_\varphi$, $x_3' = x_\nu$ at the point P represent radial and tangential directions related to the spherical surface with the radius $r = |OP|$, respectively, where $x_\varphi \parallel x_1x_2$. Finally, the infinitesimal part of a solid continuum at the point P is represented by an infinitesimal spherical cap with the parameters dr, $d\varphi$, $d\nu$ (see Figures 3.3, 3.4), where dr is a dimension of the infinitesimal spherical cap along the axis x_1'.

The surface $S_r = A_1B_1C_1D_1$ of the infinitesimal spherical cap at the radius $r = \left|\overline{OP}\right|$ has the dimensions $|A_1D_1| = |B_1C_1| = r \times d\varphi$ and $|A_1B_1| = |C_1D_1| = r \times d\nu$ along the axes x_2' and x_3', respectively.

The surface $S_{r+dr} = A_2B_2C_2D_2$ of the infinitesimal spherical cap at the radius $r + dr$ has the dimensions $|A_2D_2| = |B_2C_2| = (r+dr) \times d\varphi$ and $|A_2B_2| = |C_2D_2| = (r+dr) \times d\nu$ along the axes x_2' and x_3', respectively [1].

The axis x_1' represents a normal of the surfaces at the radii r and $r + dr$, where $|A_1A_2| = |B_1B_2| = |C_1C_2| = |D_1D_2| = dr$.

---

[1] The 'conventional approach' [45] considers the dimensions $|A_1D_1| = |B_1C_1| = r \sin\nu\, d\varphi$ and $|A_2D_2| = |B_2C_2| = (r+dr)\sin\nu\, d\varphi$ along the axis x_2' (see Figure 3.1) at the radii r and $r+dr$, respectively.

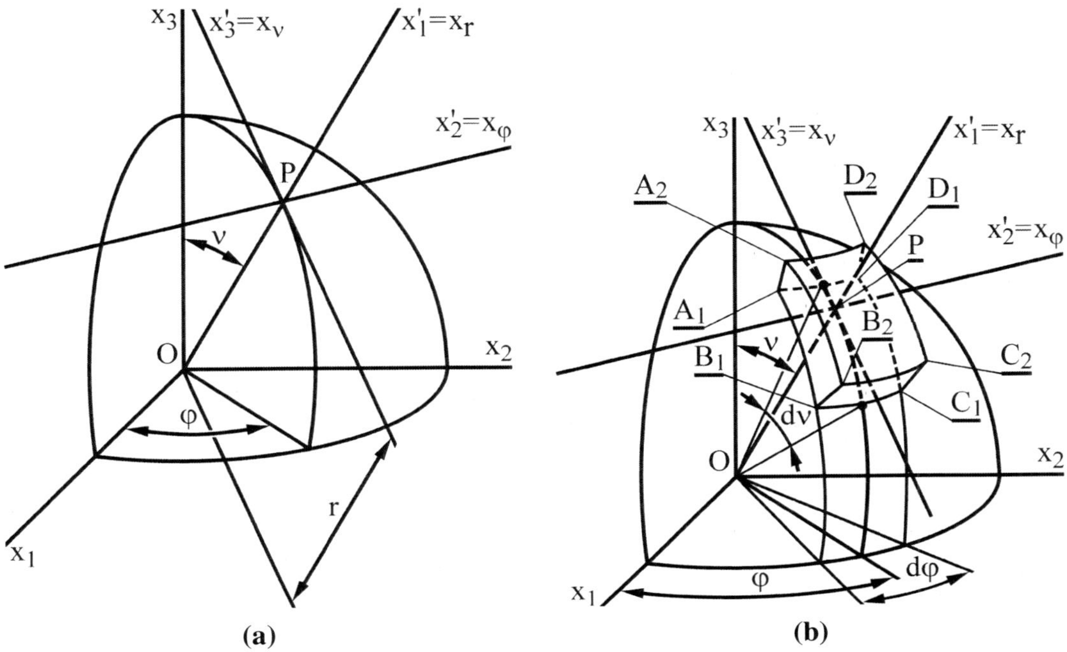

Figure 3.1. (a) The arbitrary point P with a position determined by the spherical coordinates (r, φ, ν) regarding the Cartesian system $(Ox_1x_2x_3)$, where O is a centre of the spherical particle (see Figure 2.1), and $r = |OP|$. The axes $x_1' = x_r$ and $x_2' = x_\varphi$, $x_3' = x_\nu$ thus represent radial and tangential directions, respectively, where $x_\varphi \parallel x_1x_2$. (b) The infinitesimal spherical cap at the point P with the surfaces $S_r = A_1B_1C_1D_1$ and $S_{r+dr} = A_2B_2C_2D_2$ at the radii $r = \left|\overline{OP}\right|$ and $r + dr$, respectively. Dimensions of the infinitesimal spherical cap are as follows: $|A_1A_2| = |B_1B_2| = |C_1C_2| = |D_1D_2| = dr$, $|A_1D_1| = |B_1C_1| = r \times d\varphi$, $|A_1B_1| = |C_1D_1| = r \times d\nu$, $|A_2D_2| = |B_2C_2| = (r + dr) \times d\varphi$, $|A_2B_2| = |C_2D_2| = (r + dr) \times d\nu$.

The spherical coordinates (r, φ, ν) and the infinitesimal spherical cap are considered due to the spherical shape of the particles and envelopes of the model systems (see Figure 2.1).

With regard to $x_1' = x_r$, $x_2' = x_\varphi$, $x_3' = x_\nu$ (see Figure 3.1), the conventional subscripts r, φ, ν in a connection with the spherical coordinates (r, φ, ν) are replaced by the subscripts 1, 2, 3, respectively, due to the mathematical procedures in Section 4.2. (see e.g., Eqs. (4.1)–(4.12)).

3.1.2. Intervals of Coordinates

As presented in Section 2.2., the thermal stresses in the model systems shown in Figure 2.1 are investigated within the cubic cell. Due to infinity of the matrix, formulae for the thermal stresses related to a certain cubic cell are valid for any cubic cell of the infinite matrix.

The thermal stresses are thus sufficient to be investigated within one eighth of the cubic cell (see Figure 3.2), i.e., for $\varphi \in \langle 0, \pi/2 \rangle$ and $\nu \in \langle 0, \pi/2 \rangle$. This is a consequence of symmetry of the model systems. This symmetry results from the matrix infinity and from the periodical distribution of the spherical particles and spherical envelopes.

The interval $r \in \langle 0, R_1 \rangle$ is related to the spherical particle of the multi-particle-matrix and multi-particle-envelope-matrix systems (see Figure 2.1a,b).

The interval $r \in \langle R_1, R_2 \rangle$ is related to the spherical envelope of the multi-particle-envelope-matrix system (see Figure 2.1b).

The intervals $r \in \langle R_1, r_c \rangle$ and $r \in \langle R_2, r_c \rangle$ are related to the cell matrix of the multi-particle-matrix and multi-particle-envelope-matrix systems (see Figure 2.1a,b), respectively. The distance r_c which is determined in Section 3.1.2..1 is also considered within boundary conditions for the cell matrix (see Eqs. (4.96), (4.99), (4.100)).

3.1.2.1. Determination of the Distance r_c

Section 3.1.2.1 deals with the determination of r_c which represents a distance along a radial direction from the point O to the point C_1 or C_2 on a surface of the cubic cell (see Figure 3.2), where the radial direction defined by the angles $\varphi \in \langle 0, \pi/2 \rangle$ and $\nu \in \langle 0, \pi/2 \rangle$ is given by the axis x_1' (see Figure 3.1). The points C_1 and C_2 thus represent intersections of the axis x_1' with the surface of the cubic cell for $\nu \in \langle 0, \nu^* \rangle$ and $\nu \in \langle \nu^*, \pi/2 \rangle$, respectively.

The determination of the distance r_c thus considers the intervals $\nu \in \langle 0, \nu^* \rangle$, $\nu \in \langle \nu^*, \pi/2 \rangle$. The angle ν^* (see Figure 3.2) is derived as

$$\nu^* = \arctan\left(\frac{|\overline{39}|}{|\overline{O3}|}\right) = \arctan\left(\frac{1}{c_\varphi}\right), \tag{3.1}$$

where $|\overline{O3}| = d/2$ and $|\overline{39}| = 1/(2c_\varphi)$ represent length of the abscissae $\overline{O3}$ and $\overline{39}$, respectively. The coefficient c_φ has the forms

$$c_\varphi = \cos\varphi, \quad \varphi \in \left\langle 0, \frac{\pi}{4} \right\rangle, \tag{3.2}$$

$$c_\varphi = \sin\varphi, \quad \varphi \in \left\langle \frac{\pi}{4}, \frac{\pi}{2} \right\rangle. \tag{3.3}$$

As presented in Section 2.3., the cubic cell characteristics R_1, d, v which are included in Eqs. (3.15)–(3.18) are related to the temperature $T \in \langle T_f, T_r \rangle$, where T_f is final temperature of a cooling process, and the relaxation temperature T_r is analysed in Section 3.5.1.

Interval $\nu \in \langle 0, \nu^* \rangle$. The distance $r_c = |\overline{OC_1}|$ between the particle centre O and the point C_1 on the surface *3657* (see Figure 3.2), i.e., for $\nu = \angle(\overline{OC_1}, x_3) \in \langle 0, \nu^* \rangle$, is derived by the following mathematical procedure.

Considering the sine rule [113]

$$\frac{|\overline{C_1 9}|}{\sin[\angle(\overline{OC_1}, \overline{O9})]} = \frac{|\overline{O9}|}{\sin[\angle(\overline{OC_1}, \overline{C_1 9})]} \tag{3.4}$$

and the condition

$$|\overline{O9}| = r_c \cos[\angle(\overline{OC_1}, \overline{O9})] + |\overline{C_1 9}| \cos[\angle(\overline{C_1 9}, \overline{O9})], \tag{3.5}$$

we get

$$r_c = \frac{|\overline{O9}|}{\cos[\angle(\overline{OC_1}, \overline{O9})]} \left\{ 1 - \frac{\sin[\angle(\overline{OC_1}, \overline{O9})] \cos[\angle(\overline{C_1 9}, \overline{O9})]}{\sin[\angle(\overline{OC_1}, \overline{C_1 9})]} \right\}. \tag{3.6}$$

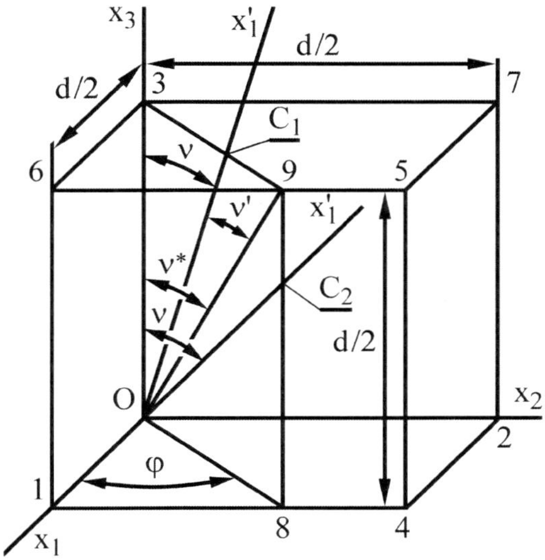

Figure 3.2. The points C_1 and C_2 as intersections of the axis x_1' (see Figure 3.1) with the surfaces *3657* and *1456* of one eighth of the cubic cell (see Figure 2.1) for $\nu \in \langle 0, \nu^* \rangle$ and $\nu \in \langle \nu^*, \pi/2 \rangle$, respectively. The x_1' represents a radial direction defined by the angles $\varphi \in \langle 0, \pi/2 \rangle$ and $\nu \in \langle 0, \pi/2 \rangle$ (see Figure 3.1).

The length $\left|\overline{O9}\right|$ of the abscissa $\overline{O9}$ and the angles $\angle\left(\overline{OC_1}, \overline{O9}\right)$, $\angle\left(\overline{C_1 9}, \overline{O9}\right)$, $\angle\left(\overline{OC_1}, \overline{C_1 9}\right)$, regarding $\left|\overline{O3}\right| = d/2$, $\left|\overline{39}\right| = 1/\left(2c_\varphi\right)$, are derived as

$$\left|\overline{O9}\right| = \sqrt{\left|\overline{O3}\right|^2 + \left|\overline{39}\right|^2} = \frac{d \times \sqrt{1 + c_\varphi^2}}{2c_\varphi}, \tag{3.7}$$

$$\angle\left(\overline{OC_1}, \overline{O9}\right) = \nu' = \nu^* - \nu, \quad \nu' \in \langle 0, \nu^* \rangle, \tag{3.8}$$

$$\angle\left(\overline{C_1 9}, \overline{O9}\right) = \frac{\pi}{2} - \nu^*, \tag{3.9}$$

$$\angle\left(\overline{OC_1}, \overline{C_1 9}\right) = \frac{\pi}{2} + \nu^* - \nu'. \tag{3.10}$$

Considering the following relationships [113]

$$\sin\left(\frac{\pi}{2} \pm \alpha\right) = \cos\alpha, \tag{3.11}$$

$$\cos\left(\frac{\pi}{2} \pm \alpha\right) = \mp \sin\alpha, \tag{3.12}$$

$$\sin(\alpha \pm \beta) = \sin\alpha\cos\beta \pm \cos\alpha\sin\beta, \tag{3.13}$$

$$\cos(\alpha \pm \beta) = \cos\alpha\cos\beta \mp \sin\alpha\sin\beta, \tag{3.14}$$

and with regard to Eqs. (2.1), the distance $r_c = \left|\overline{OC_1}\right|$ (see Figure 3.1) has the form

$$r_c = \left|\overline{OC_1}\right| = \frac{d \times \sqrt{1 + c_\varphi^2}}{2\left(\sin\nu' + c_\varphi \cos\nu'\right)} = R_1 f_c, \quad \nu' \in \langle 0, \nu^* \rangle, \tag{3.15}$$

where the function $f_c = f_c(\varphi, \nu', v)$ for $\nu' \in \langle 0, \nu^* \rangle$ is derived as

$$f_c = \frac{\sqrt{1+c_\varphi^2}}{2\left(\sin\nu' + c_\varphi \cos\nu'\right)} \left(\frac{4\pi}{3v}\right)^{1/3}, \quad \nu' \in \langle 0, \nu^* \rangle. \tag{3.16}$$

The analysis of the fact that the formula $r_c = \left|\overline{OC_1}\right| = d/(2\cos\nu)$ for $\nu \in \langle 0, \nu^* \rangle$ (see Figure 3.2) can not be considered is as follows.

As mentioned above, r_c is considered within boundary conditions for the cell matrix (see Eqs. (4.96), (4.99), (4.100)). The distance r_c is thus included in the thermal stresses and thermal strains in the cell matrix of the multi-particle-matrix and multi-particle-envelope-matrix systems.

Let the multi-particle-matrix system be considered. The thermal stresses and thermal strains in components of this system are functions of the radial stress p_1 acting at the particle-matrix boundary. The radial stress p_1 is determined by the condition (4.63) which includes the tangential strain ε'_{22m} in the cell matrix. The radial stress p_1 and consequently the thermal stresses as well as the thermal strains in components of this system include the distance r_c.

Let the multi-particle-envelope-matrix system be considered. If $\beta_p \neq \beta_e \neq \beta_m$, then the thermal stresses as well as the thermal strains in components of this system are functions of the radial stresses p_1 and p_2 acting at the particle-envelope and matrix-envelope boundaries, respectively. The radial stresses p_1, p_2 are *both* determined by the conditions (4.64), (4.65), where the condition (4.65) includes ε'_{22m}. The radial stresses p_1, p_2 and consequently the thermal stresses as well as the thermal strains in components of this system also include the distance r_c.

The coefficient β_q given by Eqs. (3.54)–(3.60) is related to the spherical particle ($q = p$), the spherical envelope ($q = e$) and the cell matrix ($q = m$). The coefficient β_q includes the thermal expansion coefficient α_q, and the phase-transformation induced radial strain $\varepsilon'_{11tq} = \varepsilon_{11tq}(\varphi, \nu)$ (see Section 3.5.2.). In case of anisotropic components we get $\alpha_q = \alpha_q(\varphi, \nu)$. In case of isotropic components, the thermal expansion coefficient α_q is not a function of the variables φ and ν, i.e., $\alpha_q \neq \alpha_q(\varphi, \nu)$.

If $\beta_p = \beta_e \neg \beta_m$, then the thermal stresses as well as the thermal strains in components of the multi-particle-envelope-matrix system are functions of the radial stress p_2 which is determined by the condition (4.65). The condition (4.65) includes ε'_{22m}. In this case, the radial stress p_2 and consequently the thermal stresses as well as the thermal strains in components of the multi-particle-envelope-matrix system also include the distance r_c.

If $\beta_p \neq \beta_e = \beta_m$, then the thermal stresses as well as the thermal strains in components of the multi-particle-envelope-matrix system are functions of the radial stress p_1 which is determined by the condition (4.64) which does not include ε'_{22m}. In this case, the radial stress p_1 and consequently the thermal stresses as well as the thermal strains in the spherical particle and spherical envelope of the multi-particle-envelope-matrix system do not include the distance r_c. However, in contrast to the spherical particle and spherical envelope, the thermal stresses and the thermal strains in the cell matrix include the distance r_c due to the boundary conditions (4.96), (4.99), (4.100).

Let the multi-particle-matrix and multi-particle-envelope-matrix systems be considered, both with isotropic components which do not exhibit phase transformations for the temperature $T \in \langle T_f, T_r \rangle$, i.e., $\varepsilon'_{11tq}(\varphi, \nu) = 0$ ($q = p,e,m$) (see Eqs. (3.54)–(3.60)). In this case, and provided that $r_c = d/(2\cos\nu)$ for $\nu \in \langle 0, \nu^* \rangle$, the thermal stresses as well as the thermal strains in components of the multi-particle-matrix system are not functions of the variable $\varphi \in \langle 0, \pi/2 \rangle$ in contrast to the interval $\nu \in \langle \nu^*, \pi/2 \rangle$ (see Eqs. (3.2), (3.3), (3.17), (3.18)). If $r_c = d/(2\cos\nu)$ for $\nu \in \langle 0, \nu^* \rangle$, then the thermal stresses and thermal strains in points of the part *O93O* of the plane *O893O* (see Figure 3.2) are independent on the variable $\varphi \in \langle 0, \pi/2 \rangle$, i.e., are independent on a position of the plane *O893O* within the cubic cell. In contrast to the part *O93O*, the thermal stresses and thermal strains in points of the part *O89* of the plane *O893O* are dependent on the variable $\varphi \in \langle 0, \pi/2 \rangle$. This paradox is not physically acceptable, and consequently the formula $r_c = |\overline{OC_1}| = d/(2\cos\nu)$ is not considered for $\nu \in \langle 0, \nu^* \rangle$.

In case of anisotropic components of the model systems, dependences of the thermal stresses and thermal strains on the variables φ, ν *also* result from the functions $\beta_q = \beta_q(\varphi, \nu)$ ($q = p,e,m$) and $s'_{ijq} = s'_{ijq}(\varphi, \nu)$, where the elastic modulus s'_{ijq} ($i, j = 1, \ldots, 6$) included in the Hooke's law is related to the anisotropic spherical particle ($q = p$), the anisotropic spherical envelope ($q = e$) and the anisotropic cell matrix ($q = m$) (see Eqs. (3.41), (3.42), (3.54)–(3.60)).

On the one hand, in case of anisotropic components, dependences of the thermal stresses and thermal strains on the variable $\varphi \in \langle 0, \pi/2 \rangle$ *also* result from $\beta_q = \beta_q(\varphi, \nu)$ and $s'_{ijq} = s'_{ijq}(\varphi, \nu)$. On the other hand, these dependences are also required to result from the function $r_c = r_c(\varphi, \nu)$.

Finally, in case of anisotropic components as well as isotropic components, the distance $r_c = |\overline{OC_1}| = r_c(\varphi, \nu)$ for $\nu \in \langle 0, \nu^* \rangle$ is determined by Eqs. (3.2), (3.3), (3.15), (3.16), where $\nu' \in \langle 0, \nu^* \rangle$. Consequently, if $\nu \in \langle 0, \nu^* \rangle$, then the variable ν included in Eq. (3.42) is required to be replaced by $\nu^* - \nu'$ for $\nu' \in \langle 0, \nu^* \rangle$. With regard to Eqs. (3.13), (3.14), the terms $\sin\nu$, $\cos\nu$ for $\nu \in \langle 0, \nu^* \rangle$ in Eq. (3.42) are required to be derived as $\sin\nu = \sin(\nu^* - \nu') = (\cos\nu' - \sin\nu'\cos\varphi)/\sqrt{1+\cos^2\varphi}$, $\cos\nu = \cos(\nu^* - \nu') = (\sin\nu' + \cos\nu'\cos\varphi)/\sqrt{1+\cos^2\varphi}$.

Interval $\nu \in \langle \nu^*, \pi/2 \rangle$. The distance $r_c = |\overline{OC_2}|$ between the particle centre O and the point C_2 on the surface *1456* (see Figure 3.2), i.e., for $\nu = \angle(\overline{OC_2}, x_3) \in \langle \nu^*, \pi/4 \rangle$, has the form

$$r_c = \frac{d}{2c_\varphi \sin\nu} = R_1 f_c, \quad \nu \in \left\langle \nu^*, \frac{\pi}{2} \right\rangle, \tag{3.17}$$

where the function $f_c = f_c(\varphi, \nu, v)$ for $\nu \in \langle \nu^*, \pi/4 \rangle$ is derived as

$$f_c = \frac{1}{2c_\varphi \sin\nu}\left(\frac{4\pi}{3v}\right)^{1/3}, \quad \nu \in \left\langle \nu^*, \frac{\pi}{2} \right\rangle, \tag{3.18}$$

and the coefficient c_φ is given by Eqs. (3.2), (3.3).

3.2. Displacement of Infinitesimal Spherical Cap. Thermal-Stress Induced Radial Displacement

Let the multi-particle-matrix system be considered. The thermal stresses in the multi-particle-matrix system (see Figure 2.1a) originate as a consequence of the condition $\beta_p \neq \beta_m$. A detailed analysis of the coefficient β_q (see Eqs. (3.54)–(3.60)) is presented in Section 3.5.1., where the subscript $q = p$ and $q = m$ is related to the spherical particle and cell matrix, respectively. The thermal stresses originate at the temperature $T \in \langle T_f, T_r \rangle$, where T_f is final temperature of a cooling process, and the relaxation temperature T_r is analysed in Section 3.5.1..

If $T \in \langle T_f, T_r \rangle$ and $\beta_m - \beta_p > 0$, then the cell matrix is pushed by the spherical particle, and the spherical particle is pushed by the cell matrix. If $T \in \langle T_f, T_r \rangle$ and $\beta_m - \beta_p < 0$, then the cell matrix is pulled by the spherical particle, and the spherical particle is pulled by the cell matrix.

An analysis of the pulling or pushing, i.e., an analysis of displacement of the infinitesimal spherical cap, is as follows. As presented in Section 3.1.2., the multi-particle-matrix system is symmetric, where this symmetry results from the matrix infinity and from the periodical distribution of the spherical particles. Due to this symmetry, the pulling or pushing of an arbitrary point at the particle-matrix boundary is realized along a normal to this surface (i.e., to the particle-matrix boundary) at this arbitrary point.

Let P (see Figure 3.1a) be such arbitrary point at the particle-matrix boundary, thus for $r = R_1$. The point P as well as the infinitesimal spherical cap at the point P (see Figure 3.1b) thus exhibit a displacement along a normal to the surfaces S_r and S_{r+dr} of the spherical the infinitesimal spherical cap, where S_r, S_{r+dr} are described in Section 3.1.1. The normal to the surfaces S_r, S_{r+dr} is represented by the axis x_1' which defines the radial direction regarding the Cartesian system $(Ox_1x_2x_3)$ (see Figure 3.1a). The infinitesimal spherical cap in the point P at the particle-matrix boundary exhibits a displacement along the axis x_1', i.e., a radial displacement.

The condition $\beta_p \neq \beta_m$ is a reason of the radial stress p_1 acting at the particle-matrix boundary along the axis x_1', i.e., along the radial direction. A condition for the determination of p_1 is derived in Section 3.6. (see Eqs. (3.82), (4.63)). The radial stress p_1 is a reason of the fact that this analysis concerning the radial displacement of the infinitesimal spherical cap in the point P at the particle-matrix boundary (i.e., for $r = R_1$) is also valid for each point of the axis x_1' (i.e., for $r \in \langle 0, r_c \rangle$), where r_c is given by Eqs. (3.1)–(3.3), (3.8), (3.15)–(3.18).

The same is also valid for the multi-particle-envelope-matrix system. In this case, p_1 and p_2 acting at the particle-envelope and matrix envelope boundaries are a consequence of the conditions $\beta_p \neq \beta_e$ and $\beta_e \neq \beta_m$. With regard to $\beta_p \neq \beta_e$ and $\beta_e \neq \beta_m$, the radial stresses p_1 and p_2 are determined by Eqs. (3.82) and (3.83), respectively, where β_e is a thermal expansion coefficient of the spherical envelope (see Eqs. (3.54)–(3.60)).

The infinitesimal spherical cap (see Figure 3.1b) in the arbitrary point P (see Figure 3.1a) with a position described by the spherical coordinates (r, φ, ν) exhibits the **radial displacement** $u_1' = u_r$ only, along the axis x_1' (see Figure 3.1).

Additionally, the following analysis concerning the radial displacement $u_1' = u_r$ is required to be considered. This analysis is based on a concept of imaginary separation which is also considered within the mathematical procedures which are used for the determination of the Eshelby's model [91].

Let the multi-particle-matrix system be considered. Let the spherical particles and the infinite matrix be imaginarily separated, and then spherical hollows are periodically distributed in the infinite matrix.

Let $T \in \langle T_f, T_r \rangle$ represent temperature of the separated spherical particles and of the infinite matrix with the spherical hollows. If the temperature T increases or decreases within the interval $\langle T_f, T_r \rangle$, then the components which are imaginarily separated expand or contract, respectively. The expansion and contraction result in displacements of points in the components. Due to the imaginary separation, these displacements result from the temperature change, and not from the difference $\beta_m - \beta_p \neq 0$ (see Eqs. (3.54)–(3.60)).

Let the spherical particles be embedded in the infinite matrix. Let $\Delta T = T - T_r \neq 0$ represent the temperature change. Let the condition $\beta_p = \beta_m$ be considered. Due to $\beta_p = \beta_m$, the thermal stresses do not originate in the multi-particle-matrix system, and the infinitesimal spherical cap is thus shifted due to the temperature change.

Let $R_{1p} = R_{1p}(T)$ and $R_{1m} = R_{1m}(T)$ represent temperature-dependent functions of radii of these separated spherical particle and the spherical hollows, respectively. Due to $T < T_r$, we get $R_{1q}(T) < R_{1q}(T_r)$ $(q = p,m)$, where $R_{1q}(T_r) = R_{1T_r}$. The same, i.e., $d(T) < d(T_r)$, is also valid for the temperature-dependent function $d = d(T)$ of the inter-particle distance d.

Due to $\beta_q = \beta_q(\varphi, \nu)$ (see Eqs. (3.42), (3.54)–(3.60)), we get $R_{1q} = R_{1q}(\varphi, \nu, T)$. Consequently, $R_{1p} = R_{1p}(\varphi, \nu, T)$ is a distance from the particle centre to a point on a surface of the separated particle along the axis x'_1 which represents the radial direction defined by the angles φ, ν. Similarly, $R_{1m} = R_{1m}(\varphi, \nu, T)$ is a distance along the axis x'_1 from a centre of the hollow to a point on a surface of the hollow.

If $\beta_p < \beta_m$, then we get $R_{1p}(\varphi, \nu, T) > R_{1m}(\varphi, \nu, T)$. Let the distance R_{1p} be changed to R_1 for each value of the variables $\varphi, \nu \in \langle 0, \pi/2 \rangle$, where $R_1 < R_{1p}$. The change $R_{1p}(\varphi, \nu, T) \to R_1$ is caused by the radial stress $p_1 = p_1(\varphi, \nu, T)$ which acts, along the axis x'_1, on a surface of the separated particle. Due to $\beta_p < \beta_m$, the radial stress p_1 is compressive regarding the surface of the separated particle. Additionally, $\left[u'_{1p}(\varphi, \nu, T)\right]_{r=R_{1p}} = R_1 - R_{1p}$ represents a **thermal-stress induced radial displacement** on a surface of the separated particle (i.e., for $r = R_{1p}$) at a point defined by the coordinates $\varphi, \nu \in \langle 0, \pi/2 \rangle$. The radial displacement $\left(u'_{1p}\right)_{r=R_{1p}}$ along the axis $x'_1 = x_r$ (see Figure 3.1) is induced by the radial stress p_1.

The separated particle can be put into the hollow provided that the distance R_{1m} is also changed to R_1 for each value of the variables $\varphi, \nu \in \langle 0, \pi/2 \rangle$, where $R_1 > R_{1m}$. The change $R_{1m}(\varphi, \nu, T) \to R_1(\varphi, \nu, T)$ is also caused by the radial stress $p_1 = p_1(\varphi, \nu, T)$ which acts on a surface of the hollow in the matrix. Due to $\beta_p < \beta_m$, the radial stress p_1 is also compressive regarding the surface of the hollow. Additionally, $[u'_{1m}(\varphi, \nu, T)]_{r=R_{1m}} = R_1 - R_{1m}$ represents a **thermal-stress induced radial displacement** on a surface of the hollow (i.e., for $r = R_{1m}$) at a point defined by the coordinates $\varphi, \nu \in \langle 0, \pi/2 \rangle$. The radial displacement $(u'_{1m})_{r=R_{1m}}$ along the axis $x'_1 = x_r$ (see Figure 3.1) is also induced by the radial stress p_1.

After the embedding of the separated particle with the radius R_{1p} in the hollow with the radius R_{1m}, a surface of the particle with the radius R_1 is pushed by a surface of the matrix, where $R_{1p} > R_1 > R_{1m}$. Similarly, a surface of the matrix is pushed by a surface of the particle. The same is also valid in case of the pulling for $\beta_p > \beta_m$. This analysis of the pushing or pulling which considers a concept of the imaginarily separated components is also valid for the multi-particle-envelope-matrix system.

An analysis of a relationship between the radial displacements $\left(u'_{1p}\right)_{r=R_{1p}}$, $\left(u'_{1q}\right)_{r=R_{iq}}$ and the coefficient $\beta_q = \beta_q(\varphi,\nu)$ $(i=1,2;\ q=e,m)$ is presented in Section 3.6.. This analysis results in the determination of the formulae Eqs. (3.81)–(3.83). In case of the multi-particle-matrix system (see Figure 2.1a), the formula Eq. (3.81) is considered for the determination of the radial stresses $p_1 = p_1(\varphi,\nu)$ acting at the particle-matrix boundary. In case of the multi-particle-envelope-matrix system (see Figure 2.1b), the formulae (3.82) and (3.83) are considered for the determination of the radial stresses $p_1 = p_1(\varphi,\nu)$ and $p_2 = p_2(\varphi,\nu)$ acting at the particle-envelope and matrix-envelope boundary, respectively.

3.3. Fundamental Equations of Solid Continuum Mechanics

3.3.1. Notation of Subscripts

The Cauchy's equations (i.e., geometric equations), compatibility and equilibrium equations in Sections 3.3.2.–3.3.4. are determined for the infinitesimal spherical cap described in Section 3.1.1. (see Figure 3.1b). The Hooke's law for anisotropic and isotropic solid continuum is presented in Section 3.3.5.

As presented in Section 3.1.1., the infinitesimal spherical cap as well as the spherical coordinates (r,φ,ν) are used due to the fact that the particle and envelopes of the multi-particle-matrix and multi-particle-envelope-matrix systems are spherical. The thermal stresses are thus investigated in the Cartesian system $(Px'_1x'_2x'_3)$ with a position defined by $r = |OP|$, φ, ν regarding the Cartesian system $(Ox_1x_2x_3)$ (see Figure 3.1).

As presented in Figure 3.1, the notation x'_1, x'_2, x'_3 instead of the conventional notation x_r, x_φ, x_ν, respectively, is used due to the notation which is used within the mathematical procedures in Section 4.2. (see e.g., the sum symbol $\sum_{i=1}^{3}$ in Eqs. (4.1)–(4.12)). Consequently, with regard to $x'_1 = x_r$, $x'_2 = x_\varphi$, $x'_3 = x_\nu$, the conventional notation for

- the thermal-stress induced radial displacement u_r is replaced by u'_1.
- the radial stress σ_r and the radial strain ε_r, is replaced by σ'_{11} and ε'_{11}, respectively.
- the tangential stresses σ_φ, σ_ν and the tangential strains ε_φ, ε_ν are replaced by σ'_{22}, σ'_{33} and ε'_{22}, ε'_{33}, respectively.
- the shear stresses $\sigma_{r\varphi}$, $\sigma_{r\nu}$, $\sigma_{\varphi\nu}$ and the shear strains $\varepsilon_{r\varphi}$, $\varepsilon_{r\nu}$, $\varepsilon_{\varphi\nu}$ are replaced by σ'_{12}, σ'_{13}, σ'_{23} and ε'_{12}, ε'_{13}, ε'_{23}, respectively.

Additionally, the reason of the notation ε'_{ii} and σ'_{ii} instead of ε'_i and σ'_i $(i=1,2,3)$, respectively, is as follows. The strain ε'_{ij} $(i,j=1,2,3)$ which is related to the infinitesimal spherical cap surface with the normal x'_j is realized along the axis x'_i, where $\varepsilon'_{ij} = \varepsilon'_{ji}$ [114]. The stress σ'_{ij} $(i,j=1,2,3)$ along the axis x'_i acts on the infinitesimal spherical cap surface with the normal x'_j, where $\sigma'_{ij} = \sigma'_{ji}$ [114].

3.3.2. Cauchy's Equations

The Cauchy's equations which represent geometric equations define relationships between displacements and strains of an infinitesimal part of a solid continuum. In case of the infinitesimal spherical cap (see Figure 3.1b), the radial strain ε'_{11} (see Figure 3.3a,b) along a radial direction defined by the axis x'_1 (see Figure 3.1a), the tangential strain ε'_{22} (see Figure 3.3a) along a tangential direction defined by the axis x'_2, the tangential strain ε'_{33} (see Figure 3.3b) along a tangential direction defined by the axis x'_3, and the shear strains ε'_{12} (see Figure 3.4a), ε_{13} (see Figure 3.4b) are derived as

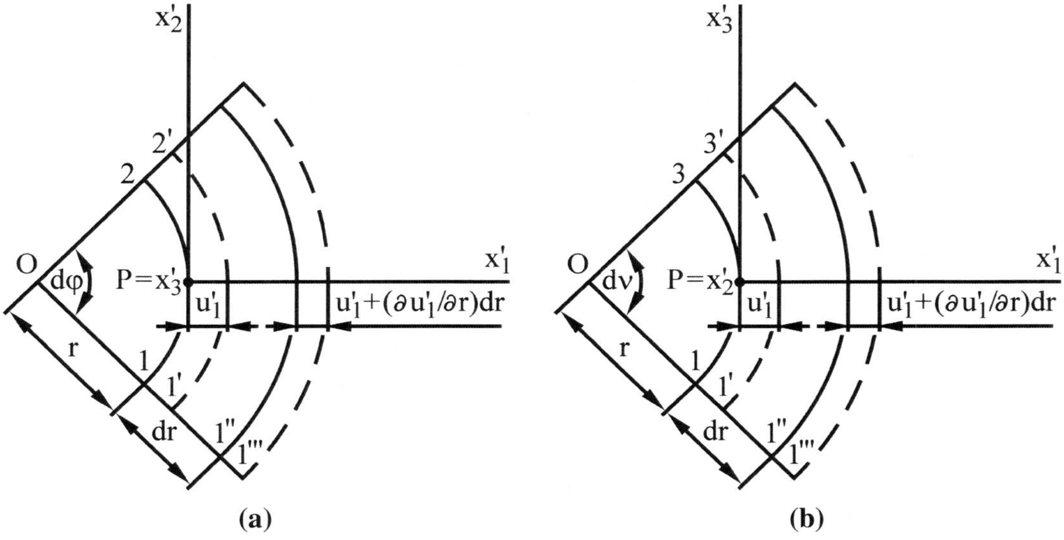

Figure 3.3. The radial displacements u'_1 and $u'_1 + (\partial u'_1/\partial r)\,dr$ of the infinitesimal spherical cap at the point P (see Figure 3.1) at the radii r and $r + dr$ in the planes (a) $x'_1x'_2$, (b) $x'_1x'_3$ (see Figure 3.1), respectively. In this case, the changes $|1'''1'| - |1''1| = (\partial u'_1/\partial r)\,dr$ and $|1'2'| - |12| = u'_1 d\varphi$, $|1'3'| - |13| = u'_1 d\nu$ are effected by the radial stress σ'_{11} and the tangential stresses σ'_{22}, σ'_{33} (see Figure 3.5), respectively. The points 1, $1'$, $1''$, $1'''$ are situated in the plane $x'_1x'_2$ as well as in the plane $x'_1x'_3$ (see Figure 3.1).

$$\varepsilon'_{11} = \frac{|1'''1'| - |1''1|}{|1''1|} = \frac{1}{dr}\left[\left(dr + \frac{\partial u'_1}{\partial r}dr\right) - dr\right] = \frac{\partial u'_1}{\partial r}, \tag{3.19}$$

$$\varepsilon'_{22} = \frac{|1'2'| - |12|}{|12|} = \frac{(u'_1 + r)\,d\varphi - r\,d\varphi}{r\,d\varphi} = \frac{u'_1}{r}, \tag{3.20}$$

$$\varepsilon'_{33} = \frac{|1'3'| - |13|}{|13|} = \frac{(u'_1 + r)\,d\nu - r\,d\nu}{r\,d\nu} = \frac{u'_1}{r}, \tag{3.21}$$

$$\varepsilon'_{12} = \tan\left[\angle\left(|12|, |12^*|\right)\right] = \frac{1}{r\,d\varphi}\left(\frac{\partial u'_1}{\partial \varphi}d\varphi\right) = \frac{1}{r}\frac{\partial u'_1}{\partial \varphi}, \tag{3.22}$$

$$\varepsilon'_{13} = \tan\left[\angle\left(|13|, |13^*|\right)\right] = \frac{1}{r\,d\nu}\left(\frac{\partial u'_1}{\partial \nu}d\nu\right) = \frac{1}{r}\frac{\partial u'_1}{\partial \nu}. \tag{3.23}$$

The Cauchy's equations (3.19)–(4.1) are related to the arbitrary point P shown in Figure 3.1.

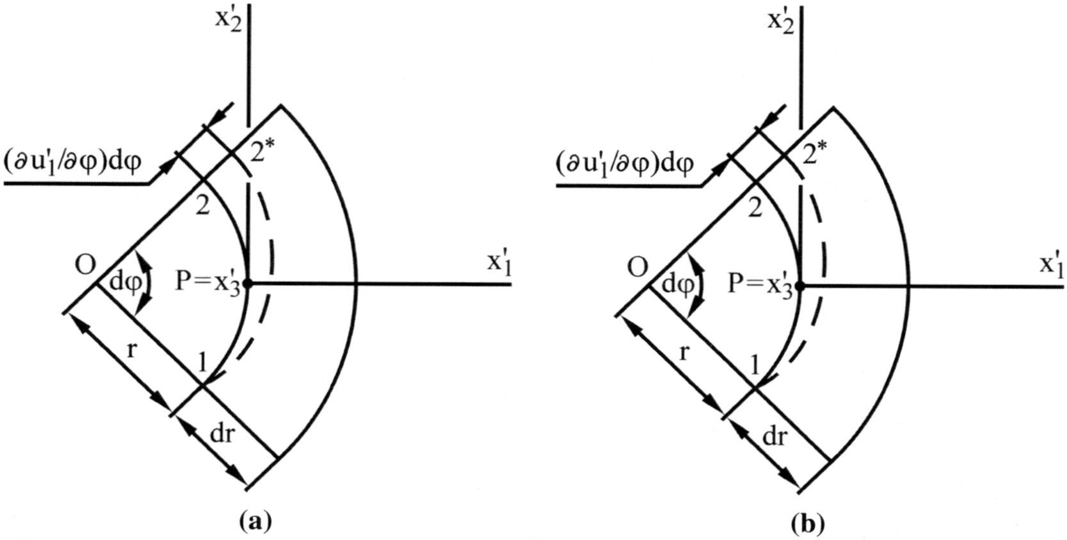

Figure 3.4. The radial displacement $u_1' = u_1'(\varphi, \nu)$ of the infinitesimal spherical cap at the point P (see Figure 3.1) in the planes (a) $x_1'x_2'$ and (b) $x_1'x_3'$ (see Figure 3.1). The changes $|22^*| = (\partial u_1'/\partial\varphi)\, d\varphi$ and $|33^*| = (\partial u_1'/\partial\nu)\, d\nu$ of the function $u_1' = u_1'(\varphi, \nu)$ which is dependent on the variables φ, ν are effected by the shear stresses σ_{12}' and σ_{13}' (see Figure 3.5), respectively.

With regard to the analysis in Section 3.2., we get the shear strain $\varepsilon_{23}' = \varepsilon_{\varphi\nu}' \propto [(\partial u_2'/\partial\nu) + (\partial u_3'/\partial\varphi)] = 0$ due to $u_2' = u_\varphi = 0$, $u_3' = u_\nu = 0$, where $u_2' = u_\varphi$ and $u_3' = u_\nu$ represent displacements of the infinitesimal spherical cap along the axes x_2' and x_3' (see Figure 3.1), respectively, i.e., along tangential directions. The analysis in Section 3.2. results in the fact that the infinitesimal spherical cap exhibits the radial displacement u_1' only, along the axis x_1' which defines a radial direction.

3.3.3. Compatibility Equations

Solid continuum mechanics considers such principle that a system which is continuous before deformation is required to exhibit this continuity after deformation as well, i.e., strains are required to be mutually compatible [1, 108]. The compatibility equations (4.2)–(4.6) for the infinitesimal spherical cap are determined in the following way.

Performing ∂Eq. (3.20) $/\partial r$ and considering Eq. (3.19), we get

$$\varepsilon_{11}' - \varepsilon_{22}' - r\frac{\partial \varepsilon_{22}'}{\partial r} = 0. \tag{3.24}$$

Performing ∂Eq. (3.19) $/\partial\varphi$, ∂Eq. (3.22) $/\partial r$, we get

$$\frac{\partial \varepsilon_{11}'}{\partial\varphi} - \varepsilon_{12}' - r\frac{\partial \varepsilon_{12}}{\partial r} = 0. \tag{3.25}$$

Performing ∂Eq. (3.19) $/\partial\nu$, ∂Eq. (4.1) $/\partial r$, we get

$$\frac{\partial \varepsilon'_{11}}{\partial \nu} - \varepsilon'_{13} - r\frac{\partial \varepsilon'_{13}}{\partial r} = 0. \tag{3.26}$$

Performing ∂Eq. (3.20) $/\partial\varphi$ and considering Eq. (3.22), we get

$$\frac{\partial \varepsilon'_{22}}{\partial \varphi} - \varepsilon'_{12} = 0. \tag{3.27}$$

Performing ∂Eq. (3.20) $/\partial\nu$ and considering Eq. (4.1), we get

$$\frac{\partial \varepsilon'_{22}}{\partial \nu} - \varepsilon'_{13} = 0. \tag{3.28}$$

3.3.4. Equilibrium Equations

Solid continuum mechanics also considers a principle of the equilibrium of forces which act within a solid continuum, strictly speaking, within on sides an infinitesimal part of the solid continuum. In case of the infinitesimal spherical cap, the equilibrium equations of the forces which act on sides of the infinitesimal spherical cap are related to the axes x'_1, x'_2, x'_3 at the arbitrary point P shown in Figure 3.1.

In case of the axis x'_1 (see Figure 3.5), we get

$$\begin{aligned}
&\left(\sigma'_{11} + \frac{\partial \sigma'_{11}}{\partial r}dr\right)(r+dr)\,d\varphi\,(r+dr)\,d\nu \\
&+\left(\sigma'_{12} + \frac{\partial \sigma'_{12}}{\partial \varphi}d\varphi\right)\cos\frac{d\varphi}{2}\,r\,d\nu\,dr \\
&+\left(\sigma'_{13} + \frac{\partial \sigma'_{13}}{\partial \nu}d\nu\right)\cos\frac{d\nu}{2}\,r\,d\varphi\,dr - \\
&\left[\sigma'_{11}\,r\,d\varphi\,r\,d\nu + \left(\sigma'_{22} + \frac{\partial \sigma'_{22}}{\partial \varphi}d\varphi\right)\sin\frac{d\varphi}{2}\,r\,d\nu\,dr\right. \\
&+\left(\sigma'_{33} + \frac{\partial \sigma'_{33}}{\partial \nu}d\nu\right)\sin\frac{d\nu}{2}\,r\,d\varphi\,dr + \sigma'_{22}\sin\frac{d\varphi}{2}\,r\,d\nu\,dr \\
&\left.+\,\sigma'_{33}\sin\frac{d\nu}{2}\,r\,d\varphi\,dr + \sigma'_{12}\cos\frac{d\varphi}{2}\,r\,d\nu\,dr + \sigma'_{13}\cos\frac{d\nu}{2}\,r\,d\varphi\,dr\right] = 0.
\end{aligned} \tag{3.29}$$

In case of the axis x'_2 (see Figure 3.5), we get

$$\begin{aligned}
&\left(\sigma'_{22} + \frac{\partial \sigma'_{22}}{\partial \varphi}d\varphi\right)\cos\frac{d\varphi}{2}\,r\,d\nu\,dr + \left(\sigma'_{21} + \frac{\partial \sigma'_{21}}{\partial r}dr\right)(r+dr)\,d\varphi\,(r+dr)\,d\nu \\
&+\left(\sigma'_{12} + \frac{\partial \sigma'_{12}}{\partial \varphi}d\varphi\right)\sin\frac{d\varphi}{2}\,r\,d\nu\,dr + \sigma'_{12}\sin\frac{d\varphi}{2}\,r\,d\nu\,dr \\
&-\left(\sigma'_{22}\cos\frac{d\varphi}{2}\,r\,d\nu\,dr + \sigma'_{21}\,r\,d\varphi\,r\,d\nu\right) = 0.
\end{aligned} \tag{3.30}$$

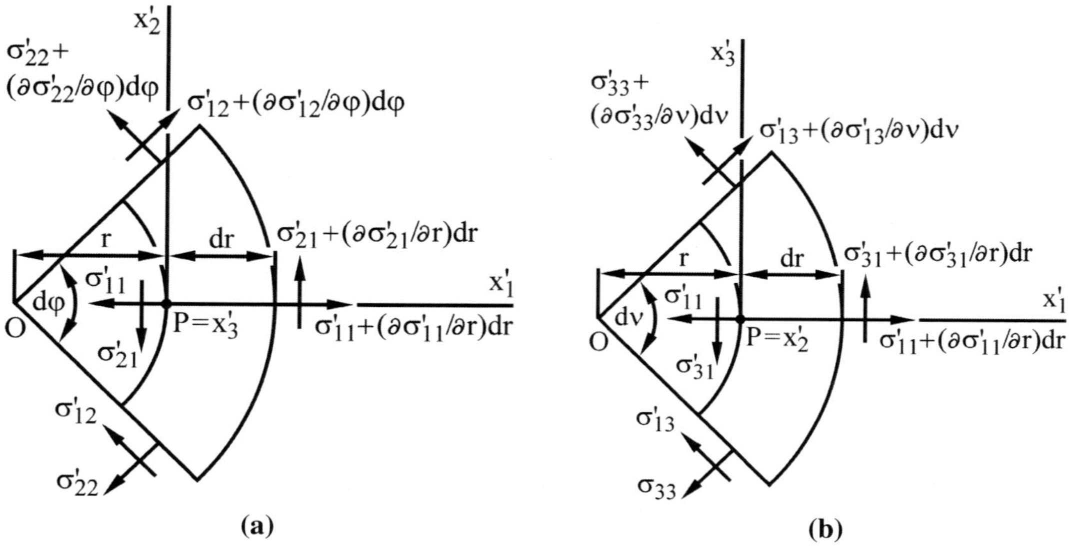

Figure 3.5. Cross-sections of the infinitesimal spherical cap in the planes (a) $x_1'x_2'$ and (b) $x_1'x_3'$. The radial stress σ_{11}', the tangential stresses σ_{22}', σ_{33}', the shear stresses $\sigma_{12}' = \sigma_{21}'$, $\sigma_{13}' = \sigma_{31}'$, along with changes of all these stresses acting on sides of the infinitesimal spherical cap at the arbitrary point P (see Figure 3.1).

In case of the axis x_3' (see Figure 3.5), we get

$$\left(\sigma_{33}' + \frac{\partial \sigma_{33}'}{\partial \nu} d\nu\right) \cos\frac{d\nu}{2} r\, d\varphi\, dr + \left(\sigma_{31}' + \frac{\partial \sigma_{31}'}{\partial r} dr\right) (r+dr)\, d\varphi\, (r+dr)\, d\nu$$
$$+ \left(\sigma_{13}' + \frac{\partial \sigma_{13}'}{\partial \nu} d\nu\right) \sin\frac{d\nu}{2} r\, d\varphi\, dr + \sigma_{13}' \sin\frac{d\nu}{2} r\, d\varphi\, dr$$
$$- \left(\sigma_{33}' \cos\frac{d\nu}{2} r\, d\varphi\, dr + \sigma_{31}'\, r\, d\varphi\, r\, d\nu\right) = 0. \tag{3.31}$$

Due to $d\varphi \approx 0$, $d\nu \approx 0$, $dr \approx 0$, we get $\sin(d\varphi/2) \approx d\varphi/2$, $\sin(d\nu/2) \approx d\nu/2$, $\cos(d\varphi/2) = \cos(d\nu/2) \approx 1$, $(dr)^2 = (d\varphi)^2 = (d\nu)^2 = 0$ [108, 115]. The equilibrium equations (4.7)–(4.9) are transformed to the forms

$$2\sigma_{11}' - \sigma_{22}' - \sigma_{33}' + r\frac{\partial \sigma_{11}'}{\partial r} + \frac{\partial \sigma_{12}'}{\partial \varphi} + \frac{\partial \sigma_{13}'}{\partial \nu} = 0, \tag{3.32}$$

$$\frac{\partial \sigma_{22}'}{\partial \varphi} + 3\sigma_{12}' + r\frac{\partial \sigma_{12}'}{\partial r} = 0, \tag{3.33}$$

$$\frac{\partial \sigma_{33}'}{\partial \nu} + 3\sigma_{13}' + r\frac{\partial \sigma_{13}'}{\partial r} = 0. \tag{3.34}$$

As analysed in Section 3.2., the infinitesimal spherical cap exhibits the thermal-stress induced radial displacement u_1' along the axis x_1' (see Figure 3.1). As analysed in Section 3.6., this radial displacement is caused by the radial stress p_1 acting at the particle-matrix and particle-envelope boundary as well as by the radial stress p_2 acting at the matrix-envelope boundary. Due to these radial stresses and radial displacement, we get $\sigma_{23}' = 0$.

The condition $\sigma'_{23} = 0$ is valid for the shear stress σ'_{23} which is determined in the Cartesian system $(Px'_1x'_2x'_3)$ (see Figure 3.1).

3.3.5. Hooke's Law

With regard to $\sigma'_{23} = 0$ (see Section 3.3.4.), the Hooke's law for an **anisotropic** component of the multi-particle-matrix and multi-particle-envelope-matrix systems is derived as [114]

$$\varepsilon'_{11} = s'_{11}\sigma'_{11} + s'_{12}\sigma'_{22} + s'_{13}\sigma'_{33} + s'_{15}\sigma'_{13} + s'_{16}\sigma'_{12}, \tag{3.35}$$

$$\varepsilon'_{22} = s'_{12}\sigma'_{11} + s'_{22}\sigma'_{22} + s'_{23}\sigma'_{33} + s'_{25}\sigma'_{13} + s'_{26}\sigma'_{12}, \tag{3.36}$$

$$\varepsilon'_{33} = s'_{13}\sigma'_{11} + s'_{23}\sigma'_{22} + s'_{33}\sigma'_{33} + s'_{35}\sigma'_{13} + s'_{36}\sigma'_{12}, \tag{3.37}$$

$$\varepsilon'_{23} = s'_{14}\sigma'_{11} + s'_{24}\sigma'_{22} + s'_{34}\sigma'_{33} + s'_{45}\sigma'_{13} + s'_{46}\sigma'_{12}, \tag{3.38}$$

$$\varepsilon'_{13} = s'_{15}\sigma'_{11} + s'_{25}\sigma'_{22} + s'_{35}\sigma'_{33} + s'_{55}\sigma'_{13} + s'_{56}\sigma'_{12}, \tag{3.39}$$

$$\varepsilon'_{12} = s'_{16}\sigma'_{11} + s'_{26}\sigma'_{22} + s'_{36}\sigma'_{33} + s'_{56}\sigma'_{13} + s'_{66}\sigma'_{12}. \tag{3.40}$$

The elastic moduli $s'_{11}, s'_{12}, \ldots, s'_{56}, s'_{66}$ in the Cartesian system with the axes x'_1, x'_2, x'_3 (see Figure 3.1) are determined by the elastic moduli $s_{11}, s_{12}, \ldots, s_{56}, s_{66}$ in the Cartesian system $(Ox_1x_2x_3)$. The determination of s'_{11}, $s'_{12}, \ldots, s'_{56}$ is as follows.

The elastic modulus s'_{ijkl} $(\equiv s'_{11}, s'_{12}, \ldots, s'_{56}, s'_{66})$ $(i,j,k,l = 1,2,3)$ in the Cartesian system with the axes x'_1, x'_2, x'_3 which is determined by s_{11}, $s_{12}, \ldots, s_{56}$, s_{66} in the Cartesian system $(Ox_1x_2x_3)$ (see Figure 3.1) has the form [114]

$$s'_{ijkl} = \sum_{r,s,t,u=1}^{3} a_{ir}\, a_{js}\, a_{kt}\, a_{lu}\, s_{rstu}, \quad i,j,k,l = 1,2,3. \tag{3.41}$$

The coefficient $a_{vw} = \cos\left[\angle\left(x'_v, x_w\right)\right]$ $(v, w = 1,2,3)$ which represents a direction cosine of an angle formed by the axes x'_v, x_w (see Figure 3.1) is derived as

$$\begin{aligned} &a_{11} = \cos\varphi\sin\nu, \quad a_{12} = \sin\varphi\sin\nu, \quad a_{13} = \cos\nu, \\ &a_{21} = -\sin\varphi, \quad a_{22} = -\cos\varphi, \quad a_{23} = 0, \\ &a_{31} = -\cos\varphi\cos\nu, \quad a_{32} = -\sin\varphi\cos\nu, \quad a_{33} = -\sin\nu. \end{aligned} \tag{3.42}$$

The transformations concerning the subscript $ij \to n$ $(i,j = 1,2,3;\ n = 1,\ldots,6)$ which are also valid for the subscript $kl \to m$ $(k,l = 1,2,3;\ m = 1,\ldots,6)$ are derived as [114]

$$\begin{aligned} &ij \equiv ji, \\ &i = j \;\; \to \;\; ij \equiv i; \\ &i \neq j \;\; \to \;\; ij = 12 \equiv 6, \;\; ij = 13 \equiv 5, \;\; ij = 23 \equiv 4. \end{aligned} \tag{3.43}$$

With regard to Eqs. (3.41)–(3.43) [115], the elastic moduli $s'_{11}, s'_{12}, \ldots, s'_{56}, s'_{66}$ determined by s_{11}, $s_{12}, \ldots, s_{56}$, s_{66} in are given by Eqs. (9.1)–(9.21).

The Hooke's law for an **isotropic** component of the multi-particle-matrix and multi-particle-envelope-matrix systems is derived as [100, 102]–[110, 116]

$$\varepsilon'_{11} = s_{11}\sigma'_{11} + s_{12}\left(\sigma'_{22} + \sigma'_{33}\right), \tag{3.44}$$

$$\varepsilon'_{22} = s_{12}\left(\sigma'_{11} + \sigma'_{33}\right) + s_{11}\sigma'_{22}, \quad (3.45)$$

$$\varepsilon'_{33} = s_{12}\left(\sigma'_{11} + \sigma'_{22}\right) + s_{11}\sigma'_{33}, \quad (3.46)$$

$$\varepsilon'_{13} = s_{44}\sigma'_{13}, \quad (3.47)$$

$$\varepsilon'_{12} = s_{44}\sigma'_{12}. \quad (3.48)$$

The elastic moduli s_{11}, s_{12}, s_{44} which represent functions of the Young's modulus E and Poisson's ratio μ have the forms [106, 113]

$$s_{11} = \frac{1}{E}, \quad s_{12} = -\frac{\mu}{E}, \quad s_{44} = \frac{2(1+\mu)}{E}, \quad (3.49)$$

As presented in [114], we get $\mu = 0.25$. In case of real materials, we get $\mu < 0.5$ [104].

The elastic moduli $s'_{11}, s'_{12}, \ldots, s'_{56}, s'_{66}$; $s_{11}, s_{12}, \ldots, s_{56}, s_{66}$; the Young' modulus E and the Poisson's ratio μ are related to the spherical particle ($q = p$), spherical envelope ($q = e$) and the cell matrix ($q = m$). Consequently, the transformations $s'_{ij} \to s'_{ijq}$, $s_{ij} \to s_{ijq}$, $E \to E_q$, $\mu \to \mu_q$ ($i, j = 1, \ldots, 6$; $q = p,e,m$) are required to be considered.

3.4. Elastic Energy

The elastic energy density w_q accumulated at the arbitrary point in the spherical particle ($q = p$), spherical envelope ($q = e$) and cell matrix ($q = m$) has the form [26, 106, 109, 110, 116]

$$w_q = \frac{1}{2}\left(\sum_{i=1}^{3} \sigma'_{iiq}\varepsilon'_{iiq} + \sum_{i,j=1;\ i\neq j}^{3} \sigma'_{ijq}\varepsilon'_{ijq}\right), \quad q = p, e, m. \quad (3.50)$$

Let the spherical coordinates (r, φ, ν) (see Figure 3.1) be considered, and then $w_q = w_q(r, \varphi, \nu)$. The elastic energy W_q which is accumulated in the volume V_q of an elastic solid continuum is derived in the form

$$W_q = \int_{V_q} w_q \, dV_q = 8 \int_0^{\pi/2} \int_0^{\pi/2} \int_{r_1}^{r_2} w_q r^2 \, dr \, d\varphi \, d\nu, \quad q = p, e, m, \quad (3.51)$$

where $dV_q = r^2 dr d\varphi d\nu$ is volume of the infinitesimal spherical cap for the angles $\varphi, \nu \in \langle 0, \pi/2 \rangle$.

In case of the multi-particle-matrix and multi-particle-envelope-matrix systems (see Section 2.2.), the integration boundaries r_1, r_2 are as follows:

1. $r_1 = 0$, $r_2 = R_1$ for the spherical particle ($q = p$) of the multi-particle-matrix and multi-particle-envelope-matrix systems,

2. $r_1 = R_1$, $r_2 = R_2$ for the spherical envelope ($q = e$),

3. $r_1 = R_1$, $r_2 = r_c$ for the cell matrix ($q = m$) of the multi-particle-matrix system,

4. $r_1 = R_2$, $r_2 = r_c$ for the cell matrix ($q = p$) of the multi-particle-envelope-matrix system.

Energy of the multi-particle-matrix and multi-particle-envelope-matrix systems which are loaded by the thermal stresses is represented by the energy W_c of the cubic cell. In this case, W_c represents thermal-stress induced elastic energy which is accumulated in the cubic cell.

In case of the multi-particle-matrix system, we get

$$W_c = W_p + W_m, \tag{3.52}$$

where W_p and W_m represent the thermal-stress induced elastic energy which is accumulated in the spherical particle and cell matrix, respectively.

In case of the multi-particle-envelope-matrix system, we get

$$W_c = W_p + W_e + W_m, \tag{3.53}$$

where W_e represent the thermal-stress induced elastic energy which is accumulated in the spherical envelope.

With regard to the analysis in Section 3.2., in case of the multi-particle-matrix system, R_1 represents a radius of the embedded spherical particle at the temperature $T \in \langle T_f, T_r \rangle$, where T_f is a final temperature of a cooling process, and the relaxation temperature T_r is determined in Section 3.5.1.1. In case of the multi-particle-envelope-matrix system, R_1, R_2 and $t = R_2 - R_1$ represents radii and thickness of the embedded spherical envelope at the temperature $T \in \langle T_f, T_r \rangle$, respectively.

As analysed in Sections 4.4., 5.7., a thermal stress-strain state in the model systems (see Figure 2.1) is determined by mutually different solutions which result in mutually different values of W_c (see Eq. (3.52), (3.53)). Due to these different values, such solution is considered to exhibit minimal value of W_c. Strictly speaking, a principle of minimal total potential energy (i.e., the tendency of an elastic solid body to exhibit minimal total potential energy) [26] is required to be considered.

As presented in [26], the total potential energy $W_t = W_d - (W_v + W_s)$ consists of the deformation energy W_d, and of the energy W_v and W_s which is induced by the volume and surface forces, $\overrightarrow{F_v}$ and $\overrightarrow{F_s}$, respectively. In case of the model systems (see Figure 2.1), we get $\overrightarrow{F_v} = \overrightarrow{F_s} = 0$, and then $W_v = W_s = 0$, $W_t = W_d$. The deformation energy W_d of the model systems is thus identical with W_c (see Eqs. (3.51)–(3.53)). Finally, in case of the model systems, the principle of minimal total potential energy is thus 'transformed' to a principle of minimal value of W_c due to $W_t = W_d = W_c$.

3.5. Reason of Thermal Stresses

3.5.1. Relaxation Temperature and Coefficients β_p, β_e, β_m

In case of the multi-particle-matrix system, the thermal stresses are a consequence of the condition $\alpha_p \neq \alpha_m$ as well as a consequence of a phase transformation which originates at the temperature T_{tq} $(q = p,m)$ at least in one component of this system, where α_q is a thermal expansion coefficient along the axis $x_1' = x_r$ (see Figure 3.1). The phase-transformation temperature T_{tq} is from the interval $\langle T_f, T_r \rangle$, where T_f is a final temperature of a cooling process, and T_r is relaxation temperature of the multi-particle-matrix system. The phase transformation at $T_{tq} \in \langle T_f, T_r \rangle$ induces the radial strain $\varepsilon'_{11tq} = \varepsilon'_{11tq}(\varphi, \nu)$ along the axis x'_1.

In case of the multi-particle-envelope-matrix system, the thermal stresses originate as a consequence of one of these conditions $\alpha_p \neq \alpha_e = \alpha_m$, $\alpha_p \neq \alpha_e \neq \alpha_m$ $\alpha_p = \alpha_e \neq \alpha_m$ as well as a consequence of the radial strain $\varepsilon'_{11tq} = \varepsilon'_{11tq}(\varphi, \nu)$ $(q = p,e,m)$ which is induced by a phase transformation at $T_{tq} \in \langle T_f, T_r \rangle$ $(q = p,e,m)$ at least in one component of this model system.

The relaxation temperature T_r of the multi-particle-matrix and multi-particle-envelope-matrix systems is analysed below.

The coefficient β_q which includes α_q and ε'_{11tq} is determined below (see Eqs. (3.54)–(3.60)). The thermal expansion coefficient α_q (along the axis x'_1) of the particle $(q = p)$, envelope $(q = p)$ and matrix $(q = m)$ is given by Eq. (3.60). The phase-transformation radial strain $\varepsilon'_{11tq} = \varepsilon'_{11tq}(\varphi, \nu)$ is induced due to the difference in dimensions of mutually transforming crystal lattices. Formulae for $\varepsilon'_{11tq} = \varepsilon'_{11tq}(\varphi, \nu)$ $(q = p,e,m)$ for anisotropic and isotropic crystal lattices are determined in Sections 3.5.2.1 and 3.5.2.2, respectively.

3.5.1.1. Relaxation Temperature

The thermal stresses originate during a cooling process. Additionally, the thermal stresses originate at the temperature $T \in \langle T_f, T_r \rangle$.

As defined in [104], the relaxation temperature T_r is such temperature below that the stress relaxation as a consequence of thermal-activated processes does not occur in a material. The relaxation temperature is defined approximately by the relationship $T_r = (0.35 - 0.4) \times T_m$ [104] and exactly by an experiment.

T_m is melting temperature of the model systems (see Section 2.2.).

The analysis of T_m for the multi-particle-matrix system is as follows. If the particles precipitate from a liquid matrix, then T_m represents a minimal value of the set $\{T_{mp}, T_{mm}\}$, where T_{mp} and T_{mm} are melting temperature of the particles and the matrix, respectively. If the particles precipitate from a solid matrix, then T_m represents melting temperature of the multi-particle-matrix system.

The analysis of T_m for the multi-particle-envelope-matrix system is as follows. If the particles and envelopes precipitate from a liquid matrix, then T_m represents a minimal value of the set $\{T_{mp}, T_{me}, T_{mm}\}$, where T_{ep} is melting temperature of the envelope. If the particles and envelopes precipitate from a solid matrix, then T_m represents melting temperature of the multi-particle-envelope-matrix system.

Let the multi-particle-matrix system be considered. If $\alpha_p = \alpha_m$, then the thermal stresses originate at the temperature T_t which represents maximal temperature of the set $\{T_{tp}, T_{tm}\}$. T_{tp} and T_{tm} thus represent temperature of a phase transformation in the spherical particle and cell matrix, where $T_{tq} \in \langle T_f, T_r \rangle$ $(q = p,m)$.

The same is also valid for multi-particle-envelope-matrix system. In this case (i.e., on the condition $\alpha_p = \alpha_e = \alpha_m$), T_t represents maximal temperature of the set $\{T_{tp}, T_{te}, T_{tm}\}$, where $T_{te} \in \langle T_f, T_r \rangle$ is temperature of a phase transformation in the spherical envelope.

3.5.1.2. Coefficients β_p, β_e, β_m

The thermal stresses in multi-component materials are a consequence of mutually different changes of dimensions of material components. These mutually different changes result in an interaction on the surface of a boundary of two material components. In case of the multi-particle-matrix system which corresponds to two-component materials (see Sections 2.1., 2.2.), this interaction is represented by the pulling or pushing of the spherical particle by the cell matrix (see Section 3.2.).

As presented in Section 3.2., the pulling or pushing at the arbitrary point P at the particle-matrix boundary is realized along a normal to an infinitesimal boundary surface at the arbitrary point P (see Figure 3.1). This fact results from the symmetry of the multi-particle-matrix and multi-particle-envelope-matrix systems. This symmetry results from the matrix infinity and from the periodical distribution of the spherical particles and spherical envelopes.

Due to the spherical shape of the particles, the particle-matrix boundary before this interaction, i.e., at the temperature T_r, is also spherical. A normal to this spherical particle-matrix boundary at the arbitrary point P is identical to the axis $x'_1 = x_r$ (see Figure 3.1). With regard to the Cartesian system $(Px'_1x'_2x'_3)$, the axis $x'_1 = x_r$ defines a radial direction.

The dimension change of a component of the multi-particle-matrix system is expressed by the coefficient β_q $(q = p,m)$ which is thus defined along the radial direction, i.e., along the axis $x'_1 = x_r$ (see Figure 3.1). Due to the spherical shape of the envelopes, the same is also valid for the multi-particle-envelope-matrix system (see Section 2.1.), and then $(q = p,e,m)$.

Let the phase transformation originate at the temperature $T_{tq} \in \langle T_f, T_r \rangle$ in the component which is related to the subscript $q = p,e,m$. Consequently, the coefficient $\beta_q = \beta_q(T)$ at the temperature $T \in \langle T_f, T_{tq} \rangle \subset \langle T_f, T_r \rangle$ has the form

$$\beta_q = \varepsilon'_{11tq} + \int_{T}^{T_{tq}} \alpha_{Iq}\, dT + \int_{T_{tq}}^{T_r} \alpha_{IIq}\, dT,$$
$$T_{tq} \in \langle T_f, T_r \rangle, \quad T \in \langle T_f, T_{tq} \rangle \subset \langle T_f, T_r \rangle, \tag{3.54}$$

where $\alpha_{Iq} = \alpha_{Iq}(T)$ and $\alpha_{IIq} = \alpha_{IIq}(T)$ (see Eq. (3.60)) represent a thermal expansion coefficient of the component at the temperature $T \leq T_{tq}$ and $T \geq T_{tq}$, respectively. If $\alpha_{Iq} \neq f(T)$ and $\alpha_{IIq} \neq f(T)$ are not functions of $T \leq T_{tq}$ and $T \geq T_{tq}$, then we get

$$\beta_q = \varepsilon'_{11tq} + \alpha_{Iq}(T_{tq} - T) + \alpha_{IIq}(T_r - T_{tq}),$$
$$T_{tq} \in \langle T_f, T_r \rangle, \quad T \in \langle T_f, T_{tq} \rangle \subset \langle T_f, T_r \rangle. \tag{3.55}$$

If $T > T_{tq}$, then the coefficient $\beta_q = \beta_q(T)$ at the temperature $T \in (T_{tq}, T_r\rangle \subset \langle T_f, T_r \rangle$ has the form

$$\beta_q = \int_{T}^{T_r} \alpha_{IIq}\, dT, \quad T_{tq} \in \langle T_f, T_r \rangle, \quad T \in (T_{tq}, T_r\rangle \subset \langle T_f, T_r \rangle. \tag{3.56}$$

If $\alpha_{IIq} \neq f(T)$ for $T \geq T_{tq}$, then we get

$$\beta_q = \alpha_{IIq}(T_r - T), \quad T \in (T_{tq}, T_r\rangle \subset \langle T_f, T_r \rangle. \tag{3.57}$$

Let the phase transformation do not originate at the temperature $T_{tq} \in \langle T_f, T_r \rangle$, i.e., $T_{tq} \notin \langle T_f, T_r \rangle$. The coefficient $\beta_q = \beta_q(T)$ at the temperature $T \in \langle T_f, T_r \rangle$ has the form

$$\beta_q = \int_T^{T_r} \alpha_q dT, \quad T_{tq} \notin \langle T_f, T_r \rangle, \quad T \in \langle T_f, T_r \rangle, \tag{3.58}$$

where $\alpha_q = \alpha_q(T)$ (see Eq. (3.60)) is a thermal expansion coefficient on the condition $T_{tq} \notin \langle T_f, T_r \rangle$. If $\alpha_q \neq f(T)$ for $T \in \langle T_f, T_r \rangle$, then we get

$$\beta_q = \alpha_q (T_r - T), \quad T_{tq} \notin \langle T_f, T_r \rangle, \quad T \in \langle T_f, T_r \rangle. \tag{3.59}$$

The thermal stresses are then a consequence of the condition $\beta_{q_1} \neq \beta_{q_2}$. In case of the multi-particle-matrix and multi-particle-envelope-matrix systems, we get $q_1, q_2 = p, m$ and $q_1, q_2 = p, e, m$, respectively, where $q_1 \neq q_2$.

As analysed in Section 3.2., the infinitesimal spherical cap exhibits a displacement along the axis $x_1' = x_r$ (see Figure 3.1), where this displacement thus represents a radial displacement. The thermal expansion coefficient along the axis $x_1' = x_r$ is determined by the following analysis.

Let α_{iq} represent a thermal expansion coefficient along the axis x_i $(i=1,2,3)$ of the Cartesian system $(Ox_1x_2x_3)$ (see Figure 3.1). The component for which one of these conditions $\alpha_{1q} \neq \alpha_{2q} = \alpha_{3q}$ or $\alpha_{1q} \neq \alpha_{2q} \neq \alpha_{3q}$ or $\alpha_{1q} = \alpha_{2q} \neq \alpha_{3q}$ is anisotropic. In case of an **anisotropic component** of the model systems (see Section 2.2.), the thermal expansion coefficient α_q along the axis x_1' is then derived as [117]

$$\alpha_q = a_{11}^2 \alpha_{1q} + a_{12}^2 \alpha_{2q} + a_{13}^2 \alpha_{3q}, \tag{3.60}$$

where the coefficient $a_{ij} = \cos[\angle(x_i', x_j)]$ $(i=1;\, i,j=1,2,3)$ is given by Eq. (3.42).

In case of an **isotropic component**, we get $\alpha_q = \alpha_{1q} = \alpha_{2q} = \alpha_{3q}$.

Due to the term α_{Iq} in Eqs. (3.54), (3.55), the terms α_{1q}, α_{2q}, α_{3q} in Eq. (3.60) are replaced by α_{1Iq}, α_{2Iq}, α_{3Iq}, respectively. Consequently, in case of α_{IIq}, the terms α_{1q}, α_{2q}, α_{3q} in Eq. (3.60) are replaced by α_{1IIq}, α_{2IIq}, α_{3IIq}, respectively.

3.5.2. Phase-transformation Induced Radial Strain

3.5.2.1. Anisotropic Crystal Lattice

The phase-transformation induced radial strain ε_{11tq}' $(q=p,e,m)$ in a component of the multi-particle-matrix and multi-particle-envelope-matrix systems is related to the *orthorhombic*, *tetragonal* and *hexagonal* crystal lattices which are anisotropic [104]. The subscripts $q=p$, $q=e$ and $q=m$ are related to a crystal lattice of the spherical particle, the spherical envelope and the cell matrix, respectively.

The subscript $n = I,II$ is used due to the following analysis for the determination of ε_{11tq}'. The subscript $n = I$ is related to the temperature $T \in \langle T_f, T_{tq} \rangle$. The subscript $n = II$ is related to the temperature $T \in \langle T_{tq}, T_r \rangle$.

The orthorhombic crystal lattice (OCL) represents a rectangular-based prism with mutually perpendicular sides. OCL is characterized by the dimensions $a_{1qn}^{(O)}$, $a_{2qn}^{(O)}$, $a_{3qn}^{(O)}$ along the axes $x_{1nq}^{(O)}$, $x_{2nq}^{(O)}$, $x_{3nq}^{(O)}$ $(n = I,II)$, respectively, where $a_{1qn}^{(O)} \neq a_{2qn}^{(O)} \neq a_{3qn}^{(O)}$ and $x_{1nq}^{(O)} \perp x_{2nq}^{(O)} \perp x_{3nq}^{(O)}$.

OCL exhibits [104]

- a simple modification which is characterized by atoms at corner points of OCL,
- a base-centered modification, which is, besides atoms at corner points of OCL, characterized by a central atom in each basis of OCL,
- a body-centered modification, which is, besides atoms at corner points of OCL, characterized by an atom at an intersection point of diagonals of OCL, i.e., at the geometrical center of OCL,
- a face-centered modification, which is, besides atoms at corner points of OCL, characterized by a central atom on each of surfaces of OCL.

The tetragonal crystal lattice (TCL) represents a square-based prism with mutually perpendicular sides. TCL is characterized by the dimensions $a_{1qn}^{(T)}$, $a_{2qn}^{(T)}$, $a_{3qn}^{(T)}$ along the axes $x_{1nq}^{(T)}$, $x_{2nq}^{(T)}$, $x_{3nq}^{(T)}$ ($n = I,II$), respectively, where $a_{1qn}^{(T)} = a_{2qn}^{(T)} \neq a_{3qn}^{(T)}$ and $x_{1nq}^{(T)} \perp x_{2nq}^{(T)} \perp x_{3nq}^{(T)}$.

The hexagonal crystal lattice (HCL) [104] represents a hexagonal-based prism with the dimension $a_{3qn}^{(H)}$ along the axis $x_{3nq}^{(H)}$, where $x_{1nq}^{(H)} \perp x_{2nq}^{(H)} \perp x_{3nq}^{(H)}$. A direction of the axis $x_{1nq}^{(H)}$ is identical with a direction of a diagonal of the basis. Bases of HCL are of a shape of the regular hexagon with the dimension $a_{1qn}^{(H)}$ along the axis $x_{1nq}^{(H)}$, and then $a_{2qn} = 2a_{1qn}\sqrt{3}/3$. HCL with atoms at corner points and with a central atom in each basis represents the most known modification.

In case of OCL, TCL and HCL, phase transformations during a cooling process are as follows:

1. the phase transformation $OCL \rightarrow OCL$, where the dimensions $a_{1qII}^{(O)}$, $a_{2qII}^{(O)}$, $a_{3qII}^{(O)}$ are then transformed to the dimensions $a_{1qI}^{(O)}$, $a_{2qI}^{(O)}$, $a_{3qI}^{(O)}$ at the temperature $T = T_{tq}$, respectively, i.e., $a_{iIIq}^{(O)} \rightarrow a_{iIq}^{(O)}$ $(i = 1,2,3)$ At least one of the conditions $a_{1qI}^{(O)} \neq a_{1qII}^{(O)}$, $a_{2qI}^{(O)} \neq a_{2qII}^{(O)}$, $a_{3qI}^{(O)} \neq a_{3qII}^{(O)}$ is valid.
2. the phase transformation $TCL \rightarrow TCL$, where $a_{iIIq}^{(T)} \rightarrow a_{iIq}^{(T)}$ $(i = 1,2,3)$ at $T = T_{tq}$. At least one of the conditions $a_{1qI}^{(T)} \neq a_{1qII}^{(T)}$, $a_{2qI}^{(T)} \neq a_{2qII}^{(T)}$, $a_{3qI}^{(T)} \neq a_{3qII}^{(T)}$ is valid.
3. the phase transformation $HCL \rightarrow HCL$, where $a_{iIIq}^{(H)} \rightarrow a_{iIq}^{(H)}$ $(i = 1,2,3)$ at $T = T_{tq}$. At least one of the conditions $a_{1qI}^{(H)} \neq a_{1qII}^{(H)}$, $a_{2qI}^{(H)} \neq a_{2qII}^{(H)}$, $a_{3qI}^{(H)} \neq a_{3qII}^{(H)}$ is valid.
4. the phase transformation $OCL \rightarrow TCL$ or $TCL \rightarrow OCL$, where $a_{iIIq}^{(O)} \rightarrow a_{iIq}^{(T)}$ or $a_{iIIq}^{(T)} \rightarrow a_{iIq}^{(O)}$ $(i = 1,2,3)$ at $T = T_{tq}$, respectively. In case of $OCL \rightarrow TCL$, at least one of the conditions $a_{1qI}^{(T)} \neq a_{1qII}^{(O)}$, $a_{2qI}^{(T)} \neq a_{2qII}^{(O)}$, $a_{3qI}^{(T)} \neq a_{3qII}^{(O)}$ is valid. In case of $TCL \rightarrow OCL$, at least one of the conditions $a_{1qI}^{(O)} \neq a_{1qII}^{(T)}$, $a_{2qI}^{(O)} \neq a_{2qII}^{(T)}$, $a_{3qI}^{(O)} \neq a_{3qII}^{(T)}$ is valid.
5. the phase transformation $OCL \rightarrow HCL$ or $HCL \rightarrow OCL$, where $a_{iIIq}^{(O)} \rightarrow a_{iIq}^{(H)}$ or $a_{iIIq}^{(H)} \rightarrow a_{iIq}^{(O)}$ $(i = 1,2,3)$ at $T = T_{tq}$, respectively. In case of $OCL \rightarrow HCL$, at least one of the

conditions $a_{1qI}^{(H)} \neq a_{1qII}^{(O)}$, $a_{2qI}^{(H)} \neq a_{2qII}^{(O)}$, $a_{3qI}^{(H)} \neq a_{3qII}^{(O)}$ is valid. In case of $HCL \to OCL$, at least one of the conditions $a_{1qI}^{(O)} \neq a_{1qII}^{(H)}$, $a_{2qI}^{(O)} \neq a_{2qII}^{(H)}$, $a_{3qI}^{(O)} \neq a_{3qII}^{(H)}$ is valid.

6. the phase transformation $TCL \to HCL$ or $HCL \to TCL$, where $a_{iIIq}^{(T)} \to a_{iIq}^{(H)}$ or $a_{iIIq}^{(H)} \to a_{iIq}^{(T)}$ ($i = 1,2,3$) at $T = T_{tq}$, respectively. In case of $TCL \to HCL$, at least one of the conditions $a_{1qI}^{(H)} \neq a_{1qII}^{(T)}$, $a_{2qI}^{(H)} \neq a_{2qII}^{(T)}$, $a_{3qI}^{(H)} \neq a_{3qII}^{(T)}$ is valid. In case of $HCL \to TCL$, at least one of the conditions $a_{1qI}^{(T)} \neq a_{1qII}^{(H)}$, $a_{2qI}^{(T)} \neq a_{2qII}^{(H)}$, $a_{3qI}^{(T)} \neq a_{3qII}^{(H)}$ is valid.

As mentioned above, the coefficient β_q is defined along the axis $x_1' = x_r$ (see Figure 3.1). Accordingly, the phase-transformation strain ε_{11tq}' which is included in β_q represents a radial strain determined along the axis $x_1' = x_r$. The determination of the strain ε_{11tq}' which is induced by a phase transformation at the temperature $T_{tq} \in \langle T_f, T_r \rangle$ results from the mathematical procedure in Section 3.1.2.1. This mathematical procedure is used for the determination of the distance r_c along the axis $x_1' = x_r$ in the cubic cell.

The determination of ε_{11tq}' in OCL, TCL, HCL considers the conditions $x_i = x_{inq}^{(O)}$, $x_i = x_{inq}^{(T)}$, $x_i = x_{inq}^{(H)}$ for $i = 1,2,3$; $n = I, II$; $q = p, e, m$;.

Orthorhombic and tetragonal crystal lattices. Figure 3.6a shows one eighth of the anisotropic crystal lattice which corresponds to OCL and TCL. The dimensions a_{iIq} and a_{iIIq}, both along the axis x_i ($i = 1,2,3$), correspond to the temperature $T \in \langle T_f, T_{tq} \rangle$ and $T \in \langle T_{tq}, T_r \rangle$, respectively. The beginning O of the Cartesian system $(Ox_1x_2x_3)$ represents a centre of the anisotropic crystal lattice. In case of OCL, we get $a_{inq} = a_{inq}^{(O)}$ for $n = I$ and/or $n = II$. In case of TCL, we get $a_{inq} = a_{inq}^{(T)}$ for $n = I$ and/or $n = II$.

The determination of ε_{11tq}' results from the determination of the distance $a_{cqn} = \left|\overline{OC_1}\right|$ and $a_{cqn} = \left|\overline{OC_2}\right|$ along the radial direction represented by the axis $x_1' = x_r$ (see Figure 3.6a) for $\nu \in \langle 0, \nu^* \rangle$ and $\nu \in \langle \nu^*, \pi/2 \rangle$, respectively. The distance $a_{cqn} = \left|\overline{OC_1}\right|$ and $a_{cqn} = \left|\overline{OC_2}\right|$ represents length of the abscissae $\overline{OC_1}$ and $\overline{OC_2}$, respectively. The points C_1 and C_2 thus represent intersections of the axis x_1' with the surface of the anisotropic crystal lattice for $\nu \in \langle 0, \nu^* \rangle$ and $\nu \in \langle \nu^*, \pi/2 \rangle$, respectively.

The angle ν_t^* (see Figure 3.6a) is derived as

$$\nu_t^* = \arctan\left(\frac{\left|\overline{39}\right|}{\left|\overline{O3}\right|}\right) = \arctan\left(\frac{1}{c_{\varphi t} a_{3qn}}\right), \quad q = p, e, m; \quad n = I, II. \tag{3.61}$$

where the coefficient c_φ along with the angle φ^* have the forms

$$c_{\varphi t} = \frac{\cos\varphi}{a_{1qn}}, \quad \varphi \in \langle 0, \varphi^* \rangle;$$
$$c_{\varphi t} = \frac{\sin\varphi}{a_{2qn}}, \quad \varphi \in \left\langle \varphi^*, \frac{\pi}{2} \right\rangle;$$
$$\varphi^* = \arctan\left(\frac{a_{2nq}}{a_{1nq}}\right), \quad q = p, e, m; \quad n = I, II. \tag{3.62}$$

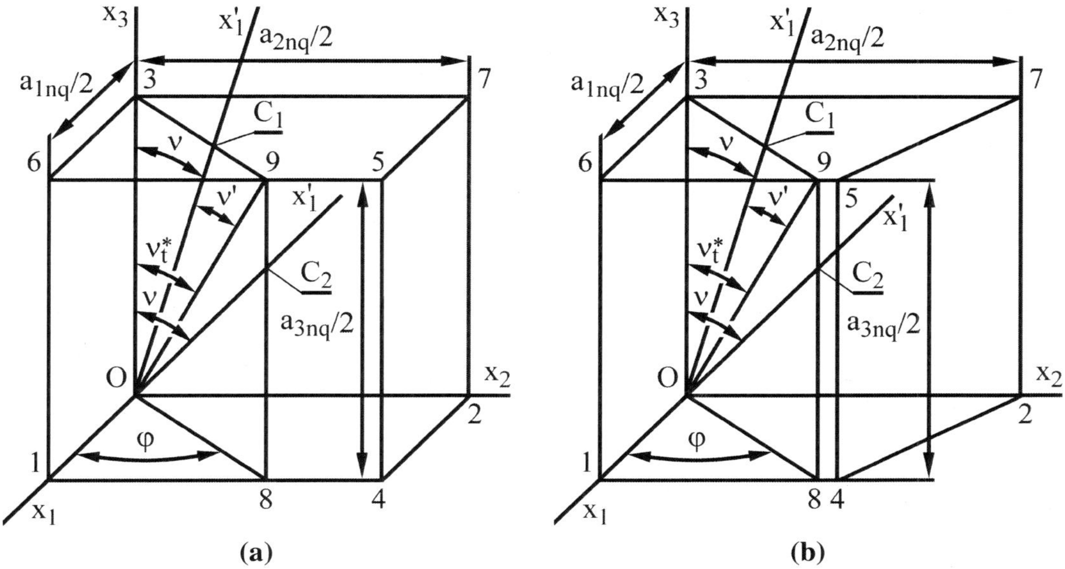

Figure 3.6. One eighth of the anisotropic crystal lattice which corresponds to (a) OCL and TCL; (b) HCL. The dimensions a_{iIq} and a_{iIIq}, both along the axis x_i ($i = 1,2,3$), correspond to the temperature $T \in \langle T_f, T_{tq} \rangle$ and $T \in \langle T_{tq}, T_r \rangle$, respectively. The beginning O of the Cartesian system $(Ox_1x_2x_3)$ represents a centre of the anisotropic crystal lattice. In case of OCL, we get $a_{inq} = a_{inq}^{(O)}$ for $n = I$ and/or $n = II$. In case of TCL, we get $a_{inq} = a_{inq}^{(T)}$ for $n = I$ and/or $n = II$. Bases of HCL are of a shape of the regular hexagon, and then $a_{2qn} = 2a_{1qn}\sqrt{3}/3$. The abscissae $\overline{OC_1}$ and $\overline{OC_2}$ represent length of the axis $x_1' = x_r$ in the anisotropic crystal lattice for $\nu \in \langle 0, \nu^* \rangle$ and $\nu \in \langle \nu^*, \pi/2 \rangle$. respectively. The axis $x_1' = x_r$ represents the radial direction (see Figure 3.1).

Interval $\nu \in \langle 0, \nu^* \rangle$. With regard to the mathematical procedure in Section 3.1.2..1, we get

$$a_{cqn} = \left|\overline{OC_1}\right| = \frac{a_{3qn} \times \sqrt{1 + \left(c_{\varphi t}\, a_{3qn}\right)^2}}{2\left(\sin\nu' + c_{\varphi t}\, a_{3qn} \cos\nu'\right)}, \quad \nu' \in \langle 0, \nu^* \rangle; \quad q = p, e, m; \quad n = I, II. \quad (3.63)$$

Interval $\nu \in \langle \nu^*, \pi/2 \rangle$. With regard to the mathematical procedure in Section 3.1.2..1, we get

$$a_{cqn} = \left|\overline{OC_2}\right| = \frac{1}{2c_{\varphi t} \sin\nu}, \quad q = p, e, m; \quad n = I, II. \quad (3.64)$$

Hexagonal crystal lattice. Figure 3.6b shows one eighth of the hexagonal crystal. The dimensions a_{iIq} and a_{iIIq}, both along the axis x_i ($i = 1,2,3$), correspond to the temperature $T \in \langle T_f, T_{tq} \rangle$ and $T \in \langle T_{tq}, T_r \rangle$, respectively. The beginning O of the Cartesian system $(Ox_1x_2x_3)$ represents a centre of the hexagonal crystal lattice. Bases of the hexagonal crystal lattice are of a shape of the regular hexagon, and then $a_{2qn} = 2a_{1qn}\sqrt{3}/3$ ($n = I, II$). The distance $a_{cqn} = \left|\overline{OC_1}\right|$ and $a_{cqn} = \left|\overline{OC_2}\right|$ for $\nu \in \langle 0, \nu^* \rangle$ and $\nu \in \langle \nu^*, \pi/2 \rangle$ is given by

Eqs. (4.39)–(4.42), respectively, where the angle φ^* and the coefficient $c_{\varphi t}$ for $\varphi \in \langle \varphi^*, \pi/2 \rangle$ have the forms

$$c_{\varphi t} = \frac{\sqrt{3}\sin\varphi + 2\cos\varphi}{3a_{1qn}}, \quad \varphi \in \left\langle \varphi^*, \frac{\pi}{2} \right\rangle; \quad \varphi^* = \frac{\pi}{6}; \quad q = p, e, m; \quad n = I, II. \tag{3.65}$$

Determination of ε'_{11tq}. The phase-transformation induced radial strain ε'_{11tq} ($q = p,e,m$) is derived as

$$\varepsilon'_{11tq} = \frac{a_{cIq} - a_{cIIq}}{a_{cIIq}}, \quad q = p, e, m, \tag{3.66}$$

where a_{cIq} and a_{cIIq} are related to the temperature $T \in \langle T_f, T_{tq} \rangle$ and $T \in \langle T_{tq}, T_r \rangle$, respectively. In case of OCL, TCL and HCL, the dimension a_{inq} ($i = 1,2,3$; $n = I,II$) in Eqs. (4.39)–(4.42) are replaced by $a_{inq}^{(O)}$, $a_{inq}^{T()}$ and $a_{inq}^{(H)}$, respectively.

3.5.2.2. Cubic Crystal Lattice

The cubic crystal lattice (CCL) exhibits [104]

- a simple modification ($K6$) which is characterized by atoms at corner points of CCL,
- a body-centered modification ($K8$), which is, besides atoms at corner points of CCL, characterized by an atom at an intersection point of diagonals of CCL, i.e., at the geometrical center of CCL,
- a face-centered modification ($K12$), which is, besides atoms at corner points of CCL, characterized by a central atom on each of surfaces of CCL.

Figure 3.7 shows the cubic crystal lattice which corresponds to $K6$, $K8$, $K12$. In case of $n = I$, the dimension a_{qI} ($\equiv a_{qI}^{(K6)}, a_{qI}^{(K8)}, a_{qI}^{(K12)}$) of the cubic crystal lattice along the axis x_{iqI} ($i = 1,2,3$; $q = p,e,m$) of the Cartesian system $(Ox_{1qI}x_{2qI}x_{3qI})$ is related to the temperature $T \in \langle T_f, T_{tq} \rangle$. In case of $n = II$, the dimension a_{qII} ($\equiv a_{qII}^{(K6)}, a_{qII}^{(K8)}, a_{qII}^{(K12)}$) along x_{iqII} ($i = 1,2,3$; $q = p,e,m$) of $(Ox_{1qII}x_{2qII}x_{3qII})$ is related to $T \in \langle T_{tq}, T_r \rangle$. Consequently, the same phase transformations of OCL, TCL, HCL which are defined in Items 1–6, Section 3.5.2..1 are also valid for $K6$, $K8$, $K12$. In contrast to the anisotropic crystal lattice (see Figure 3.6), the beginning O of the Cartesian system $(Ox_{1qn}x_{2qn}x_{3qn})$ $n = I,II$ represents a corner of the cubic crystal lattice.

The determination of the phase-transformation induced radial strain ε'_{11tq} is based on such condition that beginnings of the Cartesian systems $(Ox_{1qI}x_{2qI}x_{3qI})$, $(Ox_{1qII}x_{2qII}x_{3qII})$, $(Ox_1x_2x_3)$, $(Px'_1x'_2x'_3)$ represent the same corner point of the cubic crystal lattice. In case of $(Px'_1x'_2x'_3)$ (see Figure 3.1), this condition is valid for $r = \overline{OP} = 0$, and then $P = O$. The axis $x'_1 = x_r$ of $(Ox'_1x'_2x'_3)$ represents the radial direction regarding the spherical coordinates (r, φ, ν).

The determination of ε'_{11tq} results from the determination of the length $\left|\vec{x'_{1qn}}\right| = \left|\overline{OP_{qn}}\right|$ of the vector $\vec{x'_{1qn}} = \vec{OP}_{qn}$ along x'_1. The length $\left|\vec{x'_{1qn}}\right| = \left|\overline{OP_{qn}}\right|$ thus represents length of

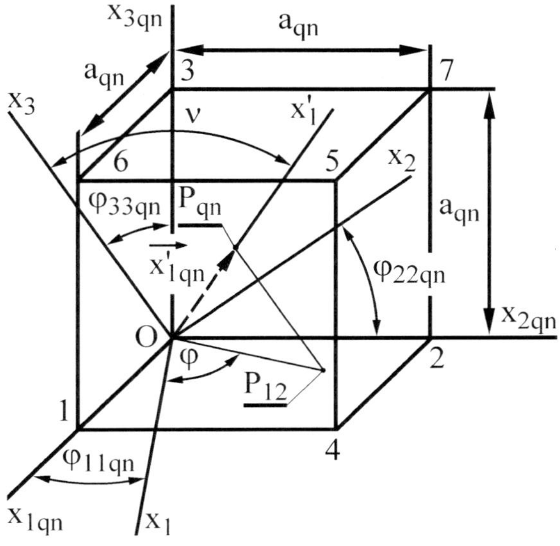

Figure 3.7. The cubic crystal lattice which corresponds to $K6$, $K8$, $K12$, where a_{qn} ($\equiv a_{qn}^{(K6)}, a_{qn}^{(K8)}, a_{qn}^{(K12)}$) represents a dimension of the cubic crystal lattice along the axis x_{iqn} ($i = 1,2,3$) at the temperature $T \in \langle T_f, T_{tq} \rangle$ for $n = I$ and $T \in \langle T_{tq}, T_r \rangle$ for $n = II$. In contrast to the anisotropic crystal lattice (see Figure 3.6), the beginning O of the Cartesian system $(Ox_{1qn}x_{2qn}x_{3qn})$ represents a corner of the cubic crystal lattice. A position of $(Ox_{1qn}x_{2qn}x_{3qn})$ $n = I,II$ regarding $(Ox_1x_2x_3)$ (see Figure 3.1) is defined by the angles $\varphi_{ijqn} = \angle(x_{iqn}, x_j)$ $(i, j = 1,2,3)$. As an example, the angles φ_{11qn}, φ_{22qn}, φ_{33qn} are shown. The points P_{qn} represents a point of intersection of x'_1 with one of the surfaces *1456*, *2754*, *3657*, and $\vec{x'_{1qn}} = \vec{OP}_{qn}$ is a vector along the axis x'_1 The axis $x'_1 = x_r$ (see Figure 3.1) represents the radial direction regarding the spherical coordinates (r, φ, ν). The point P_{12} represents a projection of P_{qn} onto the plane x_1x_2.

the axis x'_1 in the cubic crystal lattice with the dimension a_{qn} along the axis x_{iqn} ($i = 1,2,3$; $q = p,e,m$; $n = I,II$). The length $\left|\vec{x'_{1qn}}\right| = \left|\overline{OP_{qn}}\right|$ is required to be determined by the dimension a_{qn} as well as by the angles φ, ν which define an angle position of x'_1 regarding the Cartesian system $(Ox_1x_2x_3)$ (see Figure 3.1). The points P_{qn} represents a point of intersection of x'_1 with one of the surfaces *1456*, *2754*, *3657*. The point P_{12} represents a projection of P_{qn} onto the plane x_1x_2.

A position of $(Ox_{1qn}x_{2qn}x_{3qn})$ $n = I,II$ regarding the Cartesian system $(Ox_1x_2x_3)$ (see Figure 3.1) is defined by the nine coefficients a_{11qn}, $a_{12qn} \ldots, a_{32qn}$, a_{33qn}. The coefficient a_{ijqn} which is derived as

$$a_{ijqn} = \cos\varphi_{ijqn} = \cos\left[\angle(x_{iqn}, x_j)\right], \quad i, j = 1,2,3; \quad q = p,e,m; \quad n = I,II, \tag{3.67}$$

represents a direction cosine of φ_{ijqn}. The angle $\varphi_{ijqn} = \angle(x_{iqn}, x_j)$ is formed by the axes x_{iqn} and x_j ($i, j = 1,2,3$; $n = I,II$). As an example, the angles φ_{11qn}, φ_{22qn}, φ_{33qn} are shown in Figure 3.7.

Let a position of the point P_{qn} be determined by the coordinates (x_1, x_2, x_3) in $(Ox_1x_2x_3)$ as well as by the coordinates $(x_{1qn}, x_{2qn}, x_{3qn})$ in $(Ox_{1qn}x_{2qn}x_{3qn})$. The transformation formula for $x_i \rightarrow x_{iqn}$ ($i = 1,2,3$; $n = I,II$) is derived as [114]

$$x_{iqn} = \sum_{j=1}^{3} a_{ijqn} x_j, \quad i = 1,2,3; \quad q = p,e,m; \quad n = I,II, \tag{3.68}$$

where $\sum_{i=1}^{3} (x_{iqn})^2 = \sum_{i=1}^{3} x_i^2$, $\sum_{i=1}^{3} a_{ijqn} a_{ikqn} = \delta_{jk}$ ($j,k = 1,2,3$), and δ_{jk} is the Kronecker's symbol, i.e., $\delta_{jk} = 0$ and $\delta_{jk} = 1$ for $j \neq k$ and $j = k$, respectively.

Let $\vec{e_1}'$ represent a unit vector along the axis x_1'. The unit vector $\vec{e_1}'$ derived in $(Ox_1x_2x_3)$ has the form

$$\vec{e_1}' = \sum_{i=1}^{3} a_{1i} \vec{e_i}, \tag{3.69}$$

where the coefficient a_{1i} is given by Eq. (3.42), and $\vec{e_i}$ is a unit vector along the axis x_i ($i = 1,2,3$). The unit vector $\vec{e_1}'$ derived in $(Ox_{1qn}x_{2qn}x_{3qn})$ has the form

$$\vec{e_1}' = \sum_{i=1}^{3} a_{iqn}' \vec{e_{iqn}}, \quad q = p,e,m; \quad n = I,II, \tag{3.70}$$

where e_{iqn} is a unit vector along the axis x_{iqn} ($i = 1,2,3$; $n = I,II$). The coefficient a_{iqn}' which is derived as

$$a_{iqn}' = \cos\varphi_{ijgn}' = \sum_{j=1}^{3} a_{1j} a_{ijqn}, \quad i = 1,2,3; \quad q = p,e,m; \quad n = I,II, \tag{3.71}$$

represents a direction cosine of the angle φ_{ign}', and a_{1i} ($i = 1,2,3$) is given by Eq. (3.42). The angle $\varphi_{ign}' = \angle(x_{iqn}, x_1')$ is formed by the axes x_{iqn} and x_1'.

Let the intersection point P_{qn} with the coordinates $\left(x_{1qn}', x_{2qn}', x_{3qn}'\right)$ represent a point of the surface *1456*. In this case, the length $\left|\vec{x_{1qn}'}\right| = \left|\overline{OP_{qn}}\right|$ of the vector $\vec{x_{1qn}'} = \vec{OP}_{qn}$ along x_1' has the form

$$\left|\vec{x_{1qn}'}\right| = \frac{a_{qn}}{a_{1qn}'}, \quad q = p,e,m; \quad n = I,II. \tag{3.72}$$

In case of the surface *1456*, the coordinates $\left(x_{1qn}', x_{2qn}', x_{3qn}'\right)$ are as follows $x_{1qn}' = a_{qn}$, $x_{2qn}' = a_{qn} a_{2qn}' / a_{1qn}' \leq a_{qn}$, $x_{3qn}' = a_{qn} a_{3qn}' / a_{1qn}' \leq a_{qn}$. The surface *1456* with the normal x_{1qn} is thus characterized by each of the conditions

$$\frac{a_{2qn}'}{a_{1qn}'} \leq 1 \quad \text{and} \quad \frac{a_{3qn}'}{a_{1qn}'} \leq 1, \quad q = p,e,m; \quad n = I,II. \tag{3.73}$$

Let the intersection point P_{qn} represent a point of the surface *2754*. In this case, we get

$$\left|\vec{x_{1qn}'}\right| = \frac{a_{qn}}{a_{2qn}'}, \quad q = p,e,m; \quad n = I,II, \tag{3.74}$$

where $x'_{1qn} = a_{qn} a'_{1qn}/a'_{2qn} \leq a_{qn}$, $x'_{2qn} = a_{qn}$, $x'_{3qn} = a_{qn} a'_{3qn}/a'_{2qn} \leq a_{qn}$. The surface *2754* with the normal x_{2qn} is thus characterized by each of the conditions

$$\frac{a'_{1qn}}{a'_{2qn}} \leq 1 \quad \text{and} \quad \frac{a'_{3qn}}{a'_{2qn}} \leq 1, \quad q = p,e,m; \quad n = I,II. \tag{3.75}$$

Let the intersection point P_{qn} represent a point of the surface *3657*. In this case, we get

$$\left|\vec{x'_{1qn}}\right| = \frac{a_{qn}}{a'_{3qn}}, \quad q = p,e,m; \quad n = I,II, \tag{3.76}$$

where $x'_{1qn} = a_{qn} a'_{1qn}/a'_{3qn} \leq a_{qn}$, $x'_{2qn} = a_{qn} a'_{2qn}/a'_{3qn} \leq a_{qn}$, $x'_{3qn} = a_{qn}$. The surface *3657* with the normal x_{3qn} is thus characterized by each of the conditions

$$\frac{a'_{1qn}}{a'_{3qn}} \leq 1 \quad \text{and} \quad \frac{a'_{2qn}}{a'_{3qn}} \leq 1, \quad q = p,e,m; \quad n = I,II. \tag{3.77}$$

Finally, a surface with the normal x_{iqn} is thus characterized by each of the conditions

$$\frac{a'_{jqn}}{a'_{iqn}} \leq 1 \quad \text{and} \quad \frac{a'_{kqn}}{a'_{iqn}} \leq 1,$$
$$i,j,k = 1,2,3; \quad i \neq j \neq k; \quad q = p,e,m; \quad n = I,II. \tag{3.78}$$

The phase-transformation induced radial strain ε'_{11tq} $(q = p,e,m)$ is derived as

$$\varepsilon'_{11tq} = \frac{\left|\vec{x'_{1qI}}\right| - \left|\vec{x'_{1qII}}\right|}{\left|\vec{x'_{1qII}}\right|}, \quad q = p,e,m, \tag{3.79}$$

where $\left|\vec{x'_{1qI}}\right|$ and $\left|\vec{x'_{1qII}}\right|$ are related to the temperature $T \in \langle T_f, T_{tq}\rangle$ and $T \in \langle T_{tq}, T_r\rangle$, respectively. In case of $K6$, $K8$ and $K12$, the dimension a_{nq} in Eqs. (3.72), (3.74), (4.57) is replaced by $a_{nq}^{(K6)}$, $a_{nq}^{(K8)}$ and $a_{nq}^{(K12)}$ $(q = p,e,m;\ n = I,II)$, respectively.

3.6. Formulae for Determination of the Radial Stresses p_1, p_2

As presented in Section 3.5.1., the thermal stresses originate at the temperature $T \in \langle T_f, T_r\rangle$, where T_f is final temperature of a cooling process. The relaxation temperature T_r is analysed in Section 3.5.1..

An analysis of the formulae for determination of the radial stresses p_1, p_2 results from the analysis in Section 3.2..

If $T \in \langle T_f, T_r\rangle$, then R_{1p}, R_{iq} $(i = 1,2;\ q = e,m)$ which are considered in Section 3.2. have the forms

$$R_{1p} = R_{1T_r}(1-\beta_p), \quad R_{iq} = R_{iT_r}(1-\beta_q), \quad i = 1,2; \quad q = e,m, \tag{3.80}$$

where R_{1T_r} and t_{T_r} are a radius of the spherical particle and thickness of the spherical envelope both at T_r, respectively, and $R_{2T_r} = R_{1T_r} + t_{T_r}$.

Let the multi-particle-matrix system at $T \in \langle T_f, T_r \rangle$ be then considered. The thermal-stress induced radial displacement $\left(u'_{1q}\right)_{r=R_{1q}}$ is related to the change of $R_{1q} \to R_1$ ($q = p,m$), where $\left(u'_{1q}\right)_{r=R_{1q}} = R_1 - R_{1q}$. Strictly speaking, the infinitesimal spherical cap is shifted along the axis x'_1 from the position $r = R_{1q}$ to the position $r = R_1$. Accordingly, the Cauchy's equations (3.20), (3.21) are thus related to the position $r = R_{1q}$ (see Eq. 3.80). With regard to Eqs. (3.20), (3.80), the difference $(u'_{1m})_{r=R_{1m}} - \left(u'_{1p}\right)_{r=R_{1p}}$ results in the formula [2]

$$(1-\beta_m)\left(\varepsilon'_{22m}\right)_{r=R_{1m}} - (1-\beta_p)\left(\varepsilon'_{22p}\right)_{r=R_{1p}} = \beta_m - \beta_p, \tag{3.81}$$

which is considered for the determination of the radial stress $p_1 = p_1(\varphi, \nu)$ acting at the particle-matrix boundary (i.e., for $r = R_1$) along the axis x'_1 (see Figure 3.1a).

The same analysis is also valid for the multi-particle-envelope-matrix system. In case of $\beta_p \neq \beta_e$, the formula

$$(1-\beta_e)\left(\varepsilon'_{22e}\right)_{r=R_{1e}} - (1-\beta_p)\left(\varepsilon'_{22p}\right)_{r=R_{1p}} = \beta_e - \beta_p \tag{3.82}$$

is considered for the determination of the radial stress $p_1 = p_1(\varphi, \nu)$ which acts at the particle-envelope boundary.

In case of $\beta_m \neq \beta_e$, the formula

$$(1-\beta_m)\left(\varepsilon'_{22m}\right)_{r=R_{2m}} - (1-\beta_e)\left(\varepsilon'_{22e}\right)_{r=R_{2e}} = \beta_m - \beta_e \tag{3.83}$$

is considered for the determination of the radial stress $p_2 = p_2(\varphi, \nu)$ which acts at the matrix-envelope boundary.

Let the multi-particle-matrix system be considered. Let the phase transformation originate at the temperature $T_{tq} \in \langle T_f, T_r \rangle$ ($q = p,m$) at least in one of components of this system, i.e., $q = p$ and/or $q = m$. If $\alpha_{pn} \neq \alpha_{mn}$ ($n = I,II$) and $\varepsilon'_{11tp} - \varepsilon'_{11tm} \neq 0$ (see Eqs. (3.54), (3.55)), then the thermal stresses originate due to the phase transformation at $T_{tq} \in \langle T_f, T_r \rangle$. The same is also valid for the multi-particle-envelope-matrix system.

Finally, with regard to the conditions $\beta_p = \beta_m$ and $\beta_p = \beta_e = \beta_m$, the thermal stresses in the multi-particle-matrix and multi-particle-envelope-matrix systems do not occur, respectively.

3.7. Temperature Dependence of Thermal Stresses

Temperature dependences of the thermal stresses are represented by temperature dependences of the radial stresses $p_1 = p_1(\varphi, \nu, T)$, $p_2 = p_2(\varphi, \nu, T)$.

The radial stresses $p_1 = p_1(\varphi, \nu, T)$, $p_2 = p_2(\varphi, \nu, T)$ are functions of $\beta_q = \beta_q(\varphi, \nu, T)$ ($q = p,e,m$) (see Eqs. (3.54)–(3.60)) and of the elastic modulus $s'_{ijq} = s'_{jiq} = s'_{jiq}(\varphi, \nu, T)$

[2] The work [107] presents the determination of the thermal stresses in a one-particle-matrix system with isotropic components. As analysed in [107], the Cauchy's equations (3.20), (3.21) along with Eq. (3.80) are related to $r = R_1$. Due to this simplification $R_{1p} = R_{1m} = R_1$ as well as due to this simplification $R_1 = R_{1T_r}$, the difference $(u'_{1m})_{r=R_1} - \left(u'_{1p}\right)_{r=R_1}$ in [107] results in the simplified formula $(\varepsilon'_{22m})_{r=R_1} - \left(\varepsilon'_{22p}\right)_{r=R_1} = \beta_m - \beta_p$.

$(i, j = 1, \ldots, 6)$ (see Eqs. (3.41), (3.43)). The elastic modulus s'_{ijq} is included in $\varepsilon'_{22q} = \varepsilon_{22q}(\varphi, \nu, T)$. The coefficient β_q and the tangential strain ε_{22q} are included in the formulae (4.63)–(4.65). As presented in Section 3.6., these formulae are considered for the determination of p_1, p_2.

Let the thermal expansion coefficients α_{Iq}, α_{IIq}, α_q which are included in Eqs. (3.54), (3.55) be functions of T. In this case, a temperature dependence of the coefficient $\beta_q = \beta_q(T)$ results from the functions $\alpha_{Iq} = \alpha_{Iq}(T)$, $\alpha_{IIq} = \alpha_{IIq}(T)$, $\alpha_q = \alpha_q(T)$ as well as from the difference $\Delta T = T - T_r$. If $\alpha_{Iq} \neq f(T)$, $\alpha_{IIq} \neq f(T)$, $\alpha_q \neq f(T)$, then a temperature dependence of the coefficient $\beta_q = \beta_q(T)$ results from $\Delta T = T - T_r$.

Chapter 4

Thermal Stresses in Model Systems with Anisotropic Components

4.1. Introduction

Chapter 4 deals with the determination of the thermal stresses in the multi-particle-matrix and multi-particle-envelope-matrix systems (see Section 2.2.) both with anisotropic components. This determination is as follows.

As presented in Section 4.2., the tangential stresses σ'_{22}, σ'_{33} and the shear stresses σ'_{13}, σ'_{12} (see Section 3.3.1.) are determined as functions of the radial stress σ'_{11}.

Consequently, different mathematical procedures are used to determine different partial differential equations for σ'_{11} (see Section 4.3.). These different partial differential equations thus result in different solutions for σ'_{11}. As analysed in Section 4.4., these different solutions for σ'_{11} thus result in the different sets $S_q^{(1)},\ldots,S_q^{(s_q)}$, where $S_q^{(i)} = \left\{\sigma'^{(i)}_{11q}, \sigma'^{(i)}_{22q}, \sigma'^{(i)}_{33q}, \sigma'^{(i)}_{12q}, \sigma'^{(i)}_{13q}\right\}$ for $i = 1,\ldots,s_q$. The subscript $q = p,m$ and $q = p,e,m$ is considered for the multi-particle-matrix system and multi-particle-envelope-matrix system, respectively.

In case of the multi-particle-matrix system, the set $S_p^{(i)}$ $(i = 1,\ldots,s_p)$ represents one of s_p solutions which fulfil boundary conditions for the thermal stresses in the spherical particle, where $s_p \geq 1$. Similarly, the set $S_m^{(j)}$ $(j = 1,\ldots,s_{m_1})$ represents one of s_{m_1} solutions which fulfil boundary conditions for the thermal stresses in the cell matrix of the multi-particle-matrix system, where $s_{m_1} \geq 1$. Accordingly, $s_p \times s_{m_1}$ represents a number of solutions which describe a stress-strain state in the multi-particle-matrix system.

If $s_p \times s_{m_1} > 1$, then the stress-strain state induced by the thermal stresses is described by such set $\left\{S_p^{(i)}, S_m^{(j)}\right\}$ $(i = 1,\ldots,s_p;\ j = 1,\ldots,s_{m_1})$ which results in minimal elastic energy of the multi-particle-matrix system (see Eqs. (4.28)–(4.30)). Strictly speaking, due to $s_p \times s_{m_1} > 1$, a principle of minimal total potential energy of an elastic solid body (see Section 3.4.) [26] is required to be considered (see Eq. (3.52)).

The same analysis is also considered for the multi-particle-envelope-matrix system, where $S_e^{(k)}$ $(k = 1,\ldots,s_e)$ represents one of s_e solutions which fulfil boundary conditions for the thermal stresses in the spherical envelope, and $s_e \geq 1$. Similarly, if $s_p \times s_e \times s_{m_2} >$

1, then the stress-strain state induced by the thermal stresses is described by such set $\left\{S_p^{(i)}, S_e^{(j)}, S_m^{(k)}\right\}$ ($i = 1, \ldots, s_p$; $j = 1, \ldots, s_e$; $k = 1, \ldots, s_{m_2}$) which results in minimal elastic energy of the multi-particle-envelope-matrix system (see Eqs. (4.28), (4.29), (4.31)), where s_{m_2} is a number of solutions which fulfil boundary conditions for the thermal stresses in the cell matrix of the multi-particle-envelope-matrix system, and $s_{m_2} \geq 1$. Due to $s_p \times s_e \times s_{m_2} > 1$, this energy principle [26] is also required to be considered (see Eq. (3.53)).

The analysis mentioned above is also presented in detail in Section 4.4.. This detailed analysis considers the characteristic equations (4.36), (4.41), (4.45), (4.49).

As presented in Section 3.4., deformation energy of the model system in Figure 2.1 is represented by thermal-stress induced elastic energy of the cubic cell, W_c (see Eqs. (3.52), (3.53)). The deformation energy W_c is thus required to be determined for each of the $s_p \times s_{m_1}$ or $s_p \times s_e \times s_{m_2}$ solutions. This determination can be performed numerically for a real two- or three-component material defined in Section 2.1..

The mathematical procedures in Sections 4.2.–4.4. results in the mathematical solutions presented in Section 3.3. Integration constants which are included in these mathematical solutions for the thermal stresses in the spherical particle, spherical envelope and cell matrix are determined by the boundary conditions defined in Section 4.6., respectively. The thermal stresses in the cell matrix are determined by the mandatory boundary conditions (see Section 4.6.3.1) and also by the additionally boundary conditions (see Section 4.6.3.2).

Final formulae for the thermal stresses in the multi-particle-matrix and multi-particle-envelope-matrix systems, both with anisotropic components, are presented in Sections 4.7. and 4.8., respectively.

The coefficients c_i–$c_{6525+6j}$ ($i = 1-3$; $j = 0-269$;) in Sections 4.2., 4.3. (see Eqs. (4.1)–(4.49)) are presented in Appendix (see Chapter 9).

4.2. Determination of Tangential and Shear Stresses

With regard to $\varepsilon'_{22} = \varepsilon'_{33}$, $\varepsilon'_{23} = 0$, $\sigma'_{23} = 0$ (see Eqs. (3.20), (3.21), (4.14)–(4.16)), the shear stresses σ'_{13}, σ'_{12} have the form

$$\sigma'_{13} = \sum_{i=1}^{3} c_i \sigma'_{ii}, \tag{4.1}$$

$$\sigma'_{12} = \sum_{i=1}^{3} c_{3+i} \sigma'_{ii}. \tag{4.2}$$

The conditions $\varepsilon'_{22} = \varepsilon'_{33}$, $\varepsilon'_{23} = 0$, $\sigma'_{23} = 0$ are valid for the tangential strains ε'_{22}, ε'_{33}, the shear strain ε'_{23} and for the shear stress σ'_{23} which are determined in the Cartesian system $(Px'_1x'_2x'_3)$ (see Figure 3.1).

Consequently, using Eqs. (4.1), (4.2), the Hooke's law (see Eqs. (4.13)–(4.18)) is transformed to the forms

$$\varepsilon'_{11} = \sum_{i=1}^{3} c_{6+i} \sigma'_{ii}, \tag{4.3}$$

$$\varepsilon'_{22} = \varepsilon'_{33} = \sum_{i=1}^{3} c_{9+i} \sigma'_{ii}, \tag{4.4}$$

$$\varepsilon'_{13} = \sum_{i=1}^{3} c_{12+i}\sigma'_{ii}, \tag{4.5}$$

$$\varepsilon'_{12} = \sum_{i=1}^{3} c_{15+i}\sigma'_{ii}. \tag{4.6}$$

Substituting $\partial\sigma'_{13}/\partial\eta$, $\partial\sigma'_{12}/\partial\eta$ ($\eta = \varphi, \nu$) (see Eqs. (4.1), (4.2)) into Eqs. (3.25)–(3.28), (3.33), (3.34), we get

$$\frac{\partial\sigma'_{11}}{\partial\varphi} = \sum_{i=1}^{3} c_{18+i}\sigma'_{ii} + c_{21+i} r \frac{\partial\sigma'_{ii}}{\partial r}, \tag{4.7}$$

$$\frac{\partial\sigma'_{22}}{\partial\varphi} = -\sum_{i=1}^{3} c_{3+i}\left(3\sigma'_{ii} + r\frac{\partial\sigma'_{ii}}{\partial r}\right), \tag{4.8}$$

$$\frac{\partial\sigma'_{33}}{\partial\varphi} = \sum_{i=1}^{3} c_{24+i}\sigma'_{ii} + c_{27+i} r \frac{\partial\sigma'_{ii}}{\partial r}, \tag{4.9}$$

$$\frac{\partial\sigma'_{11}}{\partial\nu} = \sum_{i=1}^{3} c_{30+i}\sigma'_{ii} + c_{33+i} r \frac{\partial\sigma'_{ii}}{\partial r}, \tag{4.10}$$

$$\frac{\partial\sigma'_{22}}{\partial\nu} = \sum_{i=1}^{3} c_{36+i}\sigma'_{ii} + c_{39+i} r \frac{\partial\sigma'_{ii}}{\partial r}, \tag{4.11}$$

$$\frac{\partial\sigma'_{33}}{\partial\nu} = -\sum_{i=1}^{3} c_{i}\left(3\sigma'_{ii} + r\frac{\partial\sigma'_{ii}}{\partial r}\right). \tag{4.12}$$

The tangential stresses σ'_{22}, σ'_{33} and then the shear stresses σ'_{12}, σ'_{13} (see Eqs. (4.1), (4.2)) are derived as a function of σ'_{11}, $r\left(\partial\sigma'_{11}/\partial r\right)$, $r^2\left(\partial^2\sigma'_{11}/\partial r^2\right)$ by the following mathematical procedure.

Performing ∂Eq. (3.25)$/\partial\nu$, ∂Eq. (3.26)$/\partial\varphi$, ∂Eq. (3.27)$/\partial\nu$, ∂Eq. (3.28)$/\partial\varphi$, the compatibility equations (3.25)–(3.28) are derived as

$$\frac{\partial}{\partial\varphi}\left(\frac{\partial\varepsilon'_{11}}{\partial\nu}\right) - \frac{\partial\varepsilon'_{12}}{\partial\nu} - r\frac{\partial}{\partial r}\left(\frac{\partial\varepsilon'_{12}}{\partial\nu}\right) = 0, \tag{4.13}$$

$$\frac{\partial}{\partial\nu}\left(\frac{\partial\varepsilon'_{11}}{\partial\varphi}\right) - \frac{\partial\varepsilon'_{13}}{\partial\varphi} - r\frac{\partial}{\partial r}\left(\frac{\partial\varepsilon'_{13}}{\partial\varphi}\right) = 0, \tag{4.14}$$

$$\frac{\partial}{\partial\varphi}\left(\frac{\partial\varepsilon'_{22}}{\partial\nu}\right) - \frac{\partial\varepsilon'_{12}}{\partial\nu} = 0, \tag{4.15}$$

$$\frac{\partial}{\partial\nu}\left(\frac{\partial\varepsilon'_{22}}{\partial\varphi}\right) - \frac{\partial\varepsilon'_{13}}{\partial\varphi} = 0. \tag{4.16}$$

where the conditions the conditions $\partial^2/\partial\varphi\partial\nu = \partial^2/\partial\nu\partial\varphi$, $\partial^2/\partial r\partial\eta = \partial^2/\partial\eta\partial r$ ($\eta = \varphi, \nu$) are considered.

Substituting Eqs. (4.3)–(4.6) into Eqs. (4.13)–(4.16), we get

$$\sum_{i=1}^{3} c_{6+i}\frac{\partial}{\partial\nu}\left(\frac{\partial\sigma'_{ii}}{\partial\varphi}\right) - c_{12+i}\left[\frac{\partial\sigma'_{ii}}{\partial\varphi} + r\frac{\partial}{\partial r}\left(\frac{\partial\sigma'_{ii}}{\partial\varphi}\right)\right] = 0, \tag{4.17}$$

$$\sum_{i=1}^{3} c_{6+i} \frac{\partial}{\partial \varphi}\left(\frac{\partial \sigma'_{ii}}{\partial \nu}\right) - c_{15+i}\left[\frac{\partial \sigma'_{ii}}{\partial \nu} + r\frac{\partial}{\partial r}\left(\frac{\partial \sigma'_{ii}}{\partial \nu}\right)\right] = 0, \tag{4.18}$$

$$\sum_{i=1}^{3} c_{9+i} \frac{\partial}{\partial \nu}\left(\frac{\partial \sigma'_{ii}}{\partial \varphi}\right) - c_{12+i}\frac{\partial \sigma'_{ii}}{\partial \varphi} = 0. \tag{4.19}$$

$$\sum_{i=1}^{3} c_{9+i} \frac{\partial}{\partial \varphi}\left(\frac{\partial \sigma'_{ii}}{\partial \nu}\right) - c_{15+i}\frac{\partial \sigma'_{ii}}{\partial \nu} = 0, \tag{4.20}$$

Substituting Eqs. (4.1), (4.2) and then Eqs. (4.7)–(4.12) into Eq. (3.32); and substituting Eqs. (4.3), (4.4) into Eq. (3.6), the compatibility equation (3.6) and equilibrium equation (3.32) are transformed to the forms

$$\sum_{i=1}^{3} \left(c_{9+i} - c_{6+i}\right)\sigma'_{ii} + c_{9+i} r \frac{\partial \sigma'_{ii}}{\partial r} = 0, \tag{4.21}$$

$$\sum_{i=1}^{3} c_{42+i}\sigma'_{ii} + c_{45+i} r \frac{\partial \sigma'_{ii}}{\partial r} = 0. \tag{4.22}$$

Substituting $\partial\sigma'_{ii}/\partial\nu$ (see Eqs. (4.10)–(4.12)) and then $\partial\sigma'_{ii}/\partial\varphi$ $(i = 1-3)$ (see Eqs. (4.7)–(4.9)) into Eqs. (4.17), (4.19); and substituting $\partial\sigma'_{ii}/\partial\varphi$ (see Eqs. (4.7)–(4.9)) and then $\partial\sigma'_{ii}/\partial\nu$ $(i = 1-3)$ (see Eqs. (4.10)–(4.12)) into Eqs. (4.18), (4.20), the transformed compatibility equations (4.17)–(4.20) are derived as

$$\sum_{i=1}^{3} c_{48+i+9j}\sigma'_{ii} + c_{51+i+9j} r \frac{\partial \sigma'_{ii}}{\partial r} + c_{54+i+9j} r^2 \frac{\partial^2 \sigma'_{ii}}{\partial r^2} = 0, \quad j = 0-3. \tag{4.23}$$

The tangential stresses σ'_{22}, σ'_{33} along with the terms $r\left(\partial\sigma'_{ii}/\partial r\right)$, $r^2\left(\partial^2\sigma'_{ii}/\partial r^2\right)$ $(i=2,3)$, which represent functions of the radial stress σ'_{11} and of the terms $r\left(\partial\sigma'_{11}/\partial r\right)$, $r^2\left(\partial^2\sigma'_{11}/\partial r^2\right)$, are determined by Eqs. (4.21)–(4.23) in the forms

$$\sigma'_{2+i2+i} = \left(\delta_{1i} - \delta_{0i}\right)\left(c_{135+i}\sigma'_{11} + c_{137+i} r \frac{\partial \sigma'_{11}}{\partial r} + c_{139+i} r^2 \frac{\partial^2 \sigma'_{11}}{\partial r^2}\right), \quad i = 0, 1, \tag{4.24}$$

$$r\frac{\partial \sigma'_{2+i2+i}}{\partial r} = c_{141+i}\sigma'_{11} + c_{143+i} r \frac{\partial \sigma'_{11}}{\partial r} + c_{145+i} r^2 \frac{\partial^2 \sigma'_{11}}{\partial r^2}, \quad i = 0, 1, \tag{4.25}$$

$$r^2\frac{\partial^2 \sigma'_{2+i2+i}}{\partial r^2} = \left(\delta_{0i} - \delta_{1i}\right)\left(c_{147+i}\sigma'_{11} + c_{149+i} r \frac{\partial \sigma'_{11}}{\partial r} + c_{151+i} r^2 \frac{\partial^2 \sigma'_{11}}{\partial r^2}\right), \ i = 0, 1, \tag{4.26}$$

where δ_{0i}, δ_{1i} $(i=0,1)$ are the Kronecker's symbols.

Consequently, with regard to regarding Eqs. (4.1), (4.2), (4.24), the shear stresses σ'_{12}, σ'_{13} are derived as

$$\sigma'_{12+i} = c_{153+i}\sigma'_{11} + c_{155+i} r \frac{\partial \sigma'_{11}}{\partial r} + c_{157+i} r^2 \frac{\partial^2 \sigma'_{11}}{\partial r^2}, \quad i = 0, 1. \tag{4.27}$$

Substituting Eq. (4.26) into $r\left[\partial \mathrm{Eq.}\,(4.26)/\partial r\right]$, we get

$$r^3\frac{\partial^3\sigma'_{2+i2+i}}{\partial r^3} = (\delta_{0i}-\delta_{1i})\times$$

$$\left[-2c_{147+i}\sigma'_{11}+(c_{147+i}-c_{149+i})\,r\frac{\partial\sigma'_{11}}{\partial r}+c_{149+i}r^2\frac{\partial^2\sigma'_{11}}{\partial r^2}+c_{151+i}r^3\frac{\partial^3\sigma'_{11}}{\partial r^3}\right],$$

$$i=0,1. \tag{4.28}$$

Substituting Eq. (4.28) into $r\left[\partial \mathrm{Eq.}\,(4.28)/\partial r\right]$, we get

$$r^4\frac{\partial^4\sigma'_{2+i2+i}}{\partial r^4} = (\delta_{0i}-\delta_{1i})\times$$

$$\left[6c_{147+i}\sigma'_{11}+2\,(c_{149+i}-2c_{147+i})\,r\frac{\partial\sigma'_{11}}{\partial r}+(c_{147+i}-2c_{149+i})\,r^2\frac{\partial^2\sigma'_{11}}{\partial r^2}\right.$$

$$\left.+\,c_{149+i}r^3\frac{\partial^3\sigma'_{11}}{\partial r^3}+c_{151+i}r^4\frac{\partial^4\sigma'_{11}}{\partial r^4}\right],\quad i=0,1. \tag{4.29}$$

Substituting Eq. (4.29) into $r\left[\partial \mathrm{Eq.}\,(4.29)/\partial r\right]$, we get

$$r^5\frac{\partial^5\sigma'_{2+i2+i}}{\partial r^5} = (\delta_{0i}-\delta_{1i})\times$$

$$\left[-24c_{147+i}\sigma'_{11}+6\,(3c_{147+i}-c_{149+i})\,r\frac{\partial\sigma'_{11}}{\partial r}+2\,(c_{147+i}-5c_{149+i})\,r^2\frac{\partial^2\sigma'_{11}}{\partial r^2}\right.$$

$$\left.+\,(c_{147+i}-3c_{149+i})\,r^3\frac{\partial^3\sigma'_{11}}{\partial r^3}+c_{149+i}r^4\frac{\partial^4\sigma'_{11}}{\partial r^4}+c_{151+i}r^5\frac{\partial^5\sigma'_{11}}{\partial r^5}\right],$$

$$i=0,1. \tag{4.30}$$

Substituting Eq. (4.30) into $r\left[\partial \mathrm{Eq.}\,(4.30)/\partial r\right]$, we get

$$r^6\frac{\partial^6\sigma'_{2+i2+i}}{\partial r^6} = (\delta_{0i}-\delta_{1i})\times$$

$$\left[120c_{147+i}\sigma'_{11}+24\,(c_{149+i}-4c_{147+i})\,r\frac{\partial\sigma'_{11}}{\partial r}+12\,(c_{147+i}+2c_{149+i})\,r^2\frac{\partial^2\sigma'_{11}}{\partial r^2}\right.$$

$$-\,4c_{149+i}r^3\frac{\partial^3\sigma'_{11}}{\partial r^3}+(c_{147+i}-4c_{149+i})\,r^4\frac{\partial^4\sigma'_{11}}{\partial r^4}$$

$$\left.+\,c_{149+i}r^5\frac{\partial^5\sigma'_{11}}{\partial r^5}+c_{151+i}r^6\frac{\partial^6\sigma'_{11}}{\partial r^6}\right],\quad i=0,1. \tag{4.31}$$

4.3. Determination of Radial Stress

With regard to the spherical coordinates (r,φ,ν), the radial stress is assumed to be of the form

$$\sigma'_{11}=\sum_{i=1}^{n}C_i r^{\lambda_i}. \tag{4.32}$$

Consequently, the exponent λ_i which is required to represent a real number is thus determined by a differential equation for σ'_{11}. Such differential equation is of the order $\partial^m \sigma' / \partial r^m$ for $n \leq m$, where n is a number of boundary conditions (see Section 4.6.) for the determination of the integration constant C_i ($i = 1, \ldots, n$).

Let the mathematical operations $r(\partial/\partial r)$, $\partial/\partial\varphi$, $\partial/\partial\nu$ be applied to the transformed compatibility equation (4.21) and to the transformed equilibrium equation (4.22). Consequently, with regard to Eqs. (4.7)–(4.12), we get

$$\sum_{i=1}^{3} c_{158+i+3j} r \frac{\partial \sigma'_{ii}}{\partial r} + c_{9+i+36j} r^2 \frac{\partial^2 \sigma'_{ii}}{\partial r^2} = 0, \quad j = 0, 1, \tag{4.33}$$

$$\sum_{i=1}^{3} c_{164+i+9j} \sigma'_{ii} + c_{167+i+9j} r \frac{\partial \sigma'_{ii}}{\partial r} + c_{170+i+9j} r^2 \frac{\partial^2 \sigma'_{ii}}{\partial r^2} = 0, \quad j = 0 - 3. \tag{4.34}$$

Substituting Eqs. (4.24)–(4.26) into Eqs. (4.33), (4.34), the differential equations for σ'_{11} are derived as

$$c_{201+3i} \sigma'_{11} + c_{202+3i} r \frac{\partial \sigma'_{11}}{\partial r} + c_{203+3i} r^2 \frac{\partial^2 \sigma'_{11}}{\partial r^2} = 0, \quad i = 0 - 5. \tag{4.35}$$

Let the radial stress σ'_{11} be assumed to exhibit the form $\sigma'_{11} = \beta r^{\lambda}$. Consequently, the equation (4.35) is thus transformed to the form

$$c_{203+3i} \lambda^2 + (c_{202+3i} - c_{203+3i}) \lambda + c_{201+3i} = 0, \quad i = 0 - 5. \tag{4.36}$$

Due to $i = 0 - 5$, the equation (4.36) represents six characteristic equations with the variable $\lambda = \lambda(\varphi, \nu)$ and with the parameters φ, ν. The parameters φ, ν are included in the elastic moduli $s'_{11}, s'_{12}, \ldots, s'_{56}, s'_{66}$ (see Eqs. (4.13)–(4.18)) which are functions of the coefficient $a_{ij} = \cos[\angle(x'_i, x_j)]$ ($i, j = 1,2,3$) (see Eq. (3.42)). The elastic moduli $s'_{11}, s'_{12}, \ldots, s'_{56}, s'_{66}$ are included in the coefficients $c_i, \ldots, c_{15+i}$ (see Eqs. (9.22)–(9.27)) which are then included in the coefficients $c_{18+i}, \ldots, c_{7766+6j}$ (see Eqs. (9.28)–(9.146)) ($i = 1 - 3$; $j = 0 - 63$) which are presented in Appendix (see Chapter 9).

The exponents $\lambda_1^{(4.36)(i)} = \lambda_1^{(4.36)(i)}(\varphi, \nu)$, $\lambda_2^{(4.36)(i)} = \lambda_2^{(4.36)(i)}(\varphi, \nu)$ which represent roots of the i-th characteristic equation ($i = 0 - 5$) are derived as

$$\lambda_j^{(4.36)(i)} = \frac{1}{2c_{203+3i}} \times \left[c_{203+3i} - c_{202+3i} + (\delta_{1j} - \delta_{2j}) \sqrt{(c_{203+3i} - c_{202+3i})^2 - 4c_{201+3i} c_{203+3i}} \right], \quad i = 0 - 5, \quad j = 1, 2. \tag{4.37}$$

Due to $i = 0 - 5$, the formula (4.37) represents six sets with the exponents $\lambda_1^{(4.36)(i)}$, $\lambda_2^{(4.36)(i)}$ ($i = 0 - 5$), i.e., $\left\{\lambda_1^{(4.36)(0)}, \lambda_2^{(4.36)(0)}\right\}, \ldots, \left\{\lambda_1^{(4.36)(5)}, \lambda_2^{(4.36)(5)}\right\}$. If both exponents of each of the sets are imaginary, or none of these six sets consists of such real exponents which fulfil boundary conditions (see Section 4.6.) for a component of the model systems (see Section 2.2.), then the following mathematical procedures are considered.

Applying the mathematical operations $r\,(\partial/\partial r)$, $\partial/\partial\nu$ to Eqs. (4.33), (4.34), (4.35), and considering Eqs. (4.7)–(4.12), we get

$$\sum_{i=1}^{3} c_{218+i+9j} r \frac{\partial \sigma'_{ii}}{\partial r} + c_{221+i+9j} r^2 \frac{\partial^2 \sigma'_{ii}}{\partial r^2} + c_{224+i+9j} r^3 \frac{\partial^3 \sigma'_{ii}}{\partial r^3} = 0, \quad j = 0-13, \tag{4.38}$$

$$\sum_{i=1}^{3} c_{344+i+12j} \sigma'_{ii} + c_{347+i+12j} r \frac{\partial \sigma'_{ii}}{\partial r} + c_{350+i+12j} r^2 \frac{\partial^2 \sigma'_{ii}}{\partial r^2} + c_{354+i+12j} r^3 \frac{\partial^3 \sigma'_{ii}}{\partial r^3} = 0,$$
$$j = 0-15. \tag{4.39}$$

Substituting Eqs. (4.24)–(4.26), (4.28) into Eqs. (4.38), (4.39), we get

$$c_{537+4i} \sigma'_{11} + c_{538+4i} r \frac{\partial \sigma'_{11}}{\partial r} + c_{539+4i} r^2 \frac{\partial^2 \sigma'_{11}}{\partial r^2} + c_{540+4i} r^3 \frac{\partial^3 \sigma'_{11}}{\partial r^3} = 0, \quad i = 0-29. \tag{4.40}$$

If $\sigma'_{11} = \beta r^{\lambda}$, then the equation (4.40) is transformed to the form

$$c_{540+4i}\lambda^3 + (c_{539+4i} - 3c_{540+4i})\,\lambda^2 + (c_{538+4i} - c_{539+4i} + 2c_{540+4i})\,\lambda + c_{537+4i} = 0,$$
$$i = 0-29. \tag{4.41}$$

Due to $i = 0-29$, the equation (4.41) represents thirty characteristic equations with three roots in these thirty sets $\left\{\lambda_1^{(4.41)(0)}, \lambda_2^{(4.41)(0)}, \lambda_3^{(4.41)(0)}\right\}, \ldots, \left\{\lambda_1^{(4.41)(29)}, \lambda_2^{(4.41)(29)}, \lambda_3^{(4.41)(29)}\right\}$. The exponents $\lambda_1^{(4.41)(i)} = \lambda_1^{(4.41)(i)}(\varphi,\nu)$, $\lambda_2^{(4.41)(i)} = \lambda_2^{(4.41)(i)}(\varphi,\nu)$, $\lambda_3^{(4.41)(i)} = \lambda_3^{(4.41)(i)}(\varphi,\nu)$ represent roots of the i-th characteristic equation. The set $\left\{\lambda_1^{(4.41)(i)}, \lambda_2^{(4.41)(i)}, \lambda_3^{(4.41)(i)}\right\}$ consists either of three real roots, or of one real and two imaginary roots. If none of these thirty sets consists of such real exponents which fulfil boundary conditions (see Section 4.6.) for a component of the model systems (see Section 2.2.), then the following mathematical procedures are considered.

Applying the mathematical operations $r\,(\partial/\partial r)$, $\partial/\partial\varphi$, $\partial/\partial\nu$ to Eqs. (4.38), (4.39), and considering Eqs. (4.7)–(4.12), we get

$$\sum_{i=1}^{3} c_{656+i+12j} r \frac{\partial \sigma'_{ii}}{\partial r} + c_{659+i+12j} r^2 \frac{\partial^2 \sigma'_{ii}}{\partial r^2}$$
$$+ \sum_{i=1}^{3} c_{662+i+12j} r^3 \frac{\partial^3 \sigma'_{ii}}{\partial r^3} + c_{665+i+12j} r^4 \frac{\partial^4 \sigma'_{ii}}{\partial r^4} = 0, \quad j = 0-57, \tag{4.42}$$

$$\sum_{i=1}^{3} c_{1352+i+15j} \sigma'_{ii} + c_{1355+i+15j} r \frac{\partial \sigma'_{ii}}{\partial r} + c_{1358+i+15j} r^2 \frac{\partial^2 \sigma'_{ii}}{\partial r^2}$$
$$+ \sum_{i=1}^{3} c_{1361+i+15j} r^3 \frac{\partial^3 \sigma'_{ii}}{\partial r^3} + c_{1364+i+15j} r^4 \frac{\partial^4 \sigma'_{ii}}{\partial r^4} = 0, \quad j = 0-31. \tag{4.43}$$

Substituting Eqs. (4.24)–(4.26), (4.28), (4.29) into Eqs. (4.42), (4.43), we get

$$c_{1833+5i}\sigma'_{11} + c_{1834+5i}r\frac{\partial \sigma'_{11}}{\partial r} + c_{1835+5i}r^2\frac{\partial^2 \sigma'_{11}}{\partial r^2}$$
$$+ c_{1836+5i}r^3\frac{\partial^3 \sigma'_{11}}{\partial r^3} + c_{1837+5i}r^4\frac{\partial^4 \sigma'_{11}}{\partial r^4} = 0, \quad i = 0-89. \tag{4.44}$$

If $\sigma'_{11} = \beta r^{\lambda}$, then the equation (4.44) is transformed to the form

$$c_{1837+5i}\lambda^4 + (c_{1836+5i} - 6c_{1837+5i})\lambda^3 + (11c_{1837+5i} - 3c_{1836+5i} + c_{1835+5i})\lambda^2$$
$$+ \left[2c_{1836+5i} + c_{1834+5i} - (6c_{1837+5i} + c_{1835+5i})\right]\lambda + c_{1833+5i} = 0,$$
$$i = 0-89. \tag{4.45}$$

Due to $i = 0-89$, the equation (4.45) represents ninety characteristic equations with four roots in these ninety sets $\left\{\lambda_1^{(4.45)(0)}, \ldots, \lambda_4^{(4.45)(0)}\right\}, \ldots, \left\{\lambda_1^{(4.45)(89)}, \ldots, \lambda_4^{(4.45)(89)}\right\}$. The exponents $\lambda_1^{(4.45)(i)} = \lambda_1^{(4.45)(i)}(\varphi, \nu), \ldots, \lambda_4^{(4.45)(i)} = \lambda_4^{(4.45)(i)}(\varphi, \nu)$ represent roots of the i-th characteristic equation. The set $\left\{\lambda_1^{(4.45)(i)}, \ldots, \lambda_4^{(4.45)(i)}\right\}$ consists either of four real roots, or of two real and two imaginary roots, or of four imaginary roots. If none of these ninety sets consists of such real exponents which fulfill boundary conditions (see Section 4.6.) for a component of the model systems (see Section 2.2.), then the following mathematical procedures are considered.

Applying the mathematical operations $r(\partial/\partial r)$, $\partial/\partial\varphi$, $\partial/\partial\nu$ to Eqs. (4.42), (4.43), and considering Eqs. (4.7)–(4.12), we get

$$\sum_{i=1}^{3} c_{2282+i+15j}r\frac{\partial \sigma'_{ii}}{\partial r} + c_{2285+i+15j}r^2\frac{\partial^2 \sigma'_{ii}}{\partial r^2} + c_{2288+i+15j}r^3\frac{\partial^3 \sigma'_{ii}}{\partial r^3}$$
$$+ \sum_{i=1}^{3} c_{2291+i+15j}r^4\frac{\partial^4 \sigma'_{ii}}{\partial r^4} + c_{2294+i+15j}r^5\frac{\partial^5 \sigma'_{ii}}{\partial r^5} = 0, \quad j = 0-205, \tag{4.46}$$

$$\sum_{i=1}^{3} c_{5372+i+18j}\sigma'_{ii} + c_{5375+i+18j}r\frac{\partial \sigma'_{ii}}{\partial r} + c_{5378+i+18j}r^2\frac{\partial^2 \sigma'_{ii}}{\partial r^2} + c_{5381+i+18j}r^3\frac{\partial^3 \sigma'_{ii}}{\partial r^3}$$
$$+ \sum_{i=1}^{3} c_{5384+i+18j}r^4\frac{\partial^4 \sigma'_{ii}}{\partial r^4} + c_{5387+i+18j}r^5\frac{\partial^5 \sigma'_{ii}}{\partial r^5} = 0, \quad j = 0-63. \tag{4.47}$$

Substituting Eqs. (4.24)–(4.26), (4.28)–(4.30) into Eqs. (4.46), (4.47), we get

$$c_{6525+6i}\sigma'_{11} + c_{6526+6i}r\frac{\partial \sigma'_{11}}{\partial r} + c_{6527+6i}r^2\frac{\partial^2 \sigma'_{11}}{\partial r^2} + c_{6528+6i}r^3\frac{\partial^3 \sigma'_{11}}{\partial r^3}$$
$$+ c_{6529+6i}r^4\frac{\partial^4 \sigma'_{11}}{\partial r^4} + c_{6530+6i}r^5\frac{\partial^5 \sigma'_{11}}{\partial r^5} = 0, \quad i = 0-269. \tag{4.48}$$

If $\sigma'_{11} = \beta r^{\lambda}$, then the equation (4.48) is transformed to the form

$$c_{6530+6i}\lambda^5 + (c_{6529+6i} - 10c_{6530+6i})\lambda^4 + (c_{6528+6i} - 6c_{6529+6i} + 35c_{6530+6i})\lambda^3$$
$$+ (c_{6527+6i} - 3c_{6528+6i} + 11c_{6529+6i} - 50c_{6530+6i})\lambda^2$$
$$+ [c_{6526+6i} - c_{6527+6i} + 2(c_{6528+6i} - 3c_{6529+6i} + 12c_{6530+6i})]\lambda + c_{6525+6i} = 0,$$
$$i = 0 - 269. \tag{4.49}$$

Due to $i = 0 - 269$, the equation (4.49) represents two hundred and seventy characteristic equations with five roots in these two hundred and seventy sets $\left\{\lambda_1^{(4.49)(0)}, \ldots, \lambda_5^{(4.49)(0)}\right\}, \ldots, \left\{\lambda_1^{(4.49)(269)}, \ldots, \lambda_5^{(4.49)(269)}\right\}$. The exponents $\lambda_1^{(4.49)(i)} = \lambda_1^{(4.49)(i)}(\varphi, \nu), \ldots, \lambda_5^{(4.49)(i)} = \lambda_5^{(4.49)(i)}(\varphi, \nu)$ represent roots of the i-th characteristic equation. The set $\left\{\lambda_1^{(4.49)(i)}, \ldots, \lambda_5^{(4.49)(i)}\right\}$ consists either of five real roots, or of three real and two imaginary roots, or of one real and four imaginary roots. If none of these ninety sets consists of such real exponents which fulfill boundary conditions (see Section 4.6.) for a component of the model systems (see Section 2.2.), then the following mathematical procedures are considered.

Let the mathematical operations $r(\partial/\partial r)$, $\partial/\partial\varphi$, $s\partial/\partial\nu$ be applied to Eqs. (4.46), (4.47). Consequently, the substitution of Eqs. (4.7)–(4.12), (4.24)–(4.26), (4.28)–(4.31) along with $\sigma'_{11} = \beta r^{\lambda}$ to $r(\partial(4.46)/\partial r)$, $\partial(4.46)/\partial\varphi$, $\partial(4.46)/\partial\nu$ result in characteristic equations with the variable $\lambda = \lambda(\varphi, \nu)$ of the order λ^6.

Finally, these analyses concerning the characteristic equations (4.36), (4.41), (4.45), (4.49) are considered due to the determination of a sufficient number of the real exponent λ which is required to fulfill boundary conditions (see Section 4.6.) for a component of the model systems (see Section 2.2.).

Roots of the characteristic equations (4.41), (4.45), (4.49) are determined for a real component of the model systems (see Section 2.2.) by a suitable numerical method. Strictly speaking, numerical values of the elastic moduli $s_{11q}, s_{12q}, \ldots, s_{56q}, s_{66q}$ (see Eqs. (3.41), (3.43)) are required to be substituted to the coefficients which are included in (4.41), (4.45), (4.49), and consequently the suitable numerical method is required to be applied for the determination of the roots (the exponent $\lambda \equiv \lambda_q$) of these characteristic equation. The subscripts $q = p$, $q = p$ and $q = p$ are related to the numerical values of $s_{11q}, s_{12q}, \ldots, s_{56q}, s_{66q}$ for the spherical particle, spherical envelope and cell matrix, respectively.

4.4. Analysis of a Number of Solutions for Model Systems

Let $n_p^{(4.36)(i)}$ represent a number of such real roots of the set $\left\{\lambda_{1p}^{(4.36)(i)}, \lambda_{2p}^{(4.36)(i)}\right\}$ ($i = 0, \ldots, 5$) which fulfil boundary conditions for the spherical particle of the multi-particle-matrix system or multi-particle-envelope-matrix system (see Section 2.2.). Let n_p be a number of the boundary conditions for the spherical particle (see Section 4.6.1.), where $n_p^{(4.36)(i)} \geq n_p$.

Let $n_e^{(4.36)(i)}$ represent a number of such real roots of the set $\left\{\lambda_{1e}^{(4.36)(i)}, \lambda_{2e}^{(4.36)(i)}\right\}$ ($i = 0, \ldots, 5$) which fulfil boundary conditions for the spherical envelope of the multi-particle-envelope-matrix system. Let n_e be a number of the boundary conditions for the spherical envelope (see Section 4.6.2.), where $n_e^{(4.36)(i)} \geq n_e$.

Let $n_m^{(4.36)(i)}$ represent a number of such real roots of the set $\left\{\lambda_{1m}^{(4.36)(i)}, \lambda_{2m}^{(4.36)(i)}\right\}$ $(i=0,\ldots,5)$ which fulfil boundary conditions for the cell matrix of the multi-particle-matrix system or multi-particle-envelope-matrix system. Let n_m be a number of the boundary conditions for the cell matrix (see Section 4.6.3.), where $n_m^{(4.36)(i)} \geq n_m$. The mandatory and additional boundary conditions for the cell matrix are determined in Sections 4.6.3..1 and 4.6.3..2, respectively.

The set $\left\{\lambda_{1q}^{(4.36)(i)}, \lambda_{2q}^{(4.36)(i)}\right\}$ $(i=0,\ldots,5)$ (see Eq. (4.37)) represents the i-th set of these six sets $\left\{\lambda_1^{(4.36)(0)}, \lambda_2^{(4.36)(0)}\right\}, \ldots, \left\{\lambda_1^{(4.36)(5)}, \lambda_2^{(4.36)(5)}\right\}$ which are related to the characteristic equation (4.36).

Let $s_q^{(4.36)(i)}$ represent a number of solutions for the thermal stresses in the spherical particle $(q=p)$, in the spherical envelope $(q=p)$ and in the cell matrix $(q=m)$. The number $s_q^{(4.36)(i)}$ is related to the set $\left\{\lambda_{1q}^{(4.36)(i)}, \lambda_{2q}^{(4.36)(i)}\right\}$. The parameter $s_q^{(4.36)(i)}$ represents a number of combinations of $n_q^{(4.36)(i)}$ objects taken n_q at a time without repetitions, and then we get [113]

$$s_q^{(4.36)(i)} = \binom{n_q^{(4.36)(i)}}{n_q} = \frac{n_q^{(4.36)(i)}!}{n_q!\left(n_q^{(4.36)(i)} - n_q\right)!} = \left(n_q^{(4.36)(i)} + 1 - n_q\right)$$
$$\times \left(n_q^{(4.36)(i)} + 2 - n_q\right) \times \ldots \times \left(n_q^{(4.36)(i)} - 2\right)\left(n_q^{(4.36)(i)} - 1\right) n_q^{(4.36)(i)}$$
$$\times \left[1 \times 2 \times 3 \times \ldots \times (n_q - 2)(n_q - 1) n_q\right]^{-1}, \tag{4.50}$$

the formula $x! = 1 \times 2 \times \ldots (x-2) \times (x-1) \times x$ is considered [113].

The number $s^{(4.36)}$ of solutions which determine a stress-strain state induced by the thermal stresses in the multi-particle-matrix system is derived as

$$s^{(4.36)} = \left(\sum_i s_p^{(4.36)(i)}\right) \times \left(\sum_i s_m^{(4.36)(i)}\right), \tag{4.51}$$

where $s^{(4.36)}$ are related to the characteristic equation (4.36).

In case of the multi-particle-envelope-matrix system, we get

$$s^{(4.36)} = \left(\sum_i s_p^{(4.36)(i)}\right) \times \left(\sum_i s_e^{(4.36)(i)}\right) \times \left(\sum_i s_m^{(4.36)(i)}\right). \tag{4.52}$$

The superscript $i \subset \{0,\ldots,5\}$ concerning the sum $\sum_i$ in Eqs. (4.51), (4.52) is related to such $n_q^{(4.36)(i)}$ for which the condition $n_q^{(4.36)(i)} \geq n_q$ is valid $(q=p,e,m)$.

The same analysis is valid for $s^{(4.36)}$, $s^{(4.41)}$, $s^{(4.45)}$, $s^{(4.49)}$ which include $s_q^{(4.41)(j)}$ for $j=0,\ldots,29$, $s_q^{(4.45)(k)}$ for $k=0,\ldots,89$, $s_q^{(4.49)(l)}$ for $l=0,\ldots,269$, respectively.

The parameter $s_q^{(4.41)(j)}$ is related to $\left\{\lambda_1^{(4.41)(j)}, \lambda_2^{(4.41)(j)}, \lambda_3^{(4.41)(j)}\right\}$ which represents the j-th set of these thirty sets $\left\{\lambda_1^{(4.41)(0)}, \lambda_2^{(4.41)(0)}, \lambda_3^{(4.41)(0)}\right\}, \ldots, \left\{\lambda_1^{(4.41)(29)}, \lambda_2^{(4.41)(29)}, \lambda_3^{(4.41)(29)}\right\}$.

The parameter $s_q^{(4.45)(k)}$ is related to $\left\{\lambda_1^{(4.45)(k)},\ldots,\lambda_4^{(4.45)(k)}\right\}$ which represent the k-th set of these ninety sets $\left\{\lambda_1^{(4.45)(0)},\ldots,\lambda_4^{(4.45)(0)}\right\},\ldots,\left\{\lambda_1^{(4.45)(89)},\ldots,\lambda_4^{(4.45)(89)}\right\}$.

The parameter $s_q^{(4.49)(l)}$ is related to $\left\{\lambda_1^{(4.49)(l)},\ldots,\lambda_5^{(4.49)(l)}\right\}$ which represent the l-th set of these two hundred and seventy sets $\left\{\lambda_1^{(4.49)(0)},\ldots,\lambda_5^{(4.49)(0)}\right\},\ldots,\left\{\lambda_1^{(4.49)(269)},\ldots,\lambda_5^{(4.49)(269)}\right\}$.

As an example, let the characteristic equations (4.41), (4.49) be considered for the determination of a stress-strain state in the multi-particle-matrix and multi-particle-envelope-matrix systems (see Section 2.2.). In case of the multi-particle-matrix system, we get

$$s^{(4.41)(4.49)} = \left(\sum_j s_p^{(4.41)(i)} + \sum_l s_p^{(4.49)(i)}\right) \times \left(\sum_j s_m^{(4.41)(i)} + \sum_l s_m^{(4.49)(i)}\right). \tag{4.53}$$

In case of the multi-particle-envelope-matrix system, we get

$$\begin{aligned} s^{(4.41)(4.49)} = {} & \left(\sum_j s_p^{(4.41)(i)} + \sum_l s_p^{(4.49)(i)}\right) \times \left(\sum_j s_e^{(4.41)(i)} + \sum_l s_e^{(4.49)(i)}\right) \times \\ & \left(\sum_j s_m^{(4.41)(i)} + \sum_l s_m^{(4.49)(i)}\right). \end{aligned} \tag{4.54}$$

The superscripts $j \subset \{0,\ldots,29\}$ and $l \subset \{0,\ldots,269\}$ concerning the sums $\sum_j$, $\sum_l$ in Eqs. (4.53), (4.54) are related to such $n_q^{(4.41)(j)}$ and $n_q^{(4.49)(l)}$ for which the conditions $n_q^{(4.41)(j)} \geq n_q$ and $n_q^{(4.49)(l)} \geq n_q$ are valid ($q = p,e,m$), respectively.

As another example, let the characteristic equations (4.36), (4.41), (4.45), (4.49) be considered for the determination of a stress-strain state in the multi-particle-matrix and multi-particle-envelope-matrix systems. In case of the multi-particle-matrix system, we get

$$\begin{aligned} & s^{(4.36)(4.41)(4.45)(4.49)} = \\ & \left(\sum_i s_p^{(4.36)(i)} + \sum_j s_p^{(4.41)(i)} + \sum_k s_p^{(4.45)(i)} + \sum_l s_p^{(4.49)(i)}\right) \times \\ & \left(\sum_i s_m^{(4.36)(i)} + \sum_j s_m^{(4.41)(i)} + \sum_k s_m^{(4.45)(i)} + \sum_l s_m^{(4.49)(i)}\right). \end{aligned} \tag{4.55}$$

In case of the multi-particle-envelope-matrix system, we get

$$\begin{aligned} & s^{(4.36)(4.41)(4.45)(4.49)} = \\ & \left(\sum_i s_p^{(4.36)(i)} + \sum_j s_p^{(4.41)(i)} + \sum_k s_p^{(4.45)(i)} + \sum_l s_p^{(4.49)(i)}\right) \times \\ & \left(\sum_i s_e^{(4.36)(i)} + \sum_j s_e^{(4.41)(i)} + \sum_k s_e^{(4.45)(i)} + \sum_l s_e^{(4.49)(i)}\right) \times \\ & \left(\sum_i s_m^{(4.36)(i)} + \sum_j s_m^{(4.41)(i)} + \sum_k s_m^{(4.45)(i)} + \sum_l s_m^{(4.49)(i)}\right). \end{aligned} \tag{4.56}$$

The superscripts $i \subset \{0,\ldots,5\}$, $j \subset \{0,\ldots,29\}$, $k \subset \{0,\ldots,89\}$ and $l \subset \{0,\ldots,269\}$ concerning the sums $\sum_i$, $\sum_j$, $\sum_k$ and $\sum_l$ in Eqs. (4.55), (4.56) are related to such $n_q^{(4.36)(i)}$, $n_q^{(4.41)(j)}$, $n_q^{(4.45)(k)}$ and $n_q^{(4.49)(l)}$ for which the conditions $n_q^{(4.36)(i)} \geq n_q$, $n_q^{(4.41)(j)} \geq n_q$, $n_q^{(4.45)(k)} \geq n_q$ and $n_q^{(4.49)(l)} \geq n_q$ are valid ($q = p,e,m$), respectively.

Finally, such solutions for components of the model systems are considered to result in minimal elastic energy of the cubic cell, W_c (see Eqs. (3.52), (3.53)). As presented in Section 3.4., in case of the model systems (see Fig. 2.1), W_c is identical with total potential energy. Accordingly, a principle of minimal total potential energy [26] of the model systems is considered.

4.5. Mathematical Solutions

With regard to Eqs. (3.20), (3.21), (4.1)–(4.6), (4.24), (4.32), the stresses and strains determined by the mathematical procedures presented in Sections 4.2., 4.3. are derived as

$$\sigma'_{11q} = \sum_{i=1}^{n_q} C_{iq} r^{\lambda_{iq}}, \tag{4.57}$$

$$\sigma'_{22q} = \sum_{i=1}^{n_q} \xi_{1iq} C_{iq} r^{\lambda_{iq}}, \tag{4.58}$$

$$\sigma'_{33q} = \sum_{i=1}^{n_q} \xi_{2iq} C_{iq} r^{\lambda_{iq}}, \tag{4.59}$$

$$\sigma'_{12q} = \sum_{i=1}^{n_q} \xi_{3iq} C_{iq} r^{\lambda_{iq}}, \tag{4.60}$$

$$\sigma'_{13q} = \sum_{i=1}^{n_q} \xi_{4iq} C_{iq} r^{\lambda_{iq}}, \tag{4.61}$$

$$\varepsilon'_{11q} = \sum_{i=1}^{n_q} \xi_{5iq} C_{iq} r^{\lambda_{iq}}, \tag{4.62}$$

$$\varepsilon'_{22q} = \varepsilon'_{33q} = \sum_{i=1}^{n_q} \xi_{6iq} C_{iq} r^{\lambda_{iq}}, \tag{4.63}$$

$$\varepsilon'_{13q} = \sum_{i=1}^{n_q} \xi_{7iq} C_{iq} r^{\lambda_{iq}}, \tag{4.64}$$

$$\varepsilon'_{12q} = \sum_{i=1}^{n_q} \xi_{8iq} C_{iq} r^{\lambda_{iq}}, \tag{4.65}$$

$$u'_{1q} = r\varepsilon'_{22q} = \sum_{i=1}^{n_q} \xi_{6iq} C_{iq} r^{\lambda_{iq}+1}, \tag{4.66}$$

where n_q is a number of boundary conditions for the determination of the integration constants $C_{1q}, \ldots, C_{nq}$. The integration constants $C_{1q}, \ldots, C_{nq}$ are related to the spherical particle $(q = p)$ (see Section 4.6.1.), the spherical envelope $(q = e)$ (see Section 4.6.2.), and the cell matrix $(q = m)$ (see Section 4.6.3.).

The set $\{\lambda_{1q}, \ldots, \lambda_{nq}\}$ is required to be related to one of the characteristic equations (4.36), (4.41), (4.45), (4.49). The exponents $\lambda_{1q}, \ldots, \lambda_{nq}$ are required to be real. Finally, if $\lambda_{1q}, \ldots, \lambda_{nq}$ are real, then Equations (4.36); (4.41); (4.45); (4.49) result in $n_q = 1,2$; $n_q = 1,2,3$; $n_q = 1,\ldots,4$; $n_q = 1,\ldots,5$, respectively.

The coefficients $\xi_{1iq}, \ldots, \xi_{8iq}$ have the forms

$$\xi_{jiq} = (\delta_{2j} - \delta_{1j})\left\{c_{146+jq} + \lambda_{iq}\left[c_{148+jq} + c_{150+jq}(\lambda_{iq} - 1)\right]\right\},$$
$$i = 1, \ldots, n_q; \quad j = 1,2, \tag{4.67}$$

$$\xi_{2+jiq} = c_{152+jq} + \lambda_{iq}\left[c_{154+jq} + c_{156+jq}(\lambda_{iq} - 1)\right], \quad i = 1, \ldots, n_q; \quad j = 1,2, \tag{4.68}$$

$$\xi_{4+jiq} = c_{4+3jq} + c_{5+3jq} + c_{6+3jq}, \quad i = 1, \ldots, n_q; \quad j = 1, \ldots, 4. \tag{4.69}$$

With regard to Eqs. (4.30), (4.31), (4.57)–(4.65) and to Items 1–4 in Section 3.4., the elastic energy density w_q and the elastic energy W_q in the spherical particle $(q = p)$, the spherical envelope $(q = e)$ and the cell matrix $(q = m)$ are derived as

$$w_q = \frac{1}{2}\sum_{i,j=1}^{n_q} \omega_{ijq} C_{iq} C_{jq} r^{\lambda_{iq} + \lambda_{jq}}, \tag{4.70}$$

$$W_q = \int_0^{\pi/2}\int_0^{\pi/2} \Omega_q \, d\varphi \, d\nu. \tag{4.71}$$

The coefficients ω_{ijq} $(i, j = 1, \ldots, n_q)$, Ω_p, Ω_e and $\Omega_m = \Omega_{1+km}$ $(k = 0,1)$ have the forms

$$\omega_{ijq} = \xi_{5jq} + \xi_{6jq}(\xi_{1iq} + \xi_{2iq}) + 2(\xi_{3iq}\xi_{8jq} + \xi_{4iq}\xi_{7jq}), \quad i, j = 1, \ldots, n_q, \tag{4.72}$$

$$\Omega_p = 4\sum_{i,j=1}^{n_p} \frac{\omega_{ijp} C_{ip} C_{jp} R_1^{\lambda_{ip} + \lambda_{jp} + 3}}{\lambda_{ip} + \lambda_{jp} + 3}, \tag{4.73}$$

$$\Omega_e = 4\sum_{i,j=1}^{n_e} \frac{\omega_{ije} C_{ie} C_{je} R_2^{\lambda_{ie} + \lambda_{je} + 3}}{\lambda_{ie} + \lambda_{je} + 3}\left[1 - \left(\frac{R_1}{R_2}\right)^{\lambda_{ie} + \lambda_{je} + 3}\right], \tag{4.74}$$

$$\Omega_m = \Omega_{km} = 4\sum_{i,j=1}^{n_m} \frac{\omega_{ijm} C_{im} C_{jm} R_1^{\lambda_{im} + \lambda_{jm} + 3}}{\lambda_{im} + \lambda_{jm} + 3}\left[f_c^{\lambda_{im} + \lambda_{jm} + 3} - \left(\frac{R_k}{R_1}\right)^{\lambda_{im} + \lambda_{jm} + 3}\right],$$
$$k = 1,2, \tag{4.75}$$

where the function $f_c = f_c(\varphi, \nu, v)$ is given by Eqs. (3.1)–(3.3), (3.16), (3.18) (see Figure 3.2).

The subscript $k = 1,2$ in Eq. (4.75) is as follows:

- $k = 1$ for the multi-particle-matrix system,
- $k = 2$ for the multi-particle-envelope-matrix system.

As analysed in Section 7.3.1., the analytical model of strengthening considers the stress σ_{iq} and the elastic energy density w_{iq} ($i=1,2,3$). The stress σ_{iq} which induces the strain $\varepsilon_{iiq} = s_{iiq}\sigma_{iq}$ acts along the axis x_i ($i=1,2,3$) of the Cartesian system $(Ox_1x_2x_3)$ (see Figure 3.1). The elastic energy density w_{iq} is induced by the stress σ_{iq}.

The term $a_{ji}\sigma'_{jkq}$ represents a projection ('coordinate') of the stress σ'_{jkq} to a direction represented by the axis x_i. The coefficient a_{ji} ($i, j, k = 1,2,3$) is given by Eq. (3.42).

Consequently, the stress σ_{iq} and the elastic energy density w_{iq} ($i=1,2,3$) have the forms

$$\sigma_{iq} = \sum_{j=1}^{3} a_{ji}\sigma'_{jjq} + \sum_{j,k=1;\ j\neq k}^{3} a_{ji}\sigma'_{jkq}, \quad i = 1,2,3, \tag{4.76}$$

$$w_{iq} = \frac{\varepsilon_{iiq}\sigma_{iq}}{2} = \frac{s_{iiq}\sigma_{iq}^2}{2}, \quad i = 1,2,3. \tag{4.77}$$

With regard to Eqs. (4.57)–(4.61), (4.76), (4.77), we get

$$\sigma_{iq} = \sum_{j=1}^{n_q} \xi_{8+ijq} C_{jq} r^{\lambda_{jq}}, \quad i = 1,2,3, \tag{4.78}$$

$$w_{iq} = \frac{s_{iiq}}{2} \sum_{j,k=1}^{n_q} \xi_{8+ijq}\xi_{8+ikq} C_{jq} C_{kq} r^{\lambda_{jq}+\lambda_{kq}}, \quad i = 1,2,3, \tag{4.79}$$

where the coefficient ξ_{8+ijq} is derived as

$$\xi_{8+ijq} = a_{1i} + a_{2i}\xi_{1jq} + a_{3i}\xi_{2jq} + (a_{1i}+a_{2i})\,\xi_{3jq} + (a_{1i}+a_{3i})\,\xi_{4jq},$$
$$i = 1,2,3; \quad j = 1,\ldots,n_q. \tag{4.80}$$

4.6. Boundary Conditions

Section 4.6. deals with the determination of boundary conditions for components of the multi-particle-matrix and multi-particle-envelope-matrix systems (see Figure 2.1). In case of the multi-particle-envelope-matrix system, the conditions $\beta_p \neq \beta_e = \beta_m$, $\beta_p \neq \beta_e \neq \beta_m$, $\beta_p = \beta_e \neq \beta_m$ are considered. In case of the cell matrix (see Section 4.6.3.), mandatory and additional boundary conditions are determined.

With regard to the analysis in Section 3.2., in case of the multi-particle-matrix system, R_1 represents a radius of the embedded spherical particle at the temperature $T \in \langle T_f, T_r \rangle$, where T_f is a final temperature of a cooling process, and the relaxation temperature T_r is determined in Section 3.5.1..1. In case of the multi-particle-envelope-matrix system, R_1, R_2 and $t = R_2 - R_1$ represents radii and thickness of the embedded spherical envelope at the temperature $T \in \langle T_f, T_r \rangle$, respectively.

4.6.1. Spherical Particle

The thermal stresses σ'_{iip}, σ'_{12+jp} ($i=1,2,3$; $j=0,1$) in the Cartesian system $(Px'_1x'_2x'_3)$ (see Figure 3.1a), and the thermal-stress induced radial displacement u'_{1p} along the axis x'_1 are required to fulfil these boundary conditions

$$\left(u'_{1p}\right)_{r=0}=0, \tag{4.81}$$

$$\left(\sigma'_{iip}\right)_{r\to 0}\nrightarrow \pm\infty, \quad i=1,2,3, \tag{4.82}$$

$$\left(\sigma'_{12+ip}\right)_{r\to 0}\nrightarrow \pm\infty, \quad i=0,1, \tag{4.83}$$

$$\left(\sigma'_{11p}\right)_{r=R_1}=-p_1. \tag{4.84}$$

Equations (4.81) and (4.82)–(4.84) represent geometric and stress boundary conditions, respectively.

The boundary conditions (4.81)–(4.84) are considered for the multi-particle-matrix system on the condition $\beta_p \neq \beta_m$. The boundary conditions (4.81)–(4.84) are also considered for the multi-particle-envelope-matrix systems on the conditions $\beta_p \neq \beta_e = \beta_m$, $\beta_p \neq \beta_e \neq \beta_m$ (see Eqs. (3.54)–(3.60)).

The radial stress p_1 acting at the particle-matrix boundary is determined by the formula (4.63).

If $\beta_p \neq \beta_e = \beta_m$, then p_1 acting at the particle-envelope boundary is determined by the formula (4.64).

If $\beta_p \neq \beta_e \neq \beta_m$, then p_1 acting at the particle-envelope boundary is determined by the formulae (4.64), (4.65).

If $\beta_p = \beta_e \neq \beta_m$, then the boundary condition (4.84) is replaced by

$$\left(\varepsilon'_{22p}\right)_{r=R_1}=\left(\varepsilon'_{22e}\right)_{r=R_1}=\frac{\left(u'_{1e}\right)_{r=R_1}}{R_1}=-p_2\rho^{pe}_{1e}, \tag{4.85}$$

In this case, i.e., $\beta_p = \beta_e \neq \beta_m$, the integration constant C_{1p} includes ρ^{pe}_{1e}. The coefficient ρ^{pe}_{1e} can be thus determined after the determination of the integration constant C_{1e}. As presented in Section 4.6.2., the integration constant C_{1e} for $\beta_p = \beta_e \neq \beta_m$ is determined by the conditions (4.89), (4.91). The determination of C_{1p} for $\beta_p = \beta_e \neq \beta_m$ is then preceded by the determination of C_{1e} for $\beta_p = \beta_e \neq \beta_m$.

Equations (4.81)–(4.84) result in one integration constant only, and then we get

$$n_p = 1, \tag{4.86}$$

where n_p is a number of boundary conditions which are required for the spherical particle.

The integration constant $C_{1p} \neq 0$ is then determined by Eq. (4.84).

On each of these conditions, $\beta_p \neq \beta_m$, $\beta_p \neq \beta_e = \beta_m$, $\beta_p \neq \beta_e \neq \beta_m$, $\beta_p = \beta_e \neq \beta_m$, the absolute value $\left|u'_{1p}\right|$ is required to represent an increasing function of $r \in \langle 0, R_1 \rangle$. This increasing dependence $\left|u'_{1p}\right| - r$ exhibits a maximal value at the particle-matrix or particle-envelope boundary, thus for $r = R_1$. With regard to Eqs. (4.57)–(4.61), (4.82), (4.83), the exponent λ_{1p} is then required to fulfil the following condition

$$\lambda_{1p} > 0. \tag{4.87}$$

4.6.2. Spherical Envelope

If $\beta_p \neq \beta_e \neq \beta_m$ (see Eqs. (3.54)–(3.60)), then the following boundary conditions

$$\left(\sigma'_{11e}\right)_{r=R_1} = -p_1, \tag{4.88}$$

$$\left(\sigma'_{11e}\right)_{r=R_2} = -p_2, \tag{4.89}$$

are considered for the determination of the integration constants $C_{1e} \neq 0$ and $C_{2e} \neq 0$. In case of $\beta_p \neq \beta_e \neq \beta_m$, the real exponents λ_{1e} and λ_{2e} are not required to be defined by additional conditions which are similar to that given by Eq. (4.87). Accordingly, we get

$$n_e = 2, \tag{4.90}$$

where n_e is a number of boundary conditions which are required for the spherical envelope.

If $\beta_p \neq \beta_e = \beta_m$, then the boundary condition (4.88) is considered for the determination of the integration constant $C_{1e} \neq 0$. Accordingly, we get

$$n_e = 1. \tag{4.91}$$

In this case, i.e., $\beta_p \neq \beta_e = \beta_m$, the absolute value $|u'_{1e}|$ is required to represent a decreasing function of the variable $r \in \langle R_1, R_2 \rangle$ or, at the very most, a constant function of $r \in \langle R_1, R_2 \rangle$. This decreasing dependence $|u'_{1e}| - r$ exhibits a maximal value at the particle-envelope boundary, thus for $r = R_1$. With regard to Eq. (4.66), the exponent λ_{1e} is then required to fulfil the following condition

$$\lambda_{1e} \leq -1. \tag{4.92}$$

If $\beta_p = \beta_e \neq \beta_m$, then the boundary condition (4.89) along with Eq. (4.91) are considered for the determination of the integration constant $C_{1e} \neq 0$. The absolute value $|u'_{1e}|$ is required to represent an increasing function of $r \in \langle R_1, R_2 \rangle$ or, at the very most, a constant function of $r \in \langle R_1, R_2 \rangle$. This increasing dependence $|u'_{1e}| - r$ exhibits a maximal value at the matrix-envelope boundary, thus for $r = R_2 = R_1 + t$. With regard to Eq. (4.66), the exponent λ_{1e} is then required to fulfil the following condition

$$\lambda_{1e} \geq -1. \tag{4.93}$$

4.6.3. Cell Matrix

The thermal stresses in the cell matrix are derived regarding the standard boundary conditions (4.94)–(4.96) which are mandatory (see Section 4.6.3..1). In addition to (4.94)–(4.96), the additional boundary conditions (4.99), (4.100) (see Section 4.6.3..2) might be also considered. With regard to (4.99), the dependence $u'_{1m} - r$ is required to be extremal on the cell surface, i.e., for $r = r_c$. With regard to (4.100), the dependence $w_m - r$ is required to be extremal on the cell surface.

The mandatory boundary conditions for the multi-particle-matrix and multi-particle-envelope-matrix systems are as follows.

- In case of the multi-particle-matrix system, the conditions (4.94), (4.96) are considered.
- In case of the multi-particle-envelope-matrix system, the conditions (4.95), (4.96) are considered for $\beta_p \neq \beta_e \neq \beta_m$ or $\beta_p = \beta_e \neq \beta_m$ (see Eqs. (3.54)–(3.60)).
- In case of the multi-particle-envelope-matrix system, the conditions (4.96), (4.97) are considered for $\beta_p \neq \beta_e = \beta_m$.

Accordingly, the four following combinations of the mandatory and additional boundary conditions for the multi-particle-matrix and multi-particle-envelope-matrix systems are considered:

1. the mandatory boundary conditions (4.94), (4.96) or (4.95), (4.96).
2. the mandatory boundary conditions (4.94), (4.96) or (4.95), (4.96) along with the additional boundary condition (4.99).
3. the mandatory boundary conditions (4.94), (4.96) or (4.95), (4.96) along with the additional boundary condition (4.100).
4. the mandatory boundary conditions (4.94), (4.96) or (4.95), (4.96) along with the additional boundary conditions (4.99) and (4.100).

4.6.3.1. Mandatory Boundary Conditions

The thermal radial stress σ'_{11m} and the thermal-stress induced radial displacement u'_{1m} are required to fulfil these mandatory boundary conditions

$$\left(\sigma'_{11m}\right)_{r=R_1} = -p_1, \tag{4.94}$$

$$\left(\sigma'_{11m}\right)_{r=R_2} = -p_2, \tag{4.95}$$

$$\left(u'_{1m}\right)_{r=r_c} = 0. \tag{4.96}$$

The distance r_c is determined in Section 3.1.2..1 (see Eqs. (3.1)–(3.3), (3.8), (3.15)–(3.18)).

Equations (4.94), (4.95) and (4.96), represent stress and geometric boundary conditions, respectively.

The mandatory boundary conditions (4.94), (4.96) are considered for the multi-particle-matrix for $\beta_p \neq \beta_m$.

The mandatory boundary conditions (4.95), (4.96) are also considered for the multi-particle-envelope-matrix system on the conditions $\beta_p \neq \beta_e \neq \beta_m$ or $\beta_p = \beta_e \neq \beta_m$.

The radial stress p_1 acting at the particle-matrix boundary is determined by the formula (4.63).

If $\beta_p \neq \beta_e = \beta_m$, then p_1 acting at the particle-envelope boundary is determined by the formula (4.64).

If $\beta_p \neq \beta_e \neq \beta_m$, then the radial stresses p_1 and p_2 acting at the particle-envelope and matrix-envelope boundaries, respectively, are both determined by the formulae (4.64), (4.65).

If $\beta_p = \beta_e \neq \beta_m$, then p_2 is determined by the formula (4.65).

If $\beta_p \neq \beta_e = \beta_m$, then the boundary condition (4.95) is replaced by

$$\left(\varepsilon'_{22m}\right)_{r=R_2} = \left(\varepsilon'_{22e}\right)_{r=R_2} = \frac{\left(u'_{1e}\right)_{r=R_2}}{R_2} = -p_1\,\rho^{me}_{2e}, \tag{4.97}$$

In this case, i.e., $\beta_p \neq \beta_e = \beta_m$, the integration constants C_{1m}, C_{2m} both include ρ^{me}_{2e}. The coefficient ρ^{me}_{2e} can be thus determined after the determination of the integration constant C_{1e}. As presented in Section 4.6.2., the integration constant C_{1e} for $\beta_p \neq \beta_e = \beta_m$ is determined by the conditions (4.88), (4.92). The determination of C_{1m}, C_{2m} for $\beta_p \neq \beta_e = \beta_m$ is then preceded by the determination of C_{1e} for $\beta_p \neq \beta_e = \beta_m$.

Equations (4.94)–(4.97) result in two integration constants, and then we get

$$n_m = 2, \tag{4.98}$$

where n_m is a number of boundary conditions which are required for the cell matrix.

In case of the multi-particle-matrix system, the absolute value $|u'_{1m}|$ is required to represent a decreasing function of $r \in \langle R_1, r_c\rangle$ due to $(u'_{1m})_{r=r_c} = 0$ (see Eq. (4.96)). This decreasing dependence $|u'_{1m}| - r$ exhibits a maximal value at the particle-matrix boundary, thus for $r = R_1$. The same is also valid for the multi-particle-envelope-matrix system on each of the conditions $\beta_p \neq \beta_e = \beta_m$, $\beta_p \neq \beta_e \neq \beta_m$, $\beta_p = \beta_e \neq \beta_m$,

The decreasing course of the dependence $|u'_{1m}| - r$ in the cell matrix of the multi-particle-matrix and multi-particle-envelope-matrix systems is ensured by integration constants.

The Determination of the Mandatory Boundary Condition $(u'_{1m})_{r=r_c} = 0$. The determination of $(u'_{1m})_{r=r_c} = 0$ is based on a concept of the imaginary separation of components of the model systems (see Figure 2.1). This concept which is also considered for the determination of the Eshelby's model [91] is described in detail in Section 3.2.

Let the multi-particle-matrix system be considered. Let the components of this system be imaginarily separated at the relaxation temperature T_r. In this case, the spherical particles in the infinite matrix are replaced by spherical hollows. Let $R_1 = R_1(T)$ be a temperature dependence of a radius of these separated particles. Let $R_{1m} = R_{1m}(T)$ be a temperature dependence of a radius of the spherical hollows. Let these separated components be cooled down from the relaxation temperature T_r to the temperature T, where $T < T_r$.

A radius of the separated particle is thus changed from $R_1(T_r)$ to $R_1(T)$, where $R_1(T) < R_1(T_r)$.

A radius of the hollow is thus changed from $R_{1m}(T_r)$ to $R_{1m}(T)$, where $R_{1m}(T) < R_{1m}(T_r)$. Additionally, the condition $R_1(T_r) = R_{1m}(T_r)$ is valid.

A dimension of the cubic cell is thus changed from $d(T_r)$ to $d(T)$, where $d(T) < d(T_r)$.

Let the condition $\beta_p \lesseqgtr \beta_m$ be considered. If $\beta_p \lesseqgtr \beta_m$, then $R_1(T) \gtreqless R_{1m}(T)$ for $T < T_r$.

As analysed in Section 3.2., the infinitesimal spherical cap exhibits a displacement along the axis x'_1 (see Figure 3.1). This displacement along x'_1 results from the symmetry of the multi-particle-matrix system. The axis x'_1 of the Cartesian system $(Px'_1x'_2x'_3)$ defines a radial direction regarding the Cartesian system $(Ox_1x_2x_3)$ (see Figure 3.1).

Let the separated particle with the radius $R_1(T)$ be put into the spherical hollow with the radius $R_1(T)$ of the cubic cell with with the dimension $d(T)$. After the embedding of the separated particle in the hollow at the temperature $T < T_r$, a surface of the particle is pushed or pulled by a surface of the matrix due to $R_1(T) > R_{1m}(T)$ or $R_1(T) < R_{1m}(T)$, respectively. Similarly, a surface of the matrix is pushed or pulled by a surface of the particle due to $R_1(T) > R_{1m}(T)$ or $R_1(T) < R_{1m}(T)$, respectively. The pushing or pulling at $T < T_r$ within the spherical particle and cell matrix is expressed by the thermal-stress induced radial displacements u'_{1p} and u'_{1m} along the axis x'_1, i.e., along a radial direction.

The same is also valid for the multi-particle-envelope-matrix system as well as for the thermal-stress induced radial displacement u'_{1e} in the spherical envelope.

The pushing or pulling, which is induced in the cubic cell *after* the embedding at $T < T_r$, is related to the cell dimension $d(T)$ at the temperature $T < T_r$. This fact is required to be considered within the following analysis which concerns the determination of the mandatory boundary condition $(u'_{1m})_{r=r_c} = 0$.

An analysis of the mandatory boundary condition $(u'_{1m})_{r=r_c} = 0$ is as follows.

Let C be an arbitrary point on the surface *1234* (see Figure 4.1). The surface *1234* with the normal x_2 represents a common surface of two neighbouring cubic cells A and B with the centres O_A and O_B, respectively. These neighbouring cubic cells with the dimension $d(T)$ at the temperature $T < T_r$ are cells of the multi-particle-matrix system or the multi-particle-envelope-matrix system.

A position of the arbitrary point C on the surface *1234* of the cell A is determined by the spherical coordinates $(r_{cA}, \varphi_A, \nu_A)$ regarding the Cartesian system $(O_Ax_{1A}x_2x_{3A})$, where $r_{cA} = \left|\overline{O_AC}\right|$. Let the axis x'_{1A} (see Figures 3.1a, 4.1) define a radial displacement in the cubic cell A. Let $u'_{1mA} = u'_{1mA}(r, \varphi_A, \nu_A)$ represent the thermal-stress induced radial displacement along the axis x'_{1A} in the cell matrix of the cell A.

A position of the arbitrary point C on the surface *1234* of the cell B is determined by the spherical coordinates $(r_{cB}, \varphi_B, \nu_B)$ regarding the Cartesian system $(O_Bx_{1B}x_2x_{3B})$, where $r_{cB} = \left|\overline{O_BC}\right|$. Let the axis x'_{1B} (see Figures 3.1a, 4.1) define a radial displacement in the cubic cell B. Let $u'_{1mB} = u'_{1mB}(r, \varphi_B, \nu_B)$ represent the thermal-stress induced radial displacement along the axis x'_{1B} in the cell matrix of the cell B.

The distances $r_{cA} = r_{cA}(d, \varphi_A, \nu_A)$, $r_{cB} = r_{cB}(d, \varphi_B, \nu_B)$ are determined in Section 3.1.2.1 (see Eqs. (3.1)–(3.3), (3.8), (3.15)–(3.18)).

Due to the cubic shape of the cells A, B, we get $\varphi_A = \varphi_B = \varphi$, $\nu_A = \nu_B = \nu$, $r_{cA} = r_{cB} = r_c(d, \varphi, \nu)$, $u'_{1mA} = u'_{1mB} = u'_{1m}(r, \varphi, \nu)$.

Let the condition $(u'_{1m})_{r=r_c} < 0$ be assumed to be valid. The arbitrary point C on the surface *1234* thus tends to be shifted along the axis x'_{1A} towards the centre O_A as well as along the axis x'_{1B} towards the centre O_B.

This is possible in such case when the volume $dV_A = \left(r_{cA} - \left|(u'_{1mA})_{r=r_{cA}}\right|\right) d\varphi_A \times \left(r_{cA} - \left|(u'_{1mA})_{r=r_{cA}}\right|\right) d\nu_A \times \left|(u'_{1mA})_{r=r_{cA}}\right|$ in the cubic cell A is replaced by the vacuum, and simultaneously, the volume $dV_B = \left(r_{cB} - \left|(u'_{1mB})_{r=r_{cB}}\right|\right) d\varphi_B \times \left(r_{cA} - \left|(u'_{1mB})_{r=r_{cB}}\right|\right) d\nu_A \times \left|(u'_{1mB})_{r=r_{cB}}\right|$ in the cubic cell B is also replaced by the

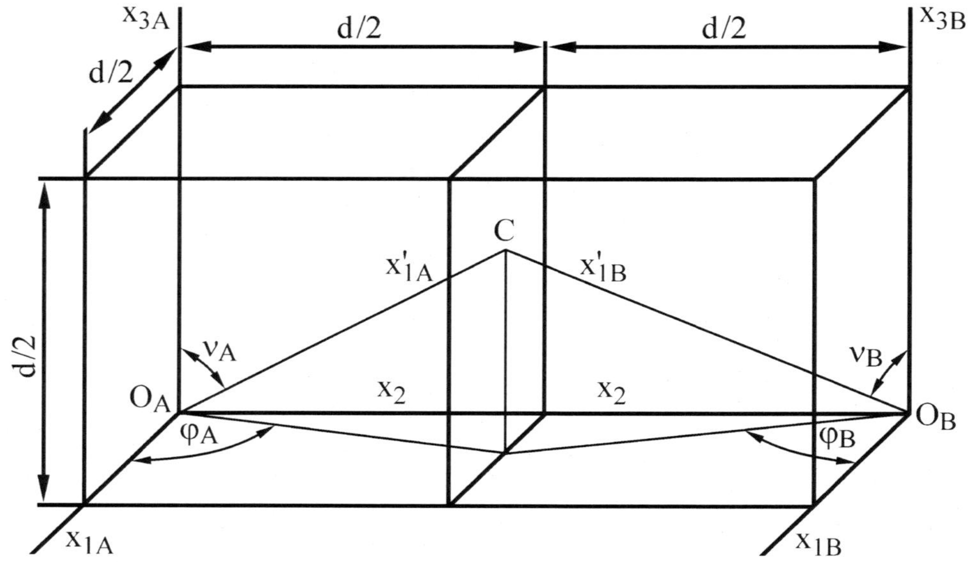

Figure 4.1. The arbitrary point C on the surface *1234* of two neighbouring cubic cells with the centres O_A and O_B both with the dimension $d(T)$ at the temperature $T < T_r$. Due to the cubic shape of the cells A, B, we get $\varphi_A = \varphi_B = \varphi$, $\nu_A = \nu_B = \nu$, $r_{cA} = \left|\overline{O_A C}\right| = r_{cB} = \left|\overline{O_B C}\right| = r_c$. The distance r_c is determined in Section 3.1.2..1 (see Eqs. (3.1)–(3.3), (3.8), (3.15)–(3.18)).

vacuum.

The volume dV_Q $(Q = A,B)$ is characterized by the length $\left|(u'_{1mA})_{r=r_{cA}}\right|$ along the axis x'_{1Q} and by the surface dS in a point on x'_{1Q} with the coordinate $r = r_{cQ} - \left|\left(u'_{1mQ}\right)_{r=r_{cQ}}\right|$. The surface dS with the normal x'_{1Q} is characterized by the dimensions $\left(r_{cQ} - \left|\left(u'_{1mQ}\right)_{r=r_{cQ}}\right|\right) d\varphi_Q$ and $\left(r_{cQ} - \left|\left(u'_{1mQ}\right)_{r=r_{cQ}}\right|\right) d\nu_Q$ $(Q = A,B)$.

Due to the general position of the point C on the surface *1234*, if $(u'_{1m})_{r=r_c} < 0$, then the matrix between neighbouring cells is replaced by the vacuum what is physically unacceptable. The same analysis is also valid for the condition $(u'_{1m})_{r=r_c} > 0$.

Accordingly, the mandatory boundary condition $(u'_{1m})_{r=r_c} = 0$ for the thermal-stress induced displacement $u'_{1m} = u'_{1m}(r, \varphi, \nu)$ in the cell matrix is thus valid.

4.6.3.2. Additional Boundary Conditions

As presented in Section 2.2., the division of the matrix the model systems into identical cubic cells is imaginary. Accordingly, the surface of the cubic cell *does not* represent a physical boundary. As an example, a physical boundary is represented by a boundary of two material components.

Due to the fact that the surface of the cubic cell is not a physical boundary, the connection of these identical functions $u'_{1mA} = u'_{1mA}(r, \varphi_A, \nu_A)$ and $u'_{1mB} = u'_{1mB}(r, \varphi_B, \nu_B)$ in the point C (i.e., for $r = r_{1cA} = r_{1cB} = r_c$) on the surface *1234* (see Figure 4.1) is assumed to be 'smooth'. Due to this assumption of the 'smooth' connection, the dependences $u'_{1mA} - r$ and $u'_{1mB} - r$ are thus required not to mutually create a singular point at the point C, i.e., for $r = r_c$ (see Section 3.1.2..1). The same analysis concerning the smooth course of the dependence $u'_{1m} - r$ for $r = r_c$ is also valid in case of the dependence $w_m - r$ for $r = r_c$, where $w_m = w_m(r, \varphi, \nu)$ is thermal-stress induced elastic energy density (see Eq. (4.28)).

To fulfil this non-singularity assumption, the functions $u'_{1m} = u'_{1m}(r, \varphi, \nu)$, $w_m = w_m(r, \varphi, \nu)$ are required to be extremal on the surface of the cubic cell, i.e., for $r = r_c$. In case of $|u'_{1m}|$ and w_m, this extreme for $r = r_c$ represents the minimum, where the dependences $|u'_{1m}| - r$ and $w_m - r$ are decreasing. If $u'_{1m} < 0$, then this extreme for $r = r_c$ represents the maximum.

With regard to Eq. (3.19), the additional boundary conditions for the cell matrix of the multi-particle-matrix and multi-particle-envelope-matrix systems (see Figure 2.1) are derived as

$$\left(\varepsilon'_{11m}\right)_{r=r_c} = \left(\frac{\partial u'_{1m}}{\partial r}\right)_{r=r_c} = 0, \tag{4.99}$$

$$\left(\frac{\partial w_m}{\partial r}\right)_{r=r_c} = 0. \tag{4.100}$$

4.7. Thermal Stresses in Multi-Particle-Matrix System

Section 4.7. deals with the determination of the thermal stresses in anisotropic components of the multi-particle-matrix system (see Figure 2.1a). This determination considers the boundary conditions in Sections 4.6.1., 4.6.3. which are applied to the mathematical solutions in Section 3.3.

With regard to the analysis in Section 3.2., R_1 represents a radius of the embedded spherical particle at the temperature $T \in \langle T_f, T_r \rangle$, where T_f is a final temperature of a cooling process, and the relaxation temperature T_r is determined in Section 3.5.1.1.

4.7.1. Thermal Stresses in Spherical Particle

With regard to the analyses in Sections 4.4., 4.5., 4.6.1., the integration constant C_{1p} is derived as

$$C_{1p} = -\frac{p_1}{R_1^{\lambda_{1p}}}, \tag{4.101}$$

where the radial stress p_1 acting at the particle-matrix boundary is given by Eq. (4.161).

Equations (4.57)–(4.65), (4.70), (4.71), (4.73), (4.78), (4.79) ($q = p$) are thus transformed to the forms

$$\sigma'_{11p} = -p_1 \left(\frac{r}{R_1}\right)^{\lambda_{1p}}, \tag{4.102}$$

$$\sigma'_{22p} = -p_1 \, \xi_{11p} \left(\frac{r}{R_1}\right)^{\lambda_{1p}}, \tag{4.103}$$

$$\sigma'_{33p} = -p_1 \xi_{21p} \left(\frac{r}{R_1}\right)^{\lambda_{1p}}, \tag{4.104}$$

$$\sigma'_{12p} = -p_1 \xi_{31p} \left(\frac{r}{R_1}\right)^{\lambda_{1p}}, \tag{4.105}$$

$$\sigma'_{13p} = -p_1 \xi_{41p} \left(\frac{r}{R_1}\right)^{\lambda_{1p}}, \tag{4.106}$$

$$\varepsilon'_{11p} = -p_1 \xi_{51p} \left(\frac{r}{R_1}\right)^{\lambda_{1p}}, \tag{4.107}$$

$$\varepsilon'_{22p} = \varepsilon'_{33p} = -p_1 \xi_{61p} \left(\frac{r}{R_1}\right)^{\lambda_{1p}}, \tag{4.108}$$

$$\varepsilon'_{13p} = -p_1 \xi_{71p} \left(\frac{r}{R_1}\right)^{\lambda_{1p}}, \tag{4.109}$$

$$\varepsilon'_{12p} = -p_1 \xi_{81p} \left(\frac{r}{R_1}\right)^{\lambda_{1p}}, \tag{4.110}$$

$$w_p = \frac{p_1^2 \omega_{11p}}{2} \left(\frac{r}{R_1}\right)^{2\lambda_{1p}}, \tag{4.111}$$

$$W_p = 4R_1^3 \int_0^{\pi/2} \int_0^{\pi/2} \frac{p_1^2 \omega_{11p}}{2\lambda_{1p}+3} \, d\varphi \, d\nu, \tag{4.112}$$

$$\sigma_{ip} = -p_1 \xi_{8+i1p} \left(\frac{r}{R_1}\right)^{\lambda_{1p}}, \quad i = 1,2,3, \tag{4.113}$$

$$w_{ip} = \frac{s_{iip} p_1^2 \xi_{8+i1p}^2}{2} \left(\frac{r}{R_1}\right)^{2\lambda_{1p}}, \quad i = 1,2,3, \tag{4.114}$$

where the coefficients $\xi_{11p}, \ldots, \xi_{81p}$, ω_{11p}, ξ_{8+i1p} ($i = 1,2,3$) are given by Eqs. (4.67)–(4.69) ($q = p$), (4.72), (4.80), respectively.

4.7.2. Thermal Stresses in Cell Matrix

As presented in Section 4.6.3., the thermal stresses in the cell matrix are determined by four combinations of boundary conditions. These four combinations are defined in Items 1–4, Section 4.6.3. (see p. 65). With regard to the analyses in Sections 4.4., 4.5., 4.6.3., the integration constant C_{im} ($i = 1, \ldots, n_m$) is derived as

$$C_{im} = -\frac{p_1 \kappa_{i1m}^{(x)}}{R_1^{\lambda_{im}}}, \quad i = 1, \ldots, n_q \tag{4.115}$$

where n_m is a number of boundary conditions for the cell matrix, and the radial stress p_1 acting at the particle-matrix boundary is given by Eq. (4.161).

The following analysis concerning the coefficient $\kappa_{i1m}^{(x)}$ and the four combinations is considered.

1. If the boundary conditions which are defined in Items 1, Section 4.6.3. are considered, then the coefficient $\kappa_{i1m}^{(x)}$ is replaced by $\kappa_{i1m}^{(1)}$ for $i = 1,2$; $n_q = 2$.
2. If the boundary conditions which are defined in Items 2, Section 4.6.3. are considered, then the coefficient $\kappa_{i1m}^{(x)}$ is replaced by $\kappa_{i1m}^{(2)}$ for $i = 1,2,3$; $n_q = 3$.
3. If the boundary conditions which are defined in Items 3, Section 4.6.3. are considered, then the coefficient $\kappa_{i1m}^{(x)}$ is replaced by $\kappa_{i1m}^{(3)}$ for $i = 1,2,3$; $n_q = 3$.
4. If the boundary conditions which are defined in Items 4, Section 4.6.3. are considered, then the coefficient $\kappa_{i1m}^{(x)}$ is replaced by $\kappa_{i1m}^{(4)}$ for $i = 1,\ldots,4$; $n_q = 4$.

Consequently, the coefficients $\kappa_{i1m}^{(1)}$, $\kappa_{i1m}^{(2)}$, $\kappa_{i1m}^{(3)}$, $\kappa_{i1m}^{(4)}$ have the forms

$$\kappa_{ijm}^{(1)} = \frac{\xi_{63-im}}{\xi_{63-im} - \xi_{60+im}\left(\frac{R_1 f_c}{R_j}\right)^{\lambda_{im}-\lambda_{3-im}}}, \quad i,j = 1,2, \tag{4.116}$$

$$\kappa_{ijm}^{(2)} = \frac{\psi_{im}}{\sum_{k=1}^{3}(-1)^{k+1}\psi_{km}\left(\frac{R_1 f_c}{R_j}\right)^{\lambda_{im}-\lambda_{km}}}, \quad i = 1,2,3; \quad j = 1,2, \tag{4.117}$$

$$\kappa_{ijm}^{(3)} = \kappa_{ijm}^{(1)} - \kappa_{3jm}^{(3)}\left[\kappa_{ijm}^{(1)} - \xi_{11+ijm}\right], \quad i,j = 1,2, \tag{4.118}$$

$$\kappa_{3jm}^{(3)} = \frac{-\xi_{15\,jm} \pm \sqrt{\xi_{15\,jm}^2 - 4\xi_{14\,jm}\xi_{16\,jm}}}{2\xi_{14\,jm}}, \quad j = 1,2, \tag{4.119}$$

$$\kappa_{ijm}^{(4)} = \xi_{16+i\,jm} - \xi_{19+i\,jm}\kappa_{4jm}^{(4)}, \quad i = 1,2,3; \quad j = 1,2, \tag{4.120}$$

$$\kappa_{4jm}^{(4)} = \frac{-\xi_{24\,jm} \pm \sqrt{\xi_{24\,jm}^2 - 4\xi_{23\,jm}\xi_{25\,jm}}}{2\xi_{23\,jm}}, \quad j = 1,2, \tag{4.121}$$

where the conditions $\xi_{15\,jm}^2 - 4\xi_{14\,jm}\xi_{16\,jm} \geq 0$, $\xi_{24\,jm}^2 - 4\xi_{23\,jm}\xi_{25\,jm} \geq 0$ (Eqs. (4.119), (4.121)) are required to be fulfilled for $\varphi, \nu \in \langle 0, \pi/2 \rangle$ due to the function $f_c = f_c(\varphi, \nu)$ (see Eqs. (3.8), (3.16), (3.18)) included in the coefficients $\xi_{13+i\,jm}$, $\xi_{22+i\,jm}$ ($i = 1,2,3$; $j = 1,2$) (see Eqs. (4.133)–(4.135), (4.142)–(4.144)). Otherwise, the additional boundary condition given by Eq. (4.100) can not be considered. Additionally, a principle of minimal total potential energy of an elastic solid body [26] is required to be considered (see Section 3.4.) due to the two solutions for each of the coefficients $\kappa_{3jm}^{(3)}$, $\kappa_{4jm}^{(4)}$ ($j = 1,2$) (see Eqs. (4.119), (4.121)).

The coefficients ψ_{im} ($i = 1,\ldots,9$); $\xi_{11+i\,jm}$ ($i = 1,\ldots,14$; $j = 1,2$) are derived as

$$\psi_{1m} = \xi_{53m}\xi_{62m} - \xi_{52m}\xi_{63m}, \tag{4.122}$$

$$\psi_{2m} = \xi_{51m}\xi_{63m} - \xi_{53m}\xi_{61m}, \tag{4.123}$$

$$\psi_{3m} = \xi_{52m}\xi_{61m} - \xi_{51m}\xi_{62m}, \tag{4.124}$$

$$\psi_{4m} = \xi_{54m}\,\xi_{63m} - \xi_{53m}\,\xi_{64m}, \tag{4.125}$$

$$\psi_{5m} = \xi_{52m}\,\xi_{64m} - \xi_{54m}\,\xi_{62m}, \tag{4.126}$$

$$\psi_{6m} = \xi_{53m}\,\xi_{64m} - \xi_{54m}\,\xi_{63m}, \tag{4.127}$$

$$\psi_{7m} = \xi_{54m}\,\xi_{61m} - \xi_{51m}\,\xi_{64m}, \tag{4.128}$$

$$\psi_{8m} = \xi_{54m}\,\xi_{62m} - \xi_{52m}\,\xi_{64m}, \tag{4.129}$$

$$\psi_{9m} = \xi_{51m}\,\xi_{64m} - \xi_{54m}\,\xi_{61m}, \tag{4.130}$$

$$\begin{aligned} \psi_{9+im} &= \psi_{1m}\left(1+\delta_{1i}\,\xi_{61m}\right)\left(\frac{R_1 f_c}{R_2}\right)^{\lambda_{2m}+\lambda_{3m}} \\ &+ \psi_{2m}\left(1+\delta_{1i}\,\xi_{62m}\right)\left(\frac{R_1 f_c}{R_2}\right)^{\lambda_{1m}+\lambda_{3m}} + \psi_{3m}\left(1+\delta_{1i}\,\xi_{63m}\right)\left(\frac{R_1 f_c}{R_2}\right)^{\lambda_{1m}+\lambda_{2m}}, \\ i &= 1,2, \end{aligned} \tag{4.131}$$

$$\xi_{11+i\,jm} = \frac{\xi_{63m}}{\xi_{63-im} - \xi_{60+im}\left(\frac{R_1 f_c}{R_j}\right)^{\lambda_{im}-\lambda_{3-im}}}\left(\frac{R_1 f_c}{R_j}\right)^{\lambda_{3m}-\lambda_{3-im}}, \quad i,j=1,2, \tag{4.132}$$

$$\begin{aligned} \xi_{14\,jm} &= 2\lambda_{1m}\,\omega_{11m}\left[\kappa_{1\,jm}^{(1)} - \xi_{12\,jm}\right]^2\left(\frac{R_1 f_c}{R_j}\right)^{2\lambda_{1m}} \\ &+ 2\lambda_{2m}\,\omega_{22m}\left[\kappa_{2\,jm}^{(1)} - \xi_{13\,jm}\right]^2\left(\frac{R_1 f_c}{R_j}\right)^{2\lambda_{2m}} + 2\lambda_{3m}\,\omega_{33m}\left(\frac{R_1 f_c}{R_j}\right)^{2\lambda_{3m}} \\ &+ \left(\lambda_{1m}+\lambda_{2m}\right)\left(\omega_{12m}+\omega_{21m}\right)\left[\kappa_{1\,jm}^{(1)} - \xi_{12\,jm}\right]\left[\kappa_{2\,jm}^{(1)} - \xi_{13\,jm}\right]\left(\frac{R_1 f_c}{R_j}\right)^{\lambda_{1m}+\lambda_{2m}} \\ &- \left(\lambda_{1m}+\lambda_{3m}\right)\left(\omega_{13m}+\omega_{31m}\right)\left(\frac{R_1 f_c}{R_j}\right)^{\lambda_{1m}+\lambda_{3m}} \\ &- \left(\lambda_{2m}+\lambda_{3m}\right)\left(\omega_{23m}+\omega_{32m}\right)\left[\kappa_{2\,jm}^{(1)} - \xi_{13\,jm}\right]\left(\frac{R_1 f_c}{R_j}\right)^{\lambda_{2m}+\lambda_{3m}}, \\ j &= 1,2, \end{aligned} \tag{4.133}$$

$$\begin{aligned} \xi_{15\,jm} &= -4\lambda_{1m}\,\omega_{11m}\,\kappa_{1\,jm}^{(1)}\left[\kappa_{1\,jm}^{(1)} - \xi_{12\,jm}\right]\left(\frac{R_1 f_c}{R_j}\right)^{2\lambda_{1m}} \\ &- 4\lambda_{2m}\,\omega_{22m}\,\kappa_{2\,jm}^{(1)}\left[\kappa_{2\,jm}^{(1)} - \xi_{13\,jm}\right]\left(\frac{R_1 f_c}{R_j}\right)^{2\lambda_{2m}} \\ &- \left(\lambda_{1m}+\lambda_{2m}\right)\left(\omega_{12m}+\omega_{21m}\right) \\ &\times\left\{\kappa_{1\,jm}^{(1)}\left[\kappa_{2\,jm}^{(1)} - \xi_{13\,jm}\right] + \kappa_{2\,jm}^{(1)}\left[\kappa_{1\,jm}^{(1)} - \xi_{12\,jm}\right]\right\}\left(\frac{R_1 f_c}{R_j}\right)^{\lambda_{1m}+\lambda_{2m}} \end{aligned}$$

$$+\kappa_{1jm}^{(1)}\left(\lambda_{1m}+\lambda_{3m}\right)\left(\omega_{13m}+\omega_{31m}\right)\left(\frac{R_1 f_c}{R_j}\right)^{\lambda_{1m}+\lambda_{3m}}$$

$$+\kappa_{2jm}^{(1)}\left(\lambda_{2m}+\lambda_{3m}\right)\left(\omega_{23m}+\omega_{32m}\right)\left(\frac{R_1 f_c}{R_j}\right)^{\lambda_{2m}+\lambda_{3m}}, \quad j=1,2, \tag{4.134}$$

$$\xi_{16\,jm}=2\lambda_{1m}\omega_{11m}\left[\kappa_{1jm}^{(1)}\right]^2\left(\frac{R_1 f_c}{R_j}\right)^{2\lambda_{1m}}+2\lambda_{2m}\omega_{22m}\left[\kappa_{2jm}^{(1)}\right]^2\left(\frac{R_1 f_c}{R_j}\right)^{2\lambda_{2m}}$$

$$+\kappa_{1jm}^{(1)}\kappa_{2jm}^{(1)}\left(\lambda_{1m}+\lambda_{2m}\right)\left(\omega_{12m}+\omega_{21m}\right)\left(\frac{R_1 f_c}{R_j}\right)^{\lambda_{1m}+\lambda_{2m}}, \quad j=1,2. \tag{4.135}$$

$$\xi_{17\,jm}=\frac{\psi_{1m}}{\psi_{10m}}\left(\frac{R_1 f_c}{R_j}\right)^{\lambda_{2m}+\lambda_{3m}}, \quad j=1,2, \tag{4.136}$$

$$\xi_{18\,jm}=\frac{\psi_{2m}}{\psi_{10m}}\left(\frac{R_1 f_c}{R_j}\right)^{\lambda_{1m}+\lambda_{3m}}, \quad j=1,2, \tag{4.137}$$

$$\xi_{19\,jm}=\frac{\psi_{3m}}{\psi_{10m}}\left(\frac{R_1 f_c}{R_j}\right)^{\lambda_{1m}+\lambda_{2m}}, \quad j=1,2, \tag{4.138}$$

$$\xi_{20\,jm}=\frac{1}{\psi_{10m}}\left[\psi_{4m}\left(\frac{R_1 f_c}{R_j}\right)^{\lambda_{3m}+\lambda_{4m}}+\psi_{5m}\left(\frac{R_1 f_c}{R_j}\right)^{\lambda_{2m}+\lambda_{4m}}+\psi_{1m}\left(\frac{R_1 f_c}{R_j}\right)^{\lambda_{2m}+\lambda_{3m}}\right], \quad j=1,2, \tag{4.139}$$

$$\xi_{21\,jm}=\frac{1}{\psi_{10m}}\left[\psi_{6m}\left(\frac{R_1 f_c}{R_j}\right)^{\lambda_{3m}+\lambda_{4m}}+\psi_{7m}\left(\frac{R_1 f_c}{R_j}\right)^{\lambda_{1m}+\lambda_{4m}}+\psi_{2m}\left(\frac{R_1 f_c}{R_j}\right)^{\lambda_{1m}+\lambda_{3m}}\right], \quad j=1,2, \tag{4.140}$$

$$\xi_{22\,jm}=\frac{1}{\psi_{10m}}\left[\psi_{8m}\left(\frac{R_1 f_c}{R_j}\right)^{\lambda_{2m}+\lambda_{4m}}+\psi_{9m}\left(\frac{R_1 f_c}{R_j}\right)^{\lambda_{1m}+\lambda_{4m}}+\psi_{3m}\left(\frac{R_1 f_c}{R_j}\right)^{\lambda_{1m}+\lambda_{2m}}\right], \quad j=1,2, \tag{4.141}$$

$$\xi_{23\,jm}=2\lambda_{1m}\omega_{11m}\xi_{20\,jm}^2\left(\frac{R_1 f_c}{R_j}\right)^{2\lambda_{1m}}+2\lambda_{2m}\omega_{22m}\xi_{21\,jm}^2\left(\frac{R_1 f_c}{R_j}\right)^{2\lambda_{2m}}$$

$$+ 2\lambda_{3m}\,\omega_{33m}\,\xi_{22\,jm}^{2}\left(\frac{R_1 f_c}{R_j}\right)^{2\lambda_{3m}} + 2\lambda_{4m}\,\omega_{44m}\left(\frac{R_1 f_c}{R_j}\right)^{2\lambda_{4m}}$$
$$+ \xi_{20\,jm}\,\xi_{21\,jm}\left(\lambda_{1m}+\lambda_{2m}\right)\left(\omega_{12m}+\omega_{21m}\right)\left(\frac{R_1 f_c}{R_j}\right)^{\lambda_{1m}+\lambda_{2m}}$$
$$+ \xi_{20\,jm}\,\xi_{22\,jm}\left(\lambda_{1m}+\lambda_{3m}\right)\left(\omega_{13m}+\omega_{31m}\right)\left(\frac{R_1 f_c}{R_j}\right)^{\lambda_{1m}+\lambda_{3m}}$$
$$+ \xi_{21\,jm}\,\xi_{22\,jm}\left(\lambda_{2m}+\lambda_{3m}\right)\left(\omega_{23m}+\omega_{32m}\right)\left(\frac{R_1 f_c}{R_j}\right)^{\lambda_{2m}+\lambda_{3m}}$$
$$- \xi_{20\,jm}\left(\lambda_{1m}+\lambda_{4m}\right)\left(\omega_{14m}+\omega_{41m}\right)\left(\frac{R_1 f_c}{R_j}\right)^{\lambda_{1m}+\lambda_{4m}}$$
$$- \xi_{21\,jm}\left(\lambda_{2m}+\lambda_{4m}\right)\left(\omega_{24m}+\omega_{42m}\right)\left(\frac{R_1 f_c}{R_j}\right)^{\lambda_{2m}+\lambda_{4m}}$$
$$- \xi_{22\,jm}\left(\lambda_{3m}+\lambda_{4m}\right)\left(\omega_{34m}+\omega_{43m}\right)\left(\frac{R_1 f_c}{R_j}\right)^{\lambda_{3m}+\lambda_{4m}}, \quad j = 1,2, \tag{4.142}$$

$$\xi_{24\,jm} = -4\lambda_{1m}\,\omega_{11m}\,\xi_{17\,jm}\,\xi_{20\,jm}\left(\frac{R_1 f_c}{R_j}\right)^{2\lambda_{1m}}$$
$$- 4\lambda_{2m}\,\omega_{22m}\,\xi_{18\,jm}\,\xi_{21\,jm}\left(\frac{R_1 f_c}{R_j}\right)^{2\lambda_{2m}}$$
$$- 4\lambda_{3m}\,\omega_{33m}\,\xi_{19\,jm}\,\xi_{22\,jm}\left(\frac{R_1 f_c}{R_j}\right)^{2\lambda_{3m}}$$
$$- \left(\lambda_{1m}+\lambda_{2m}\right)\left(\omega_{12m}+\omega_{21m}\right)\left(\xi_{17\,jm}\,\xi_{21\,jm}+\xi_{18\,jm}\,\xi_{20\,jm}\right)\left(\frac{R_1 f_c}{R_j}\right)^{\lambda_{1m}+\lambda_{2m}}$$
$$- \left(\lambda_{1m}+\lambda_{3m}\right)\left(\omega_{13m}+\omega_{31m}\right)\left(\xi_{17\,jm}\,\xi_{22\,jm}+\xi_{19\,jm}\,\xi_{20\,jm}\right)\left(\frac{R_1 f_c}{R_j}\right)^{\lambda_{1m}+\lambda_{3m}}$$
$$- \left(\lambda_{2m}+\lambda_{3m}\right)\left(\omega_{23m}+\omega_{32m}\right)\left(\xi_{18\,jm}\,\xi_{22\,jm}+\xi_{19\,jm}\,\xi_{21\,jm}\right)\left(\frac{R_1 f_c}{R_j}\right)^{\lambda_{2m}+\lambda_{3m}}$$
$$+ \xi_{17\,jm}\left(\lambda_{1m}+\lambda_{4m}\right)\left(\omega_{14m}+\omega_{41m}\right)\left(\frac{R_1 f_c}{R_j}\right)^{\lambda_{1m}+\lambda_{4m}}$$
$$+ \xi_{18\,jm}\left(\lambda_{2m}+\lambda_{4m}\right)\left(\omega_{24m}+\omega_{42m}\right)\left(\frac{R_1 f_c}{R_j}\right)^{\lambda_{2m}+\lambda_{4m}}$$
$$+ \xi_{19\,jm}\left(\lambda_{3m}+\lambda_{4m}\right)\left(\omega_{34m}+\omega_{43m}\right)\left(\frac{R_1 f_c}{R_j}\right)^{\lambda_{3m}+\lambda_{4m}}, \quad j = 1,2, \tag{4.143}$$

$$\xi_{25\,jm} = 2\lambda_{1m}\,\omega_{11m}\,\xi_{17\,jm}^{2}\left(\frac{R_1 f_c}{R_j}\right)^{2\lambda_{1m}} 2\lambda_{2m}\,\omega_{22m}\,\xi_{18\,jm}^{2}\left(\frac{R_1 f_c}{R_j}\right)^{2\lambda_{2m}}$$
$$+ 2\lambda_{3m}\,\omega_{33m}\,\xi_{19\,jm}^{2}\left(\frac{R_1 f_c}{R_j}\right)^{2\lambda_{3m}}$$

$$+\xi_{17\,jm}\xi_{18\,jm}(\lambda_{1m}+\lambda_{2m})(\omega_{12m}+\omega_{21m})\left(\frac{R_1 f_c}{R_j}\right)^{\lambda_{1m}+\lambda_{2m}}$$
$$+\xi_{17\,jm}\xi_{19\,jm}(\lambda_{1m}+\lambda_{3m})(\omega_{13m}+\omega_{31m})\left(\frac{R_1 f_c}{R_j}\right)^{\lambda_{1m}+\lambda_{3m}}$$
$$+\xi_{18\,jm}\xi_{19\,jm}(\lambda_{2m}+\lambda_{3m})(\omega_{23m}+\omega_{32m})\left(\frac{R_1 f_c}{R_j}\right)^{\lambda_{2m}+\lambda_{3m}}, \quad j=1,2, \qquad (4.144)$$

where the coefficients ξ_{5im}, ξ_{6jm} $(i,j=1,2,3)$ included in Eqs. (4.122)–(4.130) are given by Eq. (4.69) for $q=m$.

Equations (4.57)–(4.65), (4.70), (4.71), (4.75), (4.78), (4.79) $(q=m)$ along with $(\varepsilon'_{22m})_{r=R_{1m}}$ (see Eq. (3.81)) are thus transformed to the forms

$$\sigma'_{11m}=-p_1\sum_{i=1}^{n_m}\kappa_{i1m}^{(x)}\left(\frac{r}{R_1}\right)^{\lambda_{im}}, \qquad (4.145)$$

$$\sigma'_{22m}=-p_1\sum_{i=1}^{n_m}\xi_{1im}\,\kappa_{i1m}^{(x)}\left(\frac{r}{R_1}\right)^{\lambda_{im}}, \qquad (4.146)$$

$$\sigma'_{33m}=-p_1\sum_{i=1}^{n_m}\xi_{2im}\,\kappa_{i1m}^{(x)}\left(\frac{r}{R_1}\right)^{\lambda_{im}}, \qquad (4.147)$$

$$\sigma'_{12m}=-p_1\sum_{i=1}^{n_m}\xi_{3im}\,\kappa_{i1m}^{(x)}\left(\frac{r}{R_1}\right)^{\lambda_{im}}, \qquad (4.148)$$

$$\sigma'_{13m}=-p_1\sum_{i=1}^{n_m}\xi_{4im}\,\kappa_{i1m}^{(x)}\left(\frac{r}{R_1}\right)^{\lambda_{im}}, \qquad (4.149)$$

$$\varepsilon'_{11m}=-p_1\sum_{i=1}^{n_m}\xi_{5im}\,\kappa_{i1m}^{(x)}\left(\frac{r}{R_1}\right)^{\lambda_{im}}, \qquad (4.150)$$

$$\varepsilon'_{22m}=\varepsilon'_{33m}=-p_1\sum_{i=1}^{n_m}\xi_{6im}\,\kappa_{i1m}^{(x)}\left(\frac{r}{R_1}\right)^{\lambda_{im}}, \qquad (4.151)$$

$$\varepsilon'_{13m}=-p_1\sum_{i=1}^{n_m}\xi_{7im}\,\kappa_{i1m}^{(x)}\left(\frac{r}{R_1}\right)^{\lambda_{im}}, \qquad (4.152)$$

$$\varepsilon'_{12m}=-p_1\sum_{i=1}^{n_m}\xi_{8im}\,\kappa_{i1m}^{(x)}\left(\frac{r}{R_1}\right)^{\lambda_{im}}, \qquad (4.153)$$

$$w_m=\frac{p_1^2}{2}\sum_{i,j=1}^{n_m}\omega_{ijm}\,\kappa_{i1m}^{(x)}\kappa_{j1m}^{(x)}\left(\frac{r}{R_1}\right)^{\lambda_{im}+\lambda_{jm}}, \qquad (4.154)$$

$$W_m=4R_1^3\int_0^{\pi/2}\int_0^{\pi/2}p_1^2\,\Omega_{1m}\,d\varphi\,d\nu, \qquad (4.155)$$

$$\Omega_{1m} = \sum_{i,j=1}^{n_m} \frac{\omega_{ijm}\kappa_{i1m}^{(x)}\kappa_{j1m}^{(x)}}{\lambda_{im}+\lambda_{jm}+3}\left(f_c^{\lambda_{im}+\lambda_{jm}+3}-1\right), \tag{4.156}$$

$$\sigma_{im} = -p_1\sum_{j=1}^{n_m}\xi_{8+ijm}\kappa_{j1m}^{(x)}\left(\frac{r}{R_1}\right)^{\lambda_{jm}}, \quad i=1,2,3, \tag{4.157}$$

$$w_{im} = \frac{s_{iiq}p_1^2}{2}\sum_{j,k=1}^{n_m}\xi_{8+ijm}\xi_{8+ikm}\kappa_{j1m}^{(x)}\kappa_{k1m}^{(x)}\left(\frac{r}{R_1}\right)^{\lambda_{jm}+\lambda_{km}}, \quad i=1,2,3, \tag{4.158}$$

$$\left(\varepsilon_{22m}'\right)_{r=R_{1m}} = -p_1\rho_{1m}, \tag{4.159}$$

where the coefficients $\xi_{1im},\ldots,\xi_{8im}$, ω_{ijm}, ξ_{8+kim} $(i,j=1,\ldots,n_m;\ k=1,2,3)$ are given by Eqs. (4.67)–(4.69) $(q=m)$, (4.72), (4.80), respectively. The coefficient ρ_{1m} is derived as

$$\rho_{im} = (1-\beta_m)\sum_{j=1}^{n_m}\xi_{6jm}\kappa_{jim}^{(x)}\left(\frac{R_{im}}{R_1}\right)^{\lambda_{jm}}, \quad i=1,2, \tag{4.160}$$

where the radius R_{im} $(i=1,2)$ is given by Eq. (3.80).

4.7.3. Radial Stress p_1

With regard to Eqs. (3.81), (4.108), (4.159), the radial stress p_1, which is considered in Sections 4.7.1., 4.7.2. and which acts at the particle-matrix boundary, has the form

$$p_1 = \frac{\beta_m-\beta_p}{\xi_{61p}(1-\beta_p)\left(\frac{R_{1p}}{R_1}\right)^{\lambda_{1p}}-\rho_{1m}}, \tag{4.161}$$

where the radius R_{1p} and the coefficients ξ_{61p}, ρ_{1m} are given by Eqs. (3.80) and (4.69) $(q=p)$, (4.160), respectively.

4.8. Thermal Stresses in Multi-Particle-Envelope-Matrix System

Section 4.8. deals with the determination of the thermal stresses in anisotropic components of the multi-particle-envelope-matrix system (see Figure 2.1b). This determination considers the boundary conditions in Sections 4.6.1.–4.6.3. which are applied to the mathematical solutions in Section 3.3.. This determination considers the conditions $\beta_p \neq \beta_e = \beta_m$ (see Eqs. (3.54)–(3.60); see Section 4.8.1.), $\beta_p \neq \beta_e \neq \beta_m$ (see Section 4.8.2.), $\beta_p \neq \beta_e \neq \beta_m$ (see Section 4.8.3.).

With regard to the analysis in Section 3.2., R_1, R_2 and $t = R_2 - R_1$ represents radii and thickness of the embedded spherical envelope at the temperature $T \in \langle T_f, T_r\rangle$, respectively, where T_f is a final temperature of a cooling process, and the relaxation temperature T_r is determined in Section 3.5.1.1.

4.8.1. Condition $\beta_p \neq \beta_e = \beta_m$

4.8.1.1. Thermal Stresses in Spherical Particle

The integration constant C_{1p} for the spherical particle is given by Eq. (4.104). Accordingly, formulae in Section 4.7.1. are considered for the spherical particle on the condition $\beta_p \neq \beta_e = \beta_m$. The radial stress p_1 included in these formulae is given by Eq. (4.216).

4.8.1.2. Thermal Stresses in Spherical Envelope

With regard to the analyses in Sections 4.4., 4.5., 4.6.2., the integration constant C_{1e} is derived as

$$C_{1e} = -\frac{p_1}{R_1^{\lambda_{1e}}}, \tag{4.162}$$

where the radial stress p_1 acting at the particle-envelope boundary is given by Eq. (4.216).

Equations (4.57)–(4.65), (4.70), (4.71), (4.73), (4.78), (4.79) ($q = e$) along with $(\varepsilon'_{22e})_{r=R_{1e}}$ (see Eq. (3.81)) are thus transformed to the forms

$$\sigma'_{11e} = -p_1 \left(\frac{r}{R_1}\right)^{\lambda_{1e}}, \tag{4.163}$$

$$\sigma'_{22e} = -p_1 \xi_{11e} \left(\frac{r}{R_1}\right)^{\lambda_{1e}}, \tag{4.164}$$

$$\sigma'_{33e} = -p_1 \xi_{21e} \left(\frac{r}{R_1}\right)^{\lambda_{1e}}, \tag{4.165}$$

$$\sigma'_{12e} = -p_1 \xi_{31e} \left(\frac{r}{R_1}\right)^{\lambda_{1e}}, \tag{4.166}$$

$$\sigma'_{13e} = -p_1 \xi_{41e} \left(\frac{r}{R_1}\right)^{\lambda_{1e}}, \tag{4.167}$$

$$\varepsilon'_{11e} = -p_1 \xi_{51e} \left(\frac{r}{R_1}\right)^{\lambda_{1e}}, \tag{4.168}$$

$$\varepsilon'_{22e} = \varepsilon'_{33e} = -p_1 \xi_{61e} \left(\frac{r}{R_1}\right)^{\lambda_{1e}}, \tag{4.169}$$

$$\varepsilon'_{13e} = -p_1 \xi_{71e} \left(\frac{r}{R_1}\right)^{\lambda_{1e}}, \tag{4.170}$$

$$\varepsilon'_{12e} = -p_1 \xi_{81e} \left(\frac{r}{R_1}\right)^{\lambda_{1e}}, \tag{4.171}$$

$$w_e = \frac{p_1^2 \omega_{11e}}{2} \left(\frac{r}{R_1}\right)^{2\lambda_{1e}}, \tag{4.172}$$

$$W_e = 4R_1^3 \int_0^{\pi/2}\int_0^{\pi/2} \frac{p_1^2 \omega_{11e}}{2\lambda_{1e}+3}\left[\left(\frac{R_2}{R_1}\right)^{2\lambda_{1e}+3} - 1\right] d\varphi d\nu, \tag{4.173}$$

$$\sigma_{ie} = -p_1 \xi_{8+i1e}\left(\frac{r}{R_1}\right)^{\lambda_{1e}}, \quad i = 1,2,3, \tag{4.174}$$

$$w_{ie} = \frac{s_{iie} p_1^2 \xi_{8+i1e}^2}{2}\left(\frac{r}{R_1}\right)^{2\lambda_{1e}}, \quad i = 1,2,3, \tag{4.175}$$

$$\left(\varepsilon_{22e}'\right)_{r=R_{1e}} = -p_1 \rho_{1e}^{me}, \tag{4.176}$$

where the coefficients $\xi_{11e}, \ldots, \xi_{81e}$, ω_{11e}, ξ_{8+i1e} $(i = 1,2,3)$ are given by Eqs. (4.67)–(4.69) $(q = e)$, (4.72), (4.80), respectively. The coefficient ρ_{1m}^{me} is derived as

$$\rho_{ie}^{me} = \xi_{61e}(1-\beta_e)\left(\frac{R_{ie}}{R_1}\right)^{\lambda_{1e}}, \quad i = 1,2, \tag{4.177}$$

where the radius R_{ie} $(i = 1,2)$ is given by Eq. (3.80).

4.8.1.3. Thermal Stresses in Cell Matrix

As presented in Section 4.6.3., the thermal stresses in the cell matrix are determined by four combinations of boundary conditions. These four combinations are defined in Items 1–4, Section 4.6.3. (see p. 65). With regard to the analyses in Sections 4.4., 4.5., 4.6.3., the integration constant C_{im} $(i = 1, \ldots, n_m)$ is derived as

$$C_{im} = -\frac{p_1 \rho_{2e}^{me} \kappa_{i2m}^{(x)}}{R_2^{\lambda_{im}}}, \quad i = 1, \ldots, n_q, \tag{4.178}$$

where the coefficient ρ_{2e}^{me} and the radial stress p_1 acting at the particle-envelope boundary are given by Eqs. (4.177) and (4.216), respectively.

Let the analysis in Items 1–4, Section 4.7.2. be considered. Accordingly, the coefficients $\kappa_{i2m}^{(1)}$ $(i = 1,2)$, $\kappa_{i2m}^{(2)}$, $\kappa_{i2m}^{(3)}$ $(i = 1,2,3)$ have the forms

$$\kappa_{i2m}^{(1)} = \frac{1}{\xi_{60+im}\left[1 - \left(\frac{R_1 f_c}{R_2}\right)^{\lambda_{im}-\lambda_{3-im}}\right]}, \quad i = 1,2, \tag{4.179}$$

$$\kappa_{i2m}^{(2)} = \frac{\psi_{11+im}}{\sum\limits_{k=1}^{3}(-1)^{k+1}\xi_{60+km}\psi_{11+km}\left(\frac{R_1 f_c}{R_2}\right)^{\lambda_{im}-\lambda_{km}}}, \quad i = 1,2,3, \tag{4.180}$$

$$\kappa_{i2m}^{(3)} = \kappa_{i2m}^{(1)} - \kappa_{32m}^{(3)}\left[\kappa_{i2m}^{(1)} - \xi_{11+i2m}\right], \quad i = 1,2, \tag{4.181}$$

$$\kappa_{32m}^{(3)} = \frac{-\xi_{152m} \pm \sqrt{\xi_{152m}^2 - 4\xi_{142m}\xi_{162m}}}{2\xi_{142m}}, \tag{4.182}$$

where the coefficients ψ_{11+im}, ξ_{13+i2m} $(i=1,2,3)$ are given by Eq. (4.122)–(4.124), (4.133)–(4.135), respectively. The coefficient ξ_{11+i2m} $(i=1,2)$ included in Eq. (4.122)–(4.124) is derived as

$$\xi_{11+i2m} = \frac{\xi_{63m}}{\xi_{60+im}\left[1-\left(\frac{R_1 f_c}{R_2}\right)^{\lambda_{im}-\lambda_{3-im}}\right]}\left(\frac{R_1 f_c}{R_2}\right)^{\lambda_{3m}-\lambda_{3-im}}, \quad i=1,2. \tag{4.183}$$

Similarly, the coefficients $\kappa_{i2m}^{(4)}$ $(i=1,\ldots,4)$ have the forms

$$\kappa_{i2m}^{(4)} = \xi_{16+i2m} - \xi_{19+i2m}\,\kappa_{42m}^{(4)}, \quad i=1,2,3, \tag{4.184}$$

$$\kappa_{42m}^{(4)} = \frac{-\xi_{24\,2m} \pm \sqrt{\xi_{24\,2m}^2 - 4\,\xi_{23\,2m}\,\xi_{25\,2m}}}{2\,\xi_{23\,2m}}, \quad j=1,2, \tag{4.185}$$

where the coefficient ξ_{22+i2m} $(i=1,2,3)$ is given by Eq. (4.142)–(4.144). The coefficient ξ_{16+i2m} $(i=1,\ldots,6)$ included in Eq. (4.142)–(4.144) is derived as

$$\xi_{17\,2m} = \frac{\psi_{1m}}{\psi_{11m}}\left(\frac{R_1 f_c}{R_2}\right)^{\lambda_{2m}+\lambda_{3m}}, \tag{4.186}$$

$$\xi_{18\,2m} = \frac{\psi_{2m}}{\psi_{11m}}\left(\frac{R_1 f_c}{R_2}\right)^{\lambda_{1m}+\lambda_{3m}}, \tag{4.187}$$

$$\xi_{19\,2m} = \frac{\psi_{3m}}{\psi_{11m}}\left(\frac{R_1 f_c}{R_2}\right)^{\lambda_{1m}+\lambda_{2m}}, \tag{4.188}$$

$$\xi_{20\,2m} = \frac{1}{\psi_{11m}}\left[\psi_{4m}\,\xi_{62m}\left(\frac{R_1 f_c}{R_2}\right)^{\lambda_{3m}+\lambda_{4m}} + \psi_{5m}\,\xi_{63m}\left(\frac{R_1 f_c}{R_2}\right)^{\lambda_{2m}+\lambda_{4m}} + \psi_{1m}\,\xi_{64m}\left(\frac{R_1 f_c}{R_2}\right)^{\lambda_{2m}+\lambda_{3m}}\right], \tag{4.189}$$

$$\xi_{21\,2m} = \frac{1}{\psi_{11m}}\left[\psi_{6m}\,\xi_{61m}\left(\frac{R_1 f_c}{R_2}\right)^{\lambda_{3m}+\lambda_{4m}} + \psi_{7m}\,\xi_{63m}\left(\frac{R_1 f_c}{R_2}\right)^{\lambda_{1m}+\lambda_{4m}} + \psi_{2m}\,\xi_{64m}\left(\frac{R_1 f_c}{R_2}\right)^{\lambda_{1m}+\lambda_{3m}}\right], \tag{4.190}$$

$$\xi_{22\,2m} = \frac{1}{\psi_{11m}}\left[\psi_{8m}\,\xi_{61m}\left(\frac{R_1 f_c}{R_2}\right)^{\lambda_{2m}+\lambda_{4m}} + \psi_{9m}\,\xi_{62m}\left(\frac{R_1 f_c}{R_2}\right)^{\lambda_{1m}+\lambda_{4m}} + \psi_{3m}\,\xi_{64m}\left(\frac{R_1 f_c}{R_2}\right)^{\lambda_{1m}+\lambda_{2m}}\right], \tag{4.191}$$

where the coefficient ψ_{im} $(i=1,\ldots,9)$ is given by Eqs. (4.122)–(4.131).

Equations (4.57)–(4.65), (4.70), (4.71), (4.75), (4.78), (4.79) $(q=m)$ are thus transformed to the forms

$$\sigma'_{11m} = -p_1\rho_{2e}^{me}\sum_{i=1}^{n_m}\kappa_{i2m}^{(x)}\left(\frac{r}{R_2}\right)^{\lambda_{im}}, \tag{4.192}$$

$$\sigma'_{22m} = -p_1\rho_{2e}^{me}\sum_{i=1}^{n_m}\xi_{1im}\,\kappa_{i2m}^{(x)}\left(\frac{r}{R_2}\right)^{\lambda_{im}}, \tag{4.193}$$

$$\sigma'_{33m} = -p_1\rho_{2e}^{me}\sum_{i=1}^{n_m}\xi_{2im}\,\kappa_{i2m}^{(x)}\left(\frac{r}{R_2}\right)^{\lambda_{im}}, \tag{4.194}$$

$$\sigma'_{12m} = -p_1\rho_{2e}^{me}\sum_{i=1}^{n_m}\xi_{3im}\,\kappa_{i2m}^{(x)}\left(\frac{r}{R_2}\right)^{\lambda_{im}}, \tag{4.195}$$

$$\sigma'_{13m} = -p_1\rho_{2e}^{me}\sum_{i=1}^{n_m}\xi_{4im}\,\kappa_{i2m}^{(x)}\left(\frac{r}{R_2}\right)^{\lambda_{im}}, \tag{4.196}$$

$$\varepsilon'_{11m} = -p_1\rho_{2e}^{me}\sum_{i=1}^{n_m}\xi_{5im}\,\kappa_{i2m}^{(x)}\left(\frac{r}{R_2}\right)^{\lambda_{im}}, \tag{4.197}$$

$$\varepsilon'_{22m} = \varepsilon'_{33m} = -p_1\rho_{2e}^{me}\sum_{i=1}^{n_m}\xi_{6im}\,\kappa_{i2m}^{(x)}\left(\frac{r}{R_2}\right)^{\lambda_{im}}, \tag{4.198}$$

$$\varepsilon'_{13m} = -p_1\rho_{2e}^{me}\sum_{i=1}^{n_m}\xi_{7im}\,\kappa_{i2m}^{(x)}\left(\frac{r}{R_2}\right)^{\lambda_{im}}, \tag{4.199}$$

$$\varepsilon'_{12m} = -p_1\rho_{2e}^{me}\sum_{i=1}^{n_m}\xi_{8im}\,\kappa_{i2m}^{(x)}\left(\frac{r}{R_2}\right)^{\lambda_{im}}, \tag{4.200}$$

$$w_m = \frac{\left(p_1\rho_{2e}^{me}\right)^2}{2}\sum_{i,j=1}^{n_m}\omega_{ijm}\,\kappa_{i2m}^{(x)}\kappa_{j2m}^{(x)}\left(\frac{r}{R_2}\right)^{\lambda_{im}+\lambda_{jm}}, \tag{4.201}$$

$$W_m = 4\left(\rho_{2e}^{me}\right)^2 R_2^3\int_0^{\pi/2}\int_0^{\pi/2} p_1^2\,\Omega_{2m}\,d\varphi\,d\nu, \tag{4.202}$$

$$\Omega_{2m} = \sum_{i,j=1}^{n_m}\frac{\omega_{ijm}\,\kappa_{i2m}^{(x)}\kappa_{j2m}^{(x)}}{\lambda_{im}+\lambda_{jm}+3}\left[\left(\frac{R_1 f_c}{R_2}\right)^{\lambda_{im}+\lambda_{jm}+3}-1\right], \tag{4.203}$$

$$\sigma_{im} = -p_1\rho_{2e}^{me}\sum_{j=1}^{n_m}\xi_{8+ijm}\,\kappa_{j2m}^{(x)}\left(\frac{r}{R_2}\right)^{\lambda_{jm}}, \quad i=1,2,3, \tag{4.204}$$

$$w_{im} = \frac{s_{iiq}\left(p_1\rho_{2e}^{me}\right)^2}{2}\sum_{j,k=1}^{n_m}\xi_{8+ijm}\,\xi_{8+ikm}\,\kappa_{j2m}^{(x)}\kappa_{k2m}^{(x)}\left(\frac{r}{R_2}\right)^{\lambda_{jm}+\lambda_{km}}, \quad i=1,2,3, \tag{4.205}$$

where the coefficients $\xi_{1im},\ldots,\xi_{8im}$, ω_{ijm}, ξ_{8+kim} $(i,j=1,\ldots,n_m;\ k=1,2,3)$ are given by Eqs. (4.67)–(4.69) $(q=m)$, (4.72), (4.80), respectively.

4.8.1.4. Radial Stress p_1

With regard to Eqs. (3.81), (4.108), (4.176), the radial stress p_1, which is considered in Sections 4.8.1.1–4.8.1.3 and which acts at the particle-envelope boundary, has the form

$$p_1 = \frac{\beta_e - \beta_p}{\xi_{61p}(1-\beta_p)\left(\frac{R_{1p}}{R_1}\right)^{\lambda_{1p}} - \rho_{1e}^{me}}, \tag{4.206}$$

where the radius R_{1p} and the coefficients ξ_{61p}, ρ_{1e}^{me} are given by Eqs. (3.80) and (4.69) ($q = p$), (4.177), respectively.

4.8.2. Condition $\beta_p \neq \beta_e \neq \beta_m$

4.8.2.1. Thermal Stresses in Spherical Particle

The integration constant C_{1p} for the spherical particle is given by Eq. (4.101). Accordingly, formulae in Section 4.7.1. are considered for the spherical particle on the condition $\alpha'_{1p} \neq \alpha'_{1e} = \alpha'_{1m}$. The radial stress p_1 included in these formulae is given by Eq. (4.240).

4.8.2.2. Thermal Stresses in Spherical Envelope

With regard to the analyses in Sections 4.4., 4.5., 4.6.2., the integration constant C_{ie} ($i = 1,2$) is derived as

$$C_{ie} = -\sum_{j=1}^{2} \frac{p_j \kappa_{ije}}{R_j^{\lambda_{ie}}}, \quad i = 1,2. \tag{4.207}$$

where the coefficient κ_{ije} ($i, j = 1,2$) has the form

$$\kappa_{ije} = \frac{1}{1 - \left(\frac{R_i}{R_{3-i}}\right)^{\lambda_{3-je} - \lambda_{je}}}, \quad i, j = 1,2. \tag{4.208}$$

The radial stresses p_1 and p_2 acting at the particle-envelope and matrix-envelope boundaries are given by Eqs. (4.240) and (4.241), respectively.

Equations (4.57)–(4.65), (4.70), (4.71), (4.73), (4.78), (4.79) ($q = e$) along with $(\varepsilon'_{22e})_{r=R_{ie}}$ ($i = 1,2$) (see Eqs. (3.82), (3.83)) are thus transformed to the forms

$$\sigma'_{11e} = -\sum_{i,j=1}^{2} p_j \kappa_{ije} \left(\frac{r}{R_j}\right)^{\lambda_{ie}}, \tag{4.209}$$

$$\sigma'_{22e} = -\sum_{i,j=1}^{2} p_j \xi_{1ie} \kappa_{ije} \left(\frac{r}{R_j}\right)^{\lambda_{ie}}, \tag{4.210}$$

$$\sigma'_{33e} = -\sum_{i,j=1}^{2} p_j \xi_{2ie} \kappa_{ije} \left(\frac{r}{R_j}\right)^{\lambda_{ie}}, \tag{4.211}$$

$$\sigma'_{12e} = -\sum_{i,j=1}^{2} p_j \xi_{3ie} \kappa_{ije} \left(\frac{r}{R_j}\right)^{\lambda_{ie}}, \tag{4.212}$$

$$\sigma'_{13e} = -\sum_{i,j=1}^{2} p_j \xi_{4ie} \kappa_{ije} \left(\frac{r}{R_j}\right)^{\lambda_{ie}}, \tag{4.213}$$

$$\varepsilon'_{11e} = -\sum_{i,j=1}^{2} p_j \xi_{5ie} \kappa_{ije} \left(\frac{r}{R_j}\right)^{\lambda_{ie}}, \tag{4.214}$$

$$\varepsilon'_{22e} = \varepsilon'_{33e} = -\sum_{i,j=1}^{2} p_j \xi_{6ie} \kappa_{ije} \left(\frac{r}{R_j}\right)^{\lambda_{ie}}, \tag{4.215}$$

$$\varepsilon'_{13e} = -\sum_{i,j=1}^{2} p_j \xi_{7ie} \kappa_{ije} \left(\frac{r}{R_j}\right)^{\lambda_{ie}}, \tag{4.216}$$

$$\varepsilon'_{12e} = -\sum_{i,j=1}^{2} p_j \xi_{8ie} \kappa_{ije} \left(\frac{r}{R_j}\right)^{\lambda_{ie}}, \tag{4.217}$$

$$w_e = \frac{1}{2} \sum_{i,j,k,l=1}^{2} p_k p_l \omega_{ije} \kappa_{ike} \kappa_{jle} \left(\frac{r}{R_k}\right)^{\lambda_{ie}} \left(\frac{r}{R_l}\right)^{\lambda_{je}}, \tag{4.218}$$

$$W_e = 4R_2^3 \int_0^{\pi/2} \int_0^{\pi/2} \Omega_e \, d\varphi \, d\nu. \tag{4.219}$$

$$\Omega_e = \sum_{i,j,k,l=1}^{2} \frac{p_k p_l \omega_{ije} \kappa_{ike} \kappa_{jle}}{\lambda_{ie} + \lambda_{je} + 3} \left(\frac{R_2}{R_k}\right)^{\lambda_{ie}} \left(\frac{R_2}{R_l}\right)^{\lambda_{je}} \left[1 - \left(\frac{R_1}{R_2}\right)^{\lambda_{ie}+\lambda_{je}+3}\right], \tag{4.220}$$

$$\sigma_{ie} = -\sum_{j,k=1}^{2} p_k \xi_{8+ije} \kappa_{jkm} \left(\frac{r}{R_k}\right)^{\lambda_{je}}, \quad i = 1,2,3, \tag{4.221}$$

$$w_{ie} = \frac{s_{iie}}{2} \sum_{j,k,l,s=1}^{2} p_l p_s \xi_{8+ije} \xi_{8+ike} \kappa_{jlm} \kappa_{ksm} \left(\frac{r}{R_l}\right)^{\lambda_{je}} \left(\frac{r}{R_s}\right)^{\lambda_{ke}}, \quad i = 1,2,3, \tag{4.222}$$

$$\left(\varepsilon'_{22e}\right)_{r=R_{ie}} = -\sum_{j=1}^{2} p_j \rho_{ije}, \quad i = 1,2, \tag{4.223}$$

where the coefficients $\xi_{1ie}, \ldots, \xi_{8ie}$, ω_{ije}, ξ_{8+k1e} $(i,j=1,2;\ k=1,2,3)$ are given by Eqs. (4.67)–(4.69) $(q=e)$, (4.72), (4.80), respectively. The coefficient ρ_{ije} is derived as

$$\rho_{ije} = (1-\beta_e) \sum_{k=1}^{2} \xi_{6ke} \kappa_{kje} \left(\frac{R_{ie}}{R_j}\right)^{\lambda_{ke}}, \quad i,j = 1,2, \tag{4.224}$$

where the radius R_{ie} $(i=1,2)$ is given by Eq. (3.80).

4.8.2.3. Thermal Stresses in Cell Matrix

As presented in Section 4.6.3., the thermal stresses in the cell matrix are determined by four combinations of boundary conditions. These four combinations are defined in Items 1–4, Section 4.6.3. (see p. 65). With regard to the analyses in Sections 4.4., 4.5., 4.6.3., the integration constant C_{im} $(i = 1, \ldots, n_m)$ is derived as

$$C_{im} = -\frac{p_2\, \kappa_{i2m}^{(x)}}{R_2^{\lambda_{im}}}, \quad i = 1, \ldots, n_q, \tag{4.225}$$

where the radial stress p_2 acting at the matrix-envelope boundary is given by Eq. (4.241).

Let the analysis in Items 1–4, Section 4.7.2. be considered. Accordingly, the coefficients $\kappa_{i2m}^{(1)}$ $(i = 1,2)$, $\kappa_{i2m}^{(2)}$, $\kappa_{i2m}^{(3)}$ $(i = 1,2,3)$, $\kappa_{i2m}^{(4)}$ $(i = 1, \ldots, 4)$ are given by Eqs. (4.116)–(4.144).

Equations (4.57)–(4.65), (4.70), (4.71), (4.75), (4.78), (4.79) $(q = m)$ along with $(\varepsilon'_{22m})_{r=R_{2m}}$ (see Eq. (3.83)) are thus transformed to the forms

$$\sigma'_{11m} = -p_2 \sum_{i=1}^{n_m} \kappa_{i2m}^{(x)} \left(\frac{r}{R_2}\right)^{\lambda_{im}}, \tag{4.226}$$

$$\sigma'_{22m} = -p_2 \sum_{i=1}^{n_m} \xi_{1im}\, \kappa_{i2m}^{(x)} \left(\frac{r}{R_2}\right)^{\lambda_{im}}, \tag{4.227}$$

$$\sigma'_{33m} = -p_2 \sum_{i=1}^{n_m} \xi_{2im}\, \kappa_{i2m}^{(x)} \left(\frac{r}{R_2}\right)^{\lambda_{im}}, \tag{4.228}$$

$$\sigma'_{12m} = -p_2 \sum_{i=1}^{n_m} \xi_{3im}\, \kappa_{i2m}^{(x)} \left(\frac{r}{R_2}\right)^{\lambda_{im}}, \tag{4.229}$$

$$\sigma'_{13m} = -p_2 \sum_{i=1}^{n_m} \xi_{4im}\, \kappa_{i2m}^{(x)} \left(\frac{r}{R_2}\right)^{\lambda_{im}}, \tag{4.230}$$

$$\varepsilon'_{11m} = -p_2 \sum_{i=1}^{n_m} \xi_{5im}\, \kappa_{i2m}^{(x)} \left(\frac{r}{R_2}\right)^{\lambda_{im}}, \tag{4.231}$$

$$\varepsilon'_{22m} = \varepsilon'_{33m} = -p_2 \sum_{i=1}^{n_m} \xi_{6im}\, \kappa_{i2m}^{(x)} \left(\frac{r}{R_2}\right)^{\lambda_{im}}, \tag{4.232}$$

$$\varepsilon'_{13m} = -p_2 \sum_{i=1}^{n_m} \xi_{7im}\, \kappa_{i2m}^{(x)} \left(\frac{r}{R_2}\right)^{\lambda_{im}}, \tag{4.233}$$

$$\varepsilon'_{12m} = -p_2 \sum_{i=1}^{n_m} \xi_{8im}\, \kappa_{i2m}^{(x)} \left(\frac{r}{R_2}\right)^{\lambda_{im}}, \tag{4.234}$$

$$w_m = \frac{p_2^2}{2} \sum_{i,j=1}^{n_m} \omega_{ijm}\, \kappa_{i2m}^{(x)}\, \kappa_{j2m}^{(x)} \left(\frac{r}{R_2}\right)^{\lambda_{im}+\lambda_{jm}}, \tag{4.235}$$

$$W_m = 4R_2^3 \int_0^{\pi/2}\int_0^{\pi/2} p_2^2 \Omega_{2m}\, d\varphi\, d\nu, \tag{4.236}$$

$$\sigma_{im} = -p_2 \sum_{j=1}^{n_m} \xi_{8+i\,jm} \kappa_{j2m}^{(x)} \left(\frac{r}{R_2}\right)^{\lambda_{jm}}, \quad i = 1,2,3, \tag{4.237}$$

$$w_{im} = \frac{s_{iiq}\, p_2^2}{2} \sum_{j,k=1}^{n_m} \xi_{8+i\,jm} \xi_{8+ikm} \kappa_{j2m}^{(x)} \kappa_{k2m}^{(x)} \left(\frac{r}{R_2}\right)^{\lambda_{jm}+\lambda_{km}}, \quad i = 1,2,3, \tag{4.238}$$

$$\left(\varepsilon_{22m}'\right)_{r=R_{2m}} = -p_2 \rho_{2m}, \tag{4.239}$$

where the radius R_{2m} and the coefficients $\xi_{1im}, \ldots, \xi_{8im}$, ω_{ijm}, ξ_{8+kim} $(i,j = 1,\ldots,n_m;$ $k = 1,2,3)$, Ω_{2m}, ρ_{2m}, are given by Eqs. (3.80) and (4.67)–(4.69) $(q = m)$, (4.72), (4.80), (4.203), (4.160), respectively.

4.8.2.4. Radial Stresses p_1, p_2

With regard to Eqs. (4.64), (4.65), (4.108), (4.223), (4.239), the radial stresses p_1 and p_2, which act at the particle-envelope and matrix-envelope boundaries, respectively, and which are considered in Sections 4.8.2.1–4.8.2.3, have the forms

$$p_1 = \frac{\rho_{12e}(\beta_m - \beta_e) - (\rho_{2m} - \rho_{22e})(\beta_e - \beta_p)}{\rho_{12e}\rho_{21e} - (\rho_{2m} - \rho_{22e})\left[\xi_{61p}(1-\beta_p)\left(\frac{R_{1p}}{R_1}\right)^{\lambda_{1p}} - \rho_{11e}\right]}, \tag{4.240}$$

$$p_2 = \frac{\left[\xi_{61p}(1-\beta_p)\left(\frac{R_{1p}}{R_1}\right)^{\lambda_{1p}} - \rho_{11e}\right](\beta_m - \beta_e) - \rho_{21e}(\beta_e - \beta_p)}{\rho_{12e}\rho_{21e} - (\rho_{2m} - \rho_{22e})\left[\xi_{61p}(1-\beta_p)\left(\frac{R_{1p}}{R_1}\right)^{\lambda_{1p}} - \rho_{11e}\right]}, \tag{4.241}$$

where the radius R_{1p} and the coefficients ξ_{61p}, ρ_{2m}, ρ_{ije} $(i,j = 1,2)$ are given by Eqs. (3.80) and (4.69) $(q = p)$, (4.160), (4.224), respectively.

4.8.3. Condition $\beta_p = \beta_e \neq \beta_m$

4.8.3.1. Thermal Stresses in Spherical Particle

With regard to the analyses in Sections 4.4., 4.5., 4.6.1., the integration constant C_{1p} is derived as

$$C_{1p} = -\frac{p_2 \rho_{1e}^{pe}}{\xi_{61p} R_1^{\lambda_{1p}}}, \tag{4.242}$$

where the coefficients ξ_{61p}, ρ_{1e}^{pe} and the radial stress p_2 acting at the matrix-envelope boundary are given by Eqs. (4.69) $(q = p)$, (4.271) and (4.272), respectively.

Equations (4.57)–(4.65), (4.70), (4.71), (4.73), (4.78), (4.79) $(q = p)$ are thus transformed to the forms

$$\sigma_{11p}' = -\frac{p_2 \rho_{1e}^{pe}}{\xi_{61p}} \left(\frac{r}{R_1}\right)^{\lambda_{1p}}, \tag{4.243}$$

$$\sigma'_{22p} = -\frac{p_2 \rho_{1e}^{pe} \xi_{11p}}{\xi_{61p}} \left(\frac{r}{R_1}\right)^{\lambda_{1p}}, \tag{4.244}$$

$$\sigma'_{33p} = -\frac{p_2 \rho_{1e}^{pe} \xi_{21p}}{\xi_{61p}} \left(\frac{r}{R_1}\right)^{\lambda_{1p}}, \tag{4.245}$$

$$\sigma'_{12p} = -\frac{p_2 \rho_{1e}^{pe} \xi_{31p}}{\xi_{61p}} \left(\frac{r}{R_1}\right)^{\lambda_{1p}}, \tag{4.246}$$

$$\sigma'_{13p} = -\frac{p_2 \rho_{1e}^{pe} \xi_{41p}}{\xi_{61p}} \left(\frac{r}{R_1}\right)^{\lambda_{1p}}, \tag{4.247}$$

$$\varepsilon'_{11p} = -\frac{p_2 \rho_{1e}^{pe} \xi_{51p}}{\xi_{61p}} \left(\frac{r}{R_1}\right)^{\lambda_{1p}}, \tag{4.248}$$

$$\varepsilon'_{22p} = \varepsilon'_{33p} = -p_2 \rho_{1e}^{pe} \left(\frac{r}{R_1}\right)^{\lambda_{1p}}, \tag{4.249}$$

$$\varepsilon'_{13p} = -\frac{p_2 \rho_{1e}^{pe} \xi_{71p}}{\xi_{61p}} \left(\frac{r}{R_1}\right)^{\lambda_{1p}}, \tag{4.250}$$

$$\varepsilon'_{12p} = -\frac{p_2 \rho_{1e}^{pe} \xi_{81p}}{\xi_{61p}} \left(\frac{r}{R_1}\right)^{\lambda_{1p}}, \tag{4.251}$$

$$w_p = \frac{\omega_{11p}}{2} \left(\frac{p_2 \rho_{1e}^{pe}}{\xi_{61p}}\right)^2 \left(\frac{r}{R_1}\right)^{2\lambda_{1p}}, \tag{4.252}$$

$$W_p = 4R_1^3 \int_0^{\pi/2} \int_0^{\pi/2} \frac{\omega_{11p}}{2\lambda_{1p}+3} \left(\frac{p_2 \rho_{1e}^{pe}}{\xi_{61p}}\right)^2 d\varphi\, d\nu, \tag{4.253}$$

$$\sigma_{ip} = -\frac{p_2 \rho_{1e}^{pe} \xi_{8+i1p}}{\xi_{61p}} \left(\frac{r}{R_1}\right)^{\lambda_{1p}}, \quad i = 1,2,3, \tag{4.254}$$

$$w_{ip} = \frac{s_{iip} \xi_{8+i1p}^2}{2} \left(\frac{p_2 \rho_{1e}^{pe}}{\xi_{61p}}\right)^2 \left(\frac{r}{R_1}\right)^{2\lambda_{1p}}, \quad i = 1,2,3, \tag{4.255}$$

where the coefficients $\xi_{11p}, \ldots, \xi_{81p}$, ω_{11p}, ξ_{8+i1p} ($i = 1,2,3$) are given by Eqs. (4.67)–(4.69) ($q = p$), (4.72), (4.80), respectively.

4.8.3.2. Thermal Stresses in Spherical Envelope

With regard to the analyses in Sections 4.4., 4.5., 4.6.2., the integration constant C_{1e} is derived as

$$C_{1e} = -\frac{p_2}{R_2^{\lambda_{1e}}}, \tag{4.256}$$

where the radial stress p_2 acting at the matrix-envelope boundary is given by Eq. (4.272).

Equations (4.57)–(4.65), (4.70), (4.71), (4.73), (4.78), (4.79) ($q = e$) along with $(\varepsilon'_{22e})_{r=R_{2e}}$ (see Eq. (3.83)) are thus transformed to the forms

$$\sigma'_{11e} = -p_2 \left(\frac{r}{R_2}\right)^{\lambda_{1e}}, \tag{4.257}$$

$$\sigma'_{22e} = -p_2 \, \xi_{11e} \left(\frac{r}{R_2}\right)^{\lambda_{1e}}, \tag{4.258}$$

$$\sigma'_{33e} = -p_2 \, \xi_{21e} \left(\frac{r}{R_2}\right)^{\lambda_{1e}}, \tag{4.259}$$

$$\sigma'_{12e} = -p_2 \, \xi_{31e} \left(\frac{r}{R_2}\right)^{\lambda_{1e}}, \tag{4.260}$$

$$\sigma'_{13e} = -p_2 \, \xi_{41e} \left(\frac{r}{R_2}\right)^{\lambda_{1e}}, \tag{4.261}$$

$$\varepsilon'_{11e} = -p_2 \, \xi_{51e} \left(\frac{r}{R_2}\right)^{\lambda_{1e}}, \tag{4.262}$$

$$\varepsilon'_{22e} = \varepsilon'_{33e} = -p_2 \, \xi_{61e} \left(\frac{r}{R_2}\right)^{\lambda_{1e}}, \tag{4.263}$$

$$\varepsilon'_{13e} = -p_2 \, \xi_{71e} \left(\frac{r}{R_2}\right)^{\lambda_{1e}}, \tag{4.264}$$

$$\varepsilon'_{12e} = -p_2 \, \xi_{81e} \left(\frac{r}{R_2}\right)^{\lambda_{1e}}, \tag{4.265}$$

$$w_e = \frac{p_2^2 \, \omega_{11e}}{2} \left(\frac{r}{R_2}\right)^{2\lambda_{1e}}, \tag{4.266}$$

$$W_e = 4R_2^3 \int_0^{\pi/2} \int_0^{\pi/2} \frac{p_2^2 \, \omega_{11e}}{2\lambda_{1e} + 3} \left[1 - \left(\frac{R_1}{R_2}\right)^{2\lambda_{1e}+3}\right] d\varphi \, d\nu, \tag{4.267}$$

$$\sigma_{ie} = -p_2 \, \xi_{8+i1e} \left(\frac{r}{R_2}\right)^{\lambda_{1e}}, \quad i = 1,2,3, \tag{4.268}$$

$$w_{ie} = \frac{s_{iie} \, p_2^2 \, \xi_{8+i1e}^2}{2} \left(\frac{r}{R_2}\right)^{2\lambda_{1e}}, \quad i = 1,2,3, \tag{4.269}$$

$$\left(\varepsilon'_{22e}\right)_{r=R_{2e}} = -p_2 \rho_{2e}^{pe}, \tag{4.270}$$

where the coefficients $\xi_{11e}, \ldots, \xi_{81e}$, ω_{11e}, ξ_{8+i1e} ($i = 1,2,3$) are given by Eqs. (4.67)–(4.69) ($q = e$), (4.72), (4.80), respectively. The coefficient ρ_{2e}^{pe} is derived as

$$\rho_{ie}^{pe} = \xi_{61e} (1 - \beta_e) \left(\frac{R_{ie}}{R_{2e}}\right)^{\lambda_{1e}}, \quad i = 1,2, \tag{4.271}$$

where the radius R_{ie} ($i = 1,2$) is given by Eq. (3.80).

4.8.3.3. Thermal Stresses in Cell Matrix

The integration constant C_{im} $(i = 1, \ldots, n_m)$ for the cell matrix is given by Eq. (4.225). Accordingly, formulae in Section 4.8.2.3 are considered for the cell matrix on the condition $\beta_p = \beta_e \neq \beta_m$. The radial stress p_2 included in these formulae is given by Eq. (4.272).

4.8.3.4. Radial Stress p_2

With regard to Eqs. (4.65), (4.239), (4.270), the radial stress p_2, which is considered in Sections 4.8.3.1–4.8.1.3 and which acts at the matrix-envelope boundary, has the form

$$p_2 = \frac{\beta_m - \beta_e}{\rho_{2e}^{pe} - \rho_{2m}}, \tag{4.272}$$

where the coefficients ρ_{2e}^{pe}, ρ_{2m} are given by Eqs. (4.271), (4.160), respectively.

Chapter 5

Thermal Stresses in Model Systems with Isotropic Components

5.1. Introduction

Chapter 5 deals with the determination of the thermal stresses in the multi-particle-matrix and multi-particle-envelope-matrix systems (see Figure 2.1) both with isotropic components.

The determination of the thermal stresses in an isotropic component of these systems results from the Hooke's law for an isotropic continuum (see Eqs. (3.44)–(3.49)) as well as from the Cauchy's and equilibrium equations (see Eqs. (3.19)–(3.23), (3.32)–(3.34)) which are determined by the spherical coordinates (r, φ, ν) (see Figure 3.1a) for the infinitesimal spherical cap (see Figure 3.1b). The compatibility equations (see Eqs. (3.24)–(3.28)) are not considered in contrast to the determination for an anisotropic component (see Chapter 4).

As presented in Section 5.2., the connection of the Cauchy's equations and the Hooke's law results in the determination of the radial, tangential and shear stresses as functions of the thermal-stress induced radial stress u'_1 along the axis x'_1 (see Figure (3.1)), i.e., $\sigma'_{ij} = \sigma'_{ij}(u'_1)$ $(i, j = 1,2,3)$ (see Eqs. (5.1)–(5.4)). With regard to $\sigma'_{ij} = \sigma'_{ij}(u'_1)$ $(i, j = 1,2,3)$, the equilibrium equation which is related to the axis x'_1 (see Eq. (3.32)) along with a sum of the equilibrium equations which are related to the axes x'_2 and x'_2 (see Eqs. (3.33), (3.34)) represent a system of differential equations in terms of u'_1 (see Eqs. (5.6), (5.7)).

Consequently, Sections 5.3.–5.6. present four different mathematical procedures which are applied to a system of these two differential equations. This system is thus transformed to four mutually different differential equations in terms of u'_1 with zero or non-zero right sides. The Wronskian's method [113] is applied to the differential equations with non-zero right sides. These four different mathematical procedures result in four different solutions for u'_1. The determination of stresses and strains then results from the functions $\sigma'_{ij} = \sigma'_{ij}(u'_1)$ and $\varepsilon'_{ij} = \varepsilon'_{ij}(u'_1)$ $(i, j = 1,2,3)$ (see Eqs. (3.19)–(3.23), respectively.

The radial displacement $u'_{1q} = u'_{1q}(r, \varphi, \nu)$ which is related to the i-th mathematical procedure $(i = 1, \ldots, 4)$ is derived as $u'_{1q} = \sum_{j=1}^{n_q} C^{(i)}_{jq} \xi^{(i)}_{jq} u^{(i)}_{jrq}$, where n_q is a number of boundary conditions, $\xi^{(i)}_{jq}$ is a coefficient and $u^{(i)}_{jrq} = u^{(i)}_{jrq}(r)$ $(j = 1, \ldots, n_q)$ is a function of r.

The integration constants $C_{1q}^{(i)}, \ldots, C_{n_q q}^{(i)}$ are then determined by the boundary conditions for the spherical particle ($q=p$) (see Section 4.6.1.), the spherical envelope ($q=e$) (see Section 4.6.2.) and the cell matrix ($q=m$) (see Section 4.6.3.), where $C_{jq}^{(i)} = C_{jq}^{(i)}(\varphi, \nu)$ ($j=1,\ldots,n_q$). Consequently, an analysis in Section 5.7. determines such functions of the set $\left\{u_{jrq}^{(i)}, \ldots, u_{nrq}^{(i)}\right\}$ which fulfil the boundary conditions, and which are thus considered for the determination of corresponding integration constants.

In case of an *anisotropic* component, the radial displacement $u'_{1q} = u'_{1q}(r,\varphi,\nu)$ is derived as $u'_{1q} = \sum_{i=1}^{n_q} \xi_{6iq} C_{iq} r^{\lambda_{iq}+1}$ (see Eq. (4.66)), where $\lambda_{1q}, \ldots, \lambda_{n_q q}$ represent roots of one of the characteristic equations (4.36), (4.41), (4.45), (4.49), where $\lambda_{iq} = \lambda_{iq}(\varphi, \nu)$ ($i=1,\ldots,n_q$). These characteristic equations are determined by four different mathematical procedures presented in Section 4.3.. As presented in Section 4.4., the analysis of a number of solutions for the thermal stresses in an anisotropic component of the model systems (see Figure 2.1) is based on an analysis of the exponents $\lambda_{1q}, \ldots, \lambda_{n_q q}$. This analysis can be performed numerically for a real two- or three-component material defined in Section 2.1..

In contrast to an anisotropic component, a number of solutions for an isotropic component is based on an analysis of a course of the dependence $u_{jrq}^{(i)} - r$ ($i=1,\ldots,4$; $j=1,\ldots,n_q$) due to $u'_{1q} = \sum_{j=1}^{n_q} C_{jq}^{(i)} \xi_{jq}^{(i)} u_{jrq}^{(i)}$. This analysis for each isotropic component of the multi-particle-matrix and multi-particle-envelope-matrix systems (see Figure 2.1) is presented in Section 5.7..

Due to a range of this book, final formulae for the thermal stresses in isotropic components of the multi-particle-matrix and multi-particle-envelope-matrix systems (see Figure 2.1) are not presented. These final formulae can be easily determined with respect to the analysis in Section 5.7..

5.2. Differential Equations in Terms of Radial Displacement

Let the Cauchy's equations (see Eqs. (3.19)–(3.23) be substituted to the Hooke's law for an isotropic continuum (see Eqs. (3.44)–(3.49)). Consequently, the radial stress σ'_{11} acting along the axis x'_1 (see Figure 3.1), the tangential stress σ'_{22} acting along the axis x'_2, the tangential stress σ'_{33} acting along the axis x'_3, the shear stress σ'_{12} acting along the axis x'_2 and the shear stress σ'_{13} acting along the axis x'_3 have the forms

$$\sigma'_{11} = (c_1 + c_2)\frac{\partial u'_1}{\partial r} - 2c_2 \frac{u'_1}{r}, \tag{5.1}$$

$$\sigma'_{22} = \sigma'_{33} = -c_2 \frac{\partial u'_1}{\partial r} + c_1 \frac{u'_1}{r}, \tag{5.2}$$

$$\sigma'_{12} = \frac{1}{s_{44} r} \frac{\partial u'_1}{\partial \varphi}, \tag{5.3}$$

$$\sigma'_{13} = \frac{1}{s_{44} r} \frac{\partial u'_1}{\partial \nu}, \tag{5.4}$$

where the condition $\sigma'_{22} = \sigma'_{33}$ results from $\varepsilon'_{22} = \varepsilon'_{33}$ (see Eqs. (3.20), (3.21), (3.45), (3.46)). The coefficients c_1, c_2, c_3 are derived as (see Eq. (3.49))

$$c_1 = \frac{s_{11}}{s_{11}(s_{11}+s_{12}) - 2s_{12}^2} = \frac{E}{(1+\mu)(1-2\mu)},$$
$$c_2 = \frac{s_{12}c_1}{s_{11}} = -\frac{\mu E}{(1+\mu)(1-2\mu)}, \quad c_3 = \frac{s_{44}c_1 + 2}{s_{44}c_2 - 1} = -4(1-\mu). \tag{5.5}$$

In case of an elastic solid continuum, we get $\mu = 0.25$ [114]. In case of a real isotropic material, we get $\mu < 0.5$ [104], and then $c_3 < 0$. The coefficient c_3 is considered within the mathematical procedures in Sections 5.4.–5.6. (see Eqs. (5.21), (5.27), (5.33)).

Let the equations (5.1)–(5.4) be substituted to Eq. (3.32) as well as to the sum $[\partial \text{Eq.}(3.33)/\partial\varphi]$ + $[\partial \text{Eq.}(3.34)/\partial\nu]$. The equilibrium equations (3.32)–(3.34) are thus transformed to the forms

$$r^2\frac{\partial^2 u'_1}{\partial r^2} + 2r\frac{\partial u'_1}{\partial r} - 2u'_1 + \frac{U_1}{s_{44}(c_1+c_2)} = 0, \tag{5.6}$$

$$r\frac{\partial U_1}{\partial r} = c_3 U_1, \tag{5.7}$$

where the function U_1 is derived as

$$U_1 = \frac{\partial^2 u'_1}{\partial\varphi^2} + \frac{\partial^2 u'_1}{\partial\nu^2}. \tag{5.8}$$

A solution of this system of the differential equations (5.6), (5.7) is determined by the following four different mathematical procedures (see Sections 5.3.–5.6.) which result in different solutions for the radial displacement u'_1 along the axis x'_1 (see Figure 3.1).

5.3. Mathematical Procedure 1

Performing $r[\partial \text{Eq.}(5.7)/\partial r]$, we get

$$r^2\frac{\partial^2 U_1}{\partial r^2} + (1-c_3)\,r\frac{\partial U_1}{\partial r} = 0. \tag{5.9}$$

Substituting Eq. (5.7) to Eq. (5.9), we get

$$r^2\frac{\partial^2 U_1}{\partial r^2} + c_3(1-c_3)U_1 = 0. \tag{5.10}$$

If the dependence $U_1 - r$ is assumed in the form $U_1 = r^{\lambda}$, then the solution U_1 of the differential equation (5.10) has the form

$$U_1 = \frac{\partial^2 u'_1}{\partial\varphi^2} + \frac{\partial^2 u'_1}{\partial\nu^2} = \sum_{i=1}^{2} C_i r^{\lambda_i}, \tag{5.11}$$

where $C_1^{(1)}, C_2^{(1)}$ are integration constants which are related to the mathematical procedure 1. The exponent λ_i $(i = 1{,}2)$ is derived as

$$\lambda_i = \frac{1}{2}\left[1 + (\delta_{1i} - \delta_{2i})\sqrt{D}\right], \quad i = 1,2,$$
$$D = 1 - 4c_3(1 - c_3) = 1 + 16(1 - \mu)[1 + 4(1 - \mu)], \tag{5.12}$$

Due to $\mu < 0.5$ for a real isotropic material [104], and $c_3 < 0$ (see Eq. (5.5)), we get $D > 0$, and the real exponents $\lambda_1 > 3$, $\lambda_2 < -2$.

Considering Eq. (5.11), the differential equation (5.6) is transformed to the form

$$\frac{\partial^2 u_1'}{\partial r^2} + \frac{2}{r}\frac{\partial u_1'}{\partial r} - \frac{2u_1'}{r^2} = -\frac{1}{s_{44}(c_1 + c_2)}\sum_{i=1}^{2} C_i^{(1)} r^{\lambda_i - 2}. \tag{5.13}$$

Using the Wronskian's method [113], the radial displacement $u_{1q}' = u_{1q}'(r, \varphi, \nu)$ in the spherical particle $(q = p)$, the spherical envelope $(q = e)$ and the cell matrix $(q = m)$ which is determined by the mathematical procedure 1 has the form

$$u_{1q}' = \sum_{i=1}^{2} C_{iq}^{(1)} \xi_{iq}^{(1)} u_{irq}^{(1)}, \quad q = p, e, m. \tag{5.14}$$

The function $u_{irq}^{(1)} = u_{irq}^{(1)}(r)$ and the coefficient $\xi_{iq}^{(1)}$ $(i = 1{,}2;\ q = p{,}e{,}m)$ have the forms

$$u_{irq}^{(1)} = r^{\lambda_{iq}}, \quad i = 1,2; \quad \lambda_{1q} > 3, \quad \lambda_{2q} < -2; \quad q = p, e, m, \tag{5.15}$$

$$\xi_{iq}^{(1)} = \frac{1}{3s_{44}(c_1 + c_2)}\left(\frac{1}{\lambda_{iq} + 2} - \frac{1}{\lambda_{iq} - 1}\right), \quad i = 1,2; \quad q = p, e, m, \tag{5.16}$$

where $u_{1rq}^{(1)} = u_{1rq}^{(1)}(r)$ and $u_{2rq}^{(1)} = u_{2rq}^{(1)}(r)$ are increasing and decreasing function of the variable r due to $\lambda_{1q} > 3$ and $\lambda_{2q} < -2$, respectively.

5.4. Mathematical Procedure 2

Performing $\partial^2 \mathrm{Eq.}(5.7)/\partial r^2$, we get

$$r\frac{\partial^3 U_1}{\partial r^3} + (2 - c_3)\frac{\partial^2 U_1}{\partial r^2} = 0. \tag{5.17}$$

If the dependence $U_1 - r$ is assumed in the form $U_1 = r^{\lambda}$, then the solution U_1 of the differential equation (5.17) has the form

$$U_1 = \frac{\partial^2 u_1'}{\partial \varphi^2} + \frac{\partial^2 u_1'}{\partial \nu^2} = C_1^{(2)} r + C_2^{(2)} r^{c_3} + C_3^{(2)}, \tag{5.18}$$

where $C_1^{(2)}$, $C_2^{(2)}$, $C_3^{(2)}$ are integration constants which are related to the mathematical procedure 2.

Considering Eq. (5.18), the differential equation (5.6) is transformed to the form

$$\frac{\partial^2 u_1'}{\partial r^2}+\frac{2}{r}\frac{\partial u_1'}{\partial r}-\frac{2u_1'}{r^2}=-\frac{1}{s_{44}\left(c_1+c_2\right)}\left[\frac{C_1^{(2)}}{r}+C_2^{(2)}r^{c_3-2}+\frac{C_3^{(2)}}{r^2}\right]. \tag{5.19}$$

Using the Wronskian's method [113], the radial displacement $u_{1q}'=u_{1q}'(r,\varphi,\nu)$ in the spherical particle ($q=p$), the spherical envelope ($q=e$) and the cell matrix ($q=m$) which is determined by the mathematical procedure 2 has the form

$$u_{1q}'=\sum_{i=1}^{3}C_{iq}^{(2)}\xi_{iq}^{(2)}u_{irq}^{(2)},\quad q=p,e,m. \tag{5.20}$$

The function $u_{irq}^{(2)}=u_{irq}^{(2)}(r)$ and the coefficient $\xi_{iq}^{(2)}$ ($i=1,2,3$; $q=p,e,m$) have the forms

$$u_{1rq}^{(2)}=r\left(\frac{1}{3}-\ln r\right);\quad u_{2rq}^{(2)}=r^{c_{3q}},\quad c_{3q}<0;\quad u_{3rq}^{(2)}=1;\quad q=p,e,m, \tag{5.21}$$

$$\begin{aligned}\xi_{1q}^{(2)}&=\frac{1}{3s_{44q}\left(c_{1q}+c_{2q}\right)},\quad \xi_{2q}^{(2)}=\frac{1}{3s_{44q}\left(c_{1q}+c_{2q}\right)}\left(\frac{1}{c_{3q}+2}-\frac{1}{c_{3q}-1}\right),\\ \xi_{3q}^{(2)}&=\frac{1}{2s_{44q}\left(c_{1q}+c_{2q}\right)},\quad q=p,e,m,\end{aligned} \tag{5.22}$$

where $u_{2rq}^{(2)}=u_{2rq}^{(2)}(r)$ is a decreasing function of the variable r due to $c_{3q}<0$, and $u_{3rq}^{(2)}\neq f(r)$ is a constant function.

The function $u_{1rq}^{(2)}=u_{1rq}^{(2)}(r)$ increases or decreases on the condition $\partial u_{1rq}^{(2)}/\partial r>0$ or $\partial u_{1rq}^{(2)}/\partial r<0$, respectively. The conditions $\partial u_{1rq}^{(2)}/\partial r>0$ and $\partial u_{1rq}^{(2)}/\partial r<0$ result in $r<e^{-2/3}\approx 0.51342$ m and $r>e^{-2/3}\approx 0.51342$ m, respectively, where $e\approx 2.71828$ is the Euler's number [113]. In case of real two- and three-component materials , we get $R_1<0.51342$ m, $R_2<0.51342$ m, $d<0.51342$ m, and then $r<0.51342$ m. Finally, due to $r<0.51342$ m for real two- and three-component materials, the function $u_{1rq}^{(2)}=u_{1rq}^{(2)}(r)$ represents an increasing function of the variable r.

5.5. Mathematical Procedure 3

Performing $r\left[\partial\text{Eq.}(5.6)/\partial r\right]$, we get

$$r^3\frac{\partial^3 u_1'}{\partial r^3}+4r^2\frac{\partial^2 u_1'}{\partial r^2}+\frac{r}{s_{44}\left(c_1+c_2\right)}\frac{\partial U_1}{\partial r}=0. \tag{5.23}$$

Considering the right sides of Eqs. (5.7), (5.18), we get

$$r\frac{\partial U_1}{\partial r}=c_3\left[C_1^{(3)}r+C_2^{(3)}r^{c_3}+C_3^{(3)}\right], \tag{5.24}$$

where $C_1^{(3)}$, $C_2^{(3)}$, $C_3^{(3)}$ are integration constants which are related to the mathematical procedure 3.

Substituting Eq. (5.24) to Eq. (5.23), the differential equation (5.23) is transformed to the form

$$\frac{\partial^3 u_1'}{\partial r^3}+\frac{4}{r}\frac{\partial^2 u_1'}{\partial r^2}=-\frac{c_3}{s_{44}(c_1+c_2)}\left[\frac{C_1^{(3)}}{r^2}+C_2^{(3)}r^{c_3-3}+\frac{C_3^{(3)}}{r^3}\right]. \tag{5.25}$$

Using the Wronskian's method [113], the radial displacement $u_{1q}'=u_{1q}'(r,\varphi,\nu)$ in the spherical particle ($q=p$), the spherical envelope ($q=e$) and the cell matrix ($q=m$) which is determined by the mathematical procedure 3 has the form

$$u_{1q}'=\sum_{i=1}^{3}C_{iq}^{(3)}\xi_{iq}^{(3)}u_{irq}^{(3)},\quad q=p,e,m. \tag{5.26}$$

The function $u_{irq}^{(3)}=u_{irq}^{(3)}(r)$ and the coefficient $\xi_{iq}^{(3)}$ ($i=1,2,3$; $q=p,e,m$) have the forms

$$u_{1rq}^{(3)}=r\left(\frac{4}{3}-\ln r\right);\quad u_{2rq}^{(3)}=r^{c_{3q}},\quad c_{3q}<0;\quad u_{3rq}^{(3)}=\frac{1}{2}+\ln r;\quad q=p,e,m, \tag{5.27}$$

$$\xi_{1q}^{(3)}=\frac{c_{3q}}{3s_{44q}(c_{1q}+c_{2q})},\quad \xi_{2q}^{(3)}=\frac{c_{3q}}{s_{44q}(c_{1q}+c_{2q})}\left[\frac{1}{2c_{3q}}-\frac{1}{3(c_{3q}-1)}-\frac{1}{6(c_{3q}+2)}\right],$$
$$\xi_{3q}^{(3)}=\frac{c_{3q}}{2s_{44q}(c_{1q}+c_{2q})},\quad q=p,e,m, \tag{5.28}$$

where $u_{2rq}^{(3)}=u_{2rq}^{(3)}(r)$ is a decreasing function of the variable r due to $c_{3q}<0$. The function $u_{3rq}^{(3)}=u_{3rq}^{(3)}(r)$ which is defined for $r>0$ is increasing.

The function $u_{1rq}^{(3)}=u_{1rq}^{(3)}(r)$ increases or decreases on the condition $\partial u_{1rq}^{(3)}/\partial r>0$ or $\partial u_{1rq}^{(3)}/\partial r<0$, respectively. The conditions $\partial u_{1rq}^{(3)}/\partial r>0$ and $\partial u_{1rq}^{(3)}/\partial r<0$ result in $r<e^{1/3}\approx 1.39561$ m and $r>e^{1/3}\approx 1.39561$ m, respectively, where $e\approx 2.71828$ is the Euler's number [113]. In case of real two- and three-component materials, we get $R_1<1.39561$ m, $R_2<1.39561$ m, $d<1.39561$ m, and then $r<1.39561$ m. Finally, due to $r<1.39561$ m for real two- and three-component materials, the function $u_{1rq}^{(3)}=u_{1rq}^{(3)}(r)$ represents an increasing function of the variable r.

5.6. Mathematical Procedure 4

The differential equation (5.6) is transformed to the form

$$U_1=-s_{44}(c_1+c_2)\left(r^2\frac{\partial^2 u_1'}{\partial r^2}+2r\frac{\partial u_1'}{\partial r}-2u_1'\right). \tag{5.29}$$

Performing $r\left[\partial \text{Eq.}(5.29)/\partial r\right]$, we get

$$r\frac{\partial U_1}{\partial r}=-s_{44}(c_1+c_2)\left(r^3\frac{\partial^3 u_1'}{\partial r^3}+4r^2\frac{\partial^2 u_1'}{\partial r^2}\right). \tag{5.30}$$

Substituting Eqs. (5.29), (5.30) to Eq. (5.7), the differential equation (5.7) is transformed to the form

$$r^3\frac{\partial^3 u_1'}{\partial r^3}+(4-c_3)\,r^2\frac{\partial^2 u_1'}{\partial r^2}-2c_3\,r\frac{\partial u_1'}{\partial r}+2c_3 u_1'=0. \tag{5.31}$$

Let the dependence $u_1' - r$ be assumed in the form $u_1' = r^{\lambda}$. The radial displacement $u_{1q}' = u_{1q}'(r,\varphi,\nu)$ in the spherical particle ($q = p$), the spherical envelope ($q = e$) and the cell matrix ($q = m$) which is determined by the mathematical procedure 4 is then derived as

$$u_{1q}' = \sum_{i=1}^{3} C_{iq}^{(4)}\,\xi_{iq}^{(4)}\,u_{irq}^{(4)}, \quad q = p,e,m. \tag{5.32}$$

The function $u_{irq}^{(4)} = u_{irq}^{(4)}(r)$ and the coefficient $\xi_{iq}^{(4)}$ ($i = 1,2,3$; $q = p,e,m$) have the forms

$$u_{1rq}^{(4)} = r; \quad u_{2rq}^{(4)} = r^{c_{3q}}, \quad c_{3q} < 0; \quad u_{3rq}^{(4)} = \frac{1}{r^2}, \quad q = p,e,m, \tag{5.33}$$

$$\xi_{iq}^{(4)} = 1, \quad i = 1,2,3; \quad q = p,e,m, \tag{5.34}$$

where $u_{1rq}^{(4)} = u_{1rq}^{(4)}(r)$ is an increasing function of the variable r, and $u_{2rq}^{(4)} = u_{2rq}^{(4)}(r)$, $u_{3rq}^{(4)} = u_{3rq}^{(4)}(r)$ are decreasing functions of the variable r due to $c_{3q} < 0$.

5.7. Analysis of a Number of Solutions for Model Systems

The analysis in Sections 4.6.1.–4.6.3. which concern a course of the dependence $\left|u_{1q}'\right| - r$ is also valid for an isotropic component, where $u_{1q}' = u_{1q}'(r,\varphi,\nu)$ is a radial displacement along the axis x_1' (see Figure 3.1) in the spherical particle ($q = p$), the spherical envelope ($q = e$) and the cell matrix ($q = m$).

The course of $\left|u_{1q}'\right| - r$ results from courses of the functions $u_{irq}^{(1)} = u_{irq}^{(1)}(r)$, $u_{jrq}^{(2)} = u_{jrq}^{(2)}(r)$ $u_{jrq}^{(3)} = u_{jrq}^{(3)}(r)$ $u_{jrq}^{(4)} = u_{jrq}^{(4)}(r)$ ($i = 1,2$; $j = 1,2,3$; $q = p,e,m$) (see Eqs. (5.15), (5.21), (5.27), (5.33)) which are analysed in Sections 5.3.–5.6.. The analysis of courses of these functions regarding the boundary conditions in Sections 4.6.1.–4.6.3. is as follows.

Let s_q^{iso} represent a number of solutions which are determined by the mathematical procedures 1, 2, 3, 4 for the thermal stresses in the *isotropic* spherical particle ($q = p$), the *isotropic* spherical envelope ($q = e$) and the *isotropic* cell matrix ($q = m$).

5.7.1. Spherical Particle

In case of the multi-particle-matrix and multi-particle-envelope-matrix systems (see Figure 2.1), the dependence $\left|u_{1p}'\right| - r$ is required to represent an increasing function of the variable $r \in \langle 0, R_1 \rangle$ (see Section 4.6.1.). Consequently, we get

1. $C_{1p}^{(1)} \neq 0$, $C_{2p}^{(1)} = 0$ (see Eq. (5.15)),

2. $C_{1p}^{(2)} \neq 0, C_{2p}^{(2)} = 0, C_{3p}^{(2)} = 0$ (see Eq. (5.21)),

3. $C_{1p}^{(3)} \neq 0, C_{2p}^{(3)} = 0, C_{3p}^{(3)} = 0$ (see Eq. (5.27)),

4. $C_{1p}^{(4)} \neq 0, C_{2p}^{(4)} = 0, C_{3p}^{(4)} = 0$ (see Eq. (5.33)),

and then $s_p^{iso} = 4$ for each of the conditions $\beta_p \neq \beta_m$, $\beta_p \neq \beta_e = \beta_m$, $\beta_p \neq \beta_e \neq \beta_m$, $\beta_p = \beta_e \neq \beta_m$. In case of $\beta_p \neq \beta_e = \beta_m$, the difference $\beta_e - \beta_p \neq 0$ is required to represent a function of the variables φ and ν, i.e., $\beta_e - \beta_p = f(\varphi, \nu)$. If $\beta_e - \beta_p = f(\varphi, \nu)$, then $\varepsilon'_{11te}(\varphi, \nu) - \varepsilon'_{11tp}(\varphi, \nu) \neq 0$ due to $\alpha_q \neq f(\varphi, \nu)$ $(q = p,e,m)$ for an isotropic component of the model systems (see Fig. 2.1), where $\varepsilon'_{11tq} = \varepsilon'_{11tq}(\varphi, \nu)$ $(q = p,e)$ is a phase-transformation induced radial strain (see Section 3.5.2..2) which is included in the coefficient β_q (see Eqs. (3.54), (3.55)).

5.7.1.1. Conditions $\beta_p \neq \beta_e = \beta_m$ and $\varepsilon'_{11te} - \varepsilon'_{11tp} = 0$

If the conditions $\beta_p \neq \beta_e = \beta_m$ and $\varepsilon'_{11te} - \varepsilon'_{11tp} = 0)$ are both valid, then the following analysis is considered.

As analysed in Section 3.6., if $\beta_p \neq \beta_e \neq \beta_m$, then the radial stresses p_1 and p_2 acting at the particle-envelope and matrix-envelope boundaries, respectively, are simultaneously determined by the formulae Eqs. (3.82), (3.82). In this case, i.e., $\beta_p \neq \beta_e \neq \beta_m$, this determination considers ε'_{22m} (see Eq. (3.83)). Due to the mandatory boundary condition (4.96), the radial strain $\varepsilon'_{22m} = u'_{1m}/r$ (see Eq. (3.20)) includes r_c. The distance r_c represents a function of the variables φ and ν (see Section 3.1.2.1). Accordingly, if $\beta_p \neq \beta_e \neq \beta_m$, then the radial stresses p_1, p_2 also represent functions of φ and ν.

If $\beta_p \neq \beta_e = \beta_m$ and $\varepsilon'_{11te} - \varepsilon'_{11tp} = 0$, then the radial stress p_1 acting at the particle-envelope boundary is not a function of φ and ν. The thermal stresses, strains and thermal-stress induced displacements in the spherical particle and spherical envelope all include p_1. If $p_1 \neq f(\varphi, \nu)$, then the thermal stresses, strains and thermal-stress induced displacements in the spherical particle and spherical envelope are not functions of φ and ν. Accordingly, we get $\partial u'_{1q}/\partial\varphi = \partial u'_{1q}/\partial\nu = 0$ $(q = p,e)$, and then $U_{1q} = 0$ (see Eqs. (5.6)–(5.8)). Due to $U_{1q} = 0$, the differential equation (5.6) exhibits the following solution

$$u'_{1q} = \sum_{i=1}^{2} C_{iq} \xi_{iq} u_{irq}, \quad q = p,e. \tag{5.35}$$

The function $u_{irq} = u_{irq}(r)$ and the coefficient ξ_{iq} $(i = 1,2;\ q = p,e)$ have the forms

$$u_{1rq} = r; \quad u_{2rq} = \frac{1}{r^2}, \quad q = p,e, \tag{5.36}$$

$$\xi_{iq}^{(4)} = 1, \quad i = 1,2,3; \quad q = p,e, \tag{5.37}$$

where $u_{1rq} = u_{1rq}(r)$ and $u_{2rq} = u_{2rq}(r)$ represent increasing and decreasing functions of the variable r, respectively.

If $\beta_p \neq \beta_e = \beta_m$ and $\varepsilon'_{11te} - \varepsilon'_{11tp} = 0$, then $C_{1p} \neq 0$, $C_{2p} = 0$. Accordingly, we get $s_p^{iso} = 1$.

5.7.2. Spherical Envelope

5.7.2.1. Conditions $\beta_p \neq \beta_e = \beta_m$ and $\varepsilon'_{11te}(\varphi,\nu) - \varepsilon'_{11tp}(\varphi,\nu) \neq 0$

As analysed in Section 4.6.2., the dependence $|u'_{1e}| - r$ is required to represent a decreasing function of $r \in \langle R_1, R_2 \rangle$ or, at the very most, a constant function of $r \in \langle R_1, R_2 \rangle$. Consequently, we get

1. $C_{1e}^{(1)} = 0$, $C_{2e}^{(1)} \neq 0$ (see Eq. (5.15)),
2. $C_{1e}^{(2)} = 0$, $C_{2e}^{(2)} \neq 0$, $C_{3e}^{(2)} \neq 0$ (see Eq. (5.21)),
3. $C_{1e}^{(3)} = 0$, $C_{2e}^{(3)} \neq 0$, $C_{3e}^{(3)} = 0$ (see Eq. (5.27)),
4. $C_{1e}^{(4)} = 0$, $C_{2e}^{(4)} \neq 0$, $C_{3e}^{(4)} \neq 0$ (see Eq. (5.33)),

and then $s_e^{iso} = 4$ due to $u_{2re}^{(2)} = u_{2re}^{(3)} = u_{2re}^{(4)}$.

5.7.2.2. Conditions $\beta_p \neq \beta_e = \beta_m$ and $\varepsilon'_{11te}(\varphi,\nu) - \varepsilon'_{11tp}(\varphi,\nu) = 0$

With regard to the analysis in Section 5.7.1..1, we get $C_{1e} = 0$, $C_{2e} \neq 0$, and then $s_e^{iso} = 1$.

5.7.2.3. Condition $\beta_p = \beta_e \neq \beta_m$

As analysed in Section 4.6.2., the dependence $|u'_{1e}| - r$ is required to represent an increasing function of $r \in \langle R_1, R_2 \rangle$, or, at the very most, a constant function of $r \in \langle R_1, R_2 \rangle$. Consequently, we get

1. $C_{1e}^{(1)} \neq 0$, $C_{2e}^{(1)} = 0$ (see Eq. (5.15)),
2. $C_{1e}^{(2)} \neq 0$, $C_{2e}^{(2)} = 0$, $C_{3e}^{(2)} \neq 0$ (see Eq. (5.21)),
3. $C_{1e}^{(3)} \neq 0$, $C_{2e}^{(3)} = 0$, $C_{3e}^{(3)} \neq 0$ (see Eq. (5.27)),
4. $C_{1e}^{(4)} \neq 0$, $C_{2e}^{(4)} = 0$, $C_{3e}^{(4)} = 0$ (see Eq. (5.33)),

and then $s_e^{iso} = 6$.

5.7.2.4. Condition $\beta_p \neq \beta_e \neq \beta_m$

As analysed in Section 4.6.2., no requirements are considered for a course of the dependence $|u'_{1e}| - r$. Consequently, we get

1. $C_{1e}^{(1)} \neq 0$, $C_{2e}^{(1)} = 0$ (see Eq. (5.15)),
2. $C_{1e}^{(2)} \neq 0$, $C_{2e}^{(2)} \neq 0$, $C_{3e}^{(2)} = 0$ (see Eq. (5.21)),
3. $C_{1e}^{(2)} \neq 0$, $C_{2e}^{(2)} = 0$, $C_{3e}^{(2)} \neq 0$ (see Eq. (5.21)),

4. $C_{1e}^{(2)} = 0, C_{2e}^{(2)} \neq 0, C_{3e}^{(2)} \neq 0$ (see Eq. (5.21)),

5. $C_{1e}^{(3)} \neq 0, C_{2e}^{(3)} \neq 0, C_{3e}^{(3)} = 0$ (see Eq. (5.27)),

6. $C_{1e}^{(3)} \neq 0, C_{2e}^{(3)} = 0, C_{3e}^{(3)} \neq 0$ (see Eq. (5.27)),

7. $C_{1e}^{(3)} = 0, C_{2e}^{(3)} \neq 0, C_{3e}^{(3)} \neq 0$ (see Eq. (5.27)),

8. $C_{1e}^{(4)} \neq 0, C_{2e}^{(4)} \neq 0, C_{3e}^{(4)} = 0$ (see Eq. (5.33)),

9. $C_{1e}^{(4)} \neq 0, C_{2e}^{(4)} = 0, C_{3e}^{(4)} \neq 0$ (see Eq. (5.33)),

10. $C_{1e}^{(4)} = 0, C_{2e}^{(4)} \neq 0, C_{3e}^{(4)} \neq 0$ (see Eq. (5.33)),

and then $s_e^{iso} = 10$.

5.7.3. Cell Matrix

As analysed in Section 4.6.3., the dependence $|u'_{1m}| - r$ is required to represent a decreasing function of $r \in \langle R_1, r_c \rangle$ (see Figure 2.1a) or $r \in \langle R_2, r_c \rangle$ (see Figure 2.1b). The distance r_c is determined in Section 3.1.2.1 (see Eqs. (3.1)–(3.3), (3.8), (3.15)–(3.18)). The decreasing course of $|u'_{1m}| - r$ is ensured by integration constants (see Section 4.6.3.).

Consequently, we get

1. $C_{1m}^{(1)} \neq 0, C_{2m}^{(1)} \neq 0$ (see Eq. (5.15)),

2. $C_{1m}^{(2)} \neq 0, C_{2m}^{(2)} \neq 0, C_{3m}^{(2)} = 0$ (see Eq. (5.21)),

3. $C_{1m}^{(2)} \neq 0, C_{2m}^{(2)} = 0, C_{3m}^{(2)} \neq 0$ (see Eq. (5.21)),

4. $C_{1m}^{(2)} = 0, C_{2m}^{(2)} \neq 0, C_{3m}^{(2)} \neq 0$ (see Eq. (5.21)),

5. $C_{1m}^{(3)} \neq 0, C_{2m}^{(3)} \neq 0, C_{3m}^{(3)} = 0$ (see Eq. (5.27)),

6. $C_{1m}^{(3)} \neq 0, C_{2m}^{(3)} = 0, C_{3m}^{(3)} \neq 0$ (see Eq. (5.27)),

7. $C_{1m}^{(3)} = 0, C_{2m}^{(3)} \neq 0, C_{3m}^{(3)} \neq 0$ (see Eq. (5.27)),

8. $C_{1m}^{(4)} \neq 0, C_{2m}^{(4)} \neq 0, C_{3m}^{(4)} = 0$ (see Eq. (5.33)),

9. $C_{1m}^{(4)} \neq 0, C_{2m}^{(4)} = 0, C_{3m}^{(4)} \neq 0$ (see Eq. (5.33)),

10. $C_{1m}^{(4)} = 0, C_{2m}^{(4)} \neq 0, C_{3m}^{(4)} \neq 0$ (see Eq. (5.33)),

11. $C_{1m}^{(2)} \neq 0, C_{2m}^{(2)} \neq 0, C_{3m}^{(2)} \neq 0$ (see Eq. (5.21)),

12. $C_{1m}^{(3)} \neq 0, C_{2m}^{(3)} \neq 0, C_{3m}^{(3)} \neq 0$ (see Eq. (5.27)),

13. $C_{1m}^{(4)} \neq 0, C_{2m}^{(4)} \neq 0, C_{3m}^{(4)} \neq 0$ (see Eq. (5.33)),

and then $s_m^{iso} = 13$.

In case of the conditions 1–10, the mandatory boundary conditions are considered (see Section 4.6.3..1). In case of the conditions 11–13, the mandatory boundary conditions as well as the additional boundary condition (4.99) are considered. The additional boundary condition (4.100) is not considered due to the following analysis.

With regard to Eq. (3.50), the thermal-stress induced elastic energy density $w_m = w_m(r, \varphi, \nu) > 0$ in the cell matrix includes the shear stresses σ'_{12m} and σ'_{13m} as well as the shear strains ε'_{12m} and ε'_{13m} which are functions of $\partial u'_{1m}/\partial\varphi$ and $\partial u'_{1m}/\partial\nu$ (see Eqs. (3.22), (3.23), (3.47), (3.48)), respectively. The terms $\partial u'_{1m}/\partial\varphi$ and $\partial u'_{1m}/\partial\nu$ are functions of the terms $\partial C_{im}/\partial\varphi$ and $\partial C_{im}/\partial\nu$ ($i = 1, \ldots, $ n), respectively.

On the one hand, the additional boundary condition (4.100) concerning a course of the dependence $w_m - r$ on a surface of the cubic cell might be considered to be a reasonable assumption. Equation (4.100) thus includes the terms $\partial C_{im}/\partial\varphi$, $\partial C_{im}/\partial\nu$, where $C_{im} \equiv C_{im}^{(j)}$ ($i = 1, \ldots, 3$; $j = 2,3,4$) (see Eqs. (5.20), (5.26), (5.32)). On the other hand, no conditions exist for the determination of $\partial C_{im}/\partial\varphi$, $\partial C_{im}/\partial\nu$. Accordingly, if the multi-particle-matrix and multi-particle-envelope-matrix systems (see Figure 2.1) both consist of isotropic components, then the additional boundary condition (4.100) can not be considered for the determination of the thermal stresses in the cell matrix.

5.7.4. Number of Solutions for the Model Systems

The stress-strain state which is induced by the thermal stresses in the multi-particle-matrix system with isotropic components (see Figure 2.1a) is described by $s = s_p^{iso} \times s_m^{iso} = 4 \times 13 = 52$ mathematical solutions which fulfil the boundary conditions (see Section 4.6.). These mathematical solutions are determined by four different mathematical procedures (see Sections 5.3.–5.6.) which are applied to the differential equations (5.6), (5.7) (see Section 5.2.).

In case of the multi-particle-envelope-matrix system with isotropic components (see Figure 2.1b) for $\beta_p \neq \beta_e = \beta_m$, we get $s = 208$ and $s = 13$ for $\varepsilon'_{11e}(\varphi, \nu) - \varepsilon'_{11p}(\varphi, \nu) \neq 0$ and $\varepsilon'_{11e}(\varphi, \nu) - \varepsilon'_{11p}(\varphi, \nu) = 0$, respectively.

If $\beta_p \neq \beta_e \neq \beta_m$, then we get $s = 520$. If $\beta_p = \beta_e \neq \beta_m$, then we get $s = 312$.

Chapter 6

Thermal Stresses in Model Systems with Isotropic and Anisotropic Components

The determination of an analytical model of the thermal stresses in such model system (see Figure 2.1) which consists of anisotropic and isotropic components is as follows.

In case of an anisotropic component, formulae for the thermal stresses and strains as well as formulae for thermal-stress induced radial displacement, thermal-stress induced elastic energy density and elastic energy (see Section 3.4.) are taken from Chapter 4. In case of an isotropic component, these formulae are taken from Chapter 5. These formulae for anisotropic and isotropic components includes the radial stresses p_1, p_2.

As analysed in Section 3.6., p_1 acting at the particle-matrix boundary is determined by Eq. (3.81). The radial stresses p_1 and p_2 acting at the particle-envelope and matrix-envelope boundaries are determined by Eqs. (3.82) and (3.83), respectively. Equations (3.81)–(3.83) include tangential strains in components of the model systems (see Fig. 2.1). Consequently, a formula for the tangential strain in an anisotropic component is taken from Chapter 4. A formula for the tangential strain in an isotropic component is taken from Chapter 5.

Let s represent a number of solutions of the model system which consists of anisotropic and isotropic components. As an example, in case of the multi-particle-matrix system with anisotropic spherical particles and an isotropic matrix, we get $s = s_p^{aniso} + s_m^{iso}$. The numbers s_p^{aniso} and s_m^{iso} of solutions for the isotropic particle and cell matrix are taken form Chapters 4 and 5, respectively. In case of the multi-particle-envelope-matrix system, a formula for s can be analogically determined.

Finally, if $s > 1$, then a principle of minimal total potential energy of an elastic solid body [26] is required to be considered (see Section 3.4.).

Chapter 7

Related Phenomena

The analytical modelling of thermal-stress induced phenomena in the model systems (see Figure 2.1 includes

1. an analytical model of the crack formation (see Section 7.1.),
2. an analytical model of an energy barrier (see Section 7.2.),
3. an analytical model of micro- and macro-strengthening (see Section 7.3.),
4. analytical-computational and analytical-experimental-computational methods of life-time prediction (see Section 7.4.).

Section 7.1.1. presents a general analysis concerning the analytical modelling of the crack formation. This general analysis considers a curve integral of the energy density $w = w^{(ij)}$ $(ij = 12,23,31)$ (see Eqs. (7.1), (7.3)). As presented in Section 7.1.2., this general analysis is then applied to the multi-particle-matrix system. Consequently, as presented in Section 7.1.3., the curve integral of $w = w^{(ij)}$ (see Eq. (7.3)) is replaced by curve integrals of $w_p = w_p^{(ij)}$ and $w_m = w_m^{(ij)}$ (see Eqs. (7.15), (7.16), (7.21)–(7.23), (7.25), (7.26), (7.27), (7.29)), where $w_p = w_p^{(ij)}$ and $w_m = w_m^{(ij)}$ represents energy density in the spherical particle and the cell matrix, respectively.

This book deals with the analytical modelling of thermal stresses. Accordingly, on the one hand, $w = w^{(ij)}$, $w_p = w_p^{(ij)}$, $w_m = w_m^{(ij)}$ are considered to represent thermal-stress induced elastic energy density. In case of the thermal stresses, we get $w_q = w_q^{(ij)} \propto (\beta_m - \beta_p)^2$ for $q = p,m$.

On the other hand, the analyses in Sections 7.1.1.–7.1.3. along with the determination of the curve integrals are thus valid for the thermal-stress induced elastic energy density as well as for energy density which is induced by any stresses which act in the multi-particle-matrix system. In case of the energy density which is induced by any stresses, the condition $\beta_p(\varphi,\nu) = \beta_m(\varphi,\nu)$ which results in the function $\nu = \nu_{pm}(\varphi)$ (see Eqs. (7.18); see Fig. 7.6) is replaced by the conditions $w_p^{(ij)}(x_{ij},\varphi,x_k) = 0$ and $w_m^{(ij)}(x_{ij},\varphi,x_k) = 0$ which result in the function $x_{k0} = x_{k0}(x_{ij},\varphi)$ in the plane $x_{ij}x_k$ (see Figs. 7.1, 7.3), where $x_{ij} \in \langle 0, d/(2c_\varphi) \rangle$ (see Eqs. (3.2), (3.3)) and $\varphi \in \langle 0, \pi/2 \rangle$ are a variable and a parameter of $x_{k0} = x_{k0}(x_{ij},\varphi)$. Detailed analyses of the function $x_{k0} = x_{k0}(x_{ij},\varphi)$ are presented in [118]. These analyses of an extensive range are not presented in this book which is focused on the thermal stresses only.

Similarly, Sections 7.2.1. and 7.3.1. present general analyses concerning the analytical modelling of the energy barrier and strengthening, respectively. These general analyses considers surface integrals of the energy density w and w_i ($i = 1,2,3$) (see Eqs. (7.37)–(7.39), (7.48)). As presented in Sections 7.2.2., 7.3.2.1 and 7.2.3., 7.3.2.2, these general analyses are then applied to the multi-particle-matrix and multi-particle-envelope-matrix systems (see Eqs. (7.43)–(7.46), (7.55), (7.58), (7.64), (7.68)), respectively.

The analyses in Sections 7.2.1.–7.2.3., 7.3.1., 7.3.2. along with the determination of the surface integrals are thus valid for the thermal-stress induced elastic energy density as well as for energy density which is induced by any stresses which act in the multi-particle-matrix and multi-particle-envelope-matrix systems.

Finally, on the one hand, the methods of lifetime prediction which are presented in Section 7.4. are related to a stress-strain state which is induced by the thermal stresses. On the other hand, the analyses in Section 7.4. are also applicable to a stress-strain state with the radial stress σ'_{11q}, the tangential stresses σ'_{22q}, σ'_{33q} and the shear stresses σ'_{12q}, σ'_{13q} ($q = p,m$) which are required to be derived in the forms $\sigma'_{11p} = -p_1 f_{11p}$, $\sigma'_{11m} = -p_2 f_{11m}$; $\sigma'_{22p} = -p_1 f_{22p}$, $\sigma'_{33p} = -p_1 f_{33p}$, $\sigma'_{22m} = -p_2 f_{22m}$, $\sigma'_{33m} = -p_2 f_{33m}$; $\sigma'_{12p} = -p_1 f_{12p}$, $\sigma'_{13p} = -p_1 f_{13p}$, $\sigma'_{12m} = -p_2 f_{12m}$, $\sigma'_{13m} = -p_2 f_{13m}$, where f_{ijp} and f_{ijm} ($i, j = 1,2,3$) are functions of a position in the spherical particle and cell matrix, respectively.

7.1. Analytical Model of Crack Formation

The analytical modelling of crack formation is based on the comparison of thermal-stress induced elastic energy in the cubic cell with energy for the creation of the surface of a crack. This comparison is used e.g., in [71]–[74]. This comparison results from a general analysis (see Section 7.1.1.) which is applied to a solid continuum of a general shape (see Figure 7.1a). This analytical model includes the determination of a condition for the crack initiation (see Eq. (7.10)). Equation (7.10) represents a condition for the determination of the critical particle radius $R_{1cq}^{(ij)}$ which is a reason of the crack initiation in the spherical particle ($q = p$; for $\beta_p > \beta_m$) or cell matrix ($q = m$; for $\beta_p < \beta_m$) of the multi-particle-matrix system (see Figure 2.1a), where $i, j = 1,2,3$ and $i \neq j$. The superscript ij is related to the plane $x_i x_j$ which represents a plane of the crack formation[1]. The critical particle radius $R_{1cq}^{(ij)}$ thus defines a limit state with respect to the crack initiation in the plane $x_i x_j$ in the spherical particle ($q = p$) and cell matrix ($q = m$). The crack initiation is then followed by the crack propagation in the plane $x_i x_j$ for $R_1 > R_{1cq}^{(ij)}$.

The condition (7.11) for $R_1 > R_{1cq}^{(ij)}$ results in the determination of the crack shape in the crack formation plane $x_i x_j$. Additionally, the function $f_q^{(ij)}$ describing the crack shape in the plane $x_{ij} x_k$ (see Fig. 2.1b) which is perpendicular to the crack formation plane $x_i x_j$ is also determined, where $x_{ij} \subset x_i x_j$, and $\varphi = \angle(x_i, x_{ij}) \in \langle 0, 2\pi \rangle$.

[1]The term 'crack formation' which is used in this book includes the crack initiation (in the plane $x_i x_j$ for $R_1 = R_{1cq}^{(ij)}$) ($q = p,m$) and the crack propagation (in the plane $x_i x_j$ for $R_1 > R_{1cq}^{(ij)}$). The crack initiation is followed by the crack propagation.

The determination of these crack parameters considers the curve integral $W_{cq}^{(ij)}$ of thermal-stress induced elastic energy density (see Eq. (3.50)) along a curve within the cubic cell. The determination of the curve integrals $W_{cp}^{(ij)}$ and $W_{cm}^{(ij)}$ which are used within this analytical model of the cracking in the spherical particle and cell matrix, respectively, is presented in Sections 7.1.3.1, 7.1.3.2. This determination considers the conditions $\beta_p(\varphi,\nu) \neq \beta_m(\varphi,\nu)$ and $\beta_p(\varphi,\nu) = \beta_m(\varphi,\nu)$ which are analysed (see Section 7.1.3.2).

Additionally, an analysis of the crack formation in the spherical particle or cell matrix in spite of the condition $\beta_p < \beta_m$ or $\beta_p > \beta_m$, respectively, is also presented (see Section 7.1.3.2). This analysis which considers the condition $\beta_p(\varphi,\nu) = \beta_m(\varphi,\nu)$ can explain the paradoxical behaviour concerning the crack formation which is experimentally observed in real two-component materials [71]–[74].

The transformations concerning the cracking in the planes $x_i x_j$, $x_j x_k$, $x_k x_i$ ($i,j,k = 1,2,3$; $i \neq j \neq k$) are also determined (see Section 7.1.4.).

The analysis in Sections 7.1.1., 7.1.2. is valid disregarding the determination of the curve integral $W_{cq}^{(ij)}$ ($q = p,m$). On the one hand, the thermal stresses are modified (changed) during the crack propagation for $R_1 > R_{1cq}^{(ij)}$. Accordingly, $W_{cq}^{(ij)}$ is also modified (changed) during the crack propagation. Such modification might be analytically and/or computationally determined, and then this modification can be applied to the analysis in Sections 7.1.1., 7.1.2.

On the other hand, formulae for $W_{cq}^{(ij)}$ in Section 7.1.3. do not consider the modification of $W_{cq}^{(ij)}$ during the crack propagation. These formulae for $W_{cq}^{(ij)}$ are thus determined for a stress-strain state which is induced by the thermal stresses before the crack initiation. Consequently, these formulae are then required to be considered only in case of a two-component material with ceramic components. In general, a ceramic component exhibits a high-speed crack propagation [71]–[74] during which the modification (change) of $W_{cq}^{(ij)}$ can be assumed to be neglected.

The determination of $R_{1cq}^{(ij)}$ which considers the thermal stresses *before* the crack initiation is valid for any two-component material.

Due to a range of this book, this analytical model of the crack formation induced by the thermal stresses is determined for the multi-particle-matrix system with anisotropic and/or isotropic components (see Fig 2.1a). Results for the multi-particle-envelope-matrix system (see Fig 2.1b) can be determined analogically.

7.1.1. General Analysis

The analytical model of the crack formation is based on the following general analysis. This analysis concerns the crack formation in a solid continuum with the volume V and with a general shape as shown in Figures 7.1a,b.

Let the crack be formed in the solid continuum in the plane $x_i x_j$. Figures 7.1a and 7.1b show the solid continuum before and after the crack propagation in the plane $x_i x_j$, respectively.

The shaded area in Figure 7.1a represents cuts of the solid continuum in the planes $x_i x_j$, $x_i x_k$, $x_{ij} x_k$ ($i,j,k = 1,2,3$; $i \neq j \neq k$), where $x_{ij} \subset x_i x_j$. The curves *1, 2, 3, 4* on a surface of the solid continuum are outlines of the cuts in the planes $x_i x_k$, $x_i x_j$, $x_j x_k$, $x_{ij} x_k$, respectively. The

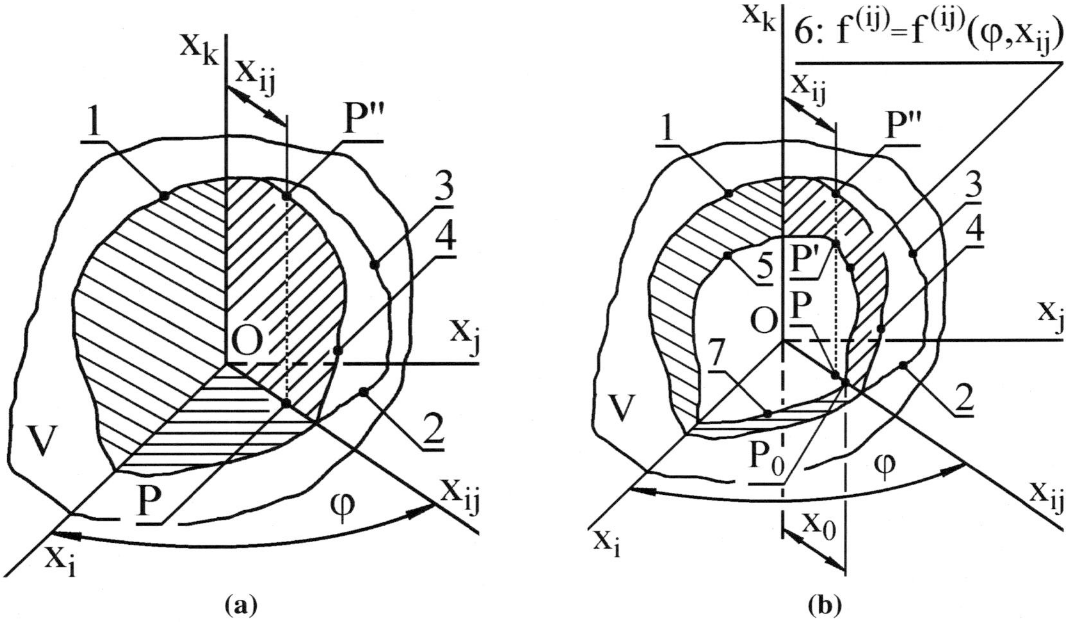

Figure 7.1. The solid continuum with a general shape with the volume V in the Cartesian system $(Ox_ix_jx_k)$ $(i, j, k = 1,2,3;\ i \neq j \neq k)$; (a) without and (b) with a crack. The crack is formed in the plane x_ix_j. The shaded area represents cuts of the solid continuum in the planes x_ix_j, x_ix_k, $x_{ij}x_k$, where $x_{ij} \subset x_ix_j$, $\varphi = \angle(x_i, x_{ij}) \in \langle 0, 2\pi \rangle$. The curves *1, 2, 3, 4* on a surface of the solid continuum are outlines of the cuts in the planes x_ix_k, x_ix_j, x_jx_k, $x_{ij}x_k$, respectively. P and P'' are points on the axis x_{ij} and the curve *4*, respectively, where $|OP| = x_{ij}$, $\overline{PP''} \parallel x_k$. The curves *5* and *6* represent the crack shape in the planes x_ix_k and $x_{ij}x_k$, respectively. The planes x_ix_k, $x_{ij}x_k$ are perpendicular to the crack formation plane x_ix_j. The curve *7* determines a position of the crack tip in the crack formation plane x_ix_j. The function $f^{(ij)} = f^{(ij)}(x_{ij}, \varphi, x_k)$ related to the curve *6* describes the crack shape in the plane $x_{ij}x_k$. The point $P_0 \in x_ix_j$ with the coordinate $x_{ij} = x_0(\varphi)$ represents the crack tip related to the plane $x_{ij}x_k$.

plane $x_{ij}x_k$ is perpendicular to the crack formation plane x_ix_j. A position of $x_{ij}x_k$ regarding the Cartesian system $(0x_ix_jx_k)$ is determined by the angle $\varphi = \angle(x_i, x_{ij}) \in \langle 0, 2\pi \rangle$, and then $x_{ij} \equiv x_i$ for $\varphi = 0$; $x_{ij} \equiv x_j$ for $\varphi = \pi/2$. The points P and P'', both with the coordinate x_{ij}, represent points on the axis x_{ij} and the curve *4*, respectively, where $\overline{PP''} \parallel x_k$.

Figure 7.1b presents a crack which is formed in the plane x_ix_j. The curves *5* and *6* represent the crack shape in the planes x_ix_k and $x_{ij}x_k$ which are perpendicular to the the crack formation plane x_ix_j, respectively. The curve *7* determines a position of the crack tip in the the crack formation plane x_ix_j. The function $f^{(ij)} = f^{(ij)}(x_{ij}, \varphi, x_k)$ related to the curve *6* (b) describes the crack shape in the plane $x_{ij}x_k$. The point $P_0 \in x_ix_j$ with the coordinate $x_{ij} = x_0(\varphi)$ represents the crack tip related to the plane $x_{ij}x_k$.

Let the thermal-stress induced elastic energy density $w = w(r, \varphi, \nu)$ (see Eq. (3.50)) be determined by the cylindrical coordinates (x_{ij}, φ, x_k), and then we get

$$w = w^{(ij)} = w^{(ij)}(x_{ij}, \varphi, x_k), \quad ij = 12,\ k = 3; \quad ij = 23,\ k = 1; \quad ij = 31,\ k = 2. \tag{7.1}$$

where $x_i x_j \perp x_k$ ($i \neq j \neq k$), and $x_i = x_{ij} \cos\varphi$, $x_j = x_{ij} \sin\varphi$. The superscript ij in Eq. (7.1) thus indicates that the energy density w is derived as a function of the cylindrical coordinates (x_{ij}, φ, x_k).

The thermal-stress induced elastic energy density $w = w^{(ij)}(x_{ij}, \varphi, x_k)$ results in the thermal-stress induced energy $W = \int_V w dV$ accumulated in the volume V. The energy W tends to be released by the crack formation in the plane $x_i x_j$ ($ij = 12$, 23, 31). The same is also valid for the energy $dW = dW\left(w^{(ij)}\right)$ (see Eq. (7.1)) which is accumulated in a volume of the infinitesimal prism in Figure 7.2, where $\left|\overline{PP''}\right|$ (see Figure 7.1b) is height, and $dS_{ij} = x_{ij}\, d\varphi\, dx_{ij}$ is a surface area of a basis in the crack formation plane $x_i x_j$. Consequently, we get

$$\begin{aligned} dW &= \int_{\overline{PP''}} w^{(ij)} dV = \int_{\overline{PP''}} w^{(ij)} dS_{ij}\, dx_k = x_{ij}\, d\varphi\, dx_{ij} \int_{\overline{PP''}} w^{(ij)} dx_k \\ &= W_c^{(ij)} x_{ij}\, d\varphi\, dx_{ij}, \end{aligned} \tag{7.2}$$

$$W_c^{(ij)} = W_c^{(ij)}(x_{ij}, \varphi) = \int_{\overline{PP''}} w^{(ij)} dx_k, \tag{7.3}$$

where $dV = dS_{ij}\, dx_k$ (see Figure 7.2). The curve integral $W_c^{(ij)} = W_c^{(ij)}(x_{ij}, \varphi)$ is a function of x_{ij}, φ due to $w^{(ij)} = w^{(ij)}(x_{ij}, \varphi, x_k)$.

The crack is formed in the plane $x_i x_j$ provided that the condition

$$\int_{\overline{PP''}} w^{(ij)}(x_{ij}, \varphi, x_k)\, dx_k = \int_{\overline{PP''(-)}} w^{(ij)}(x_{ij}, \varphi, -x_k)\, dx_k, \quad \varphi \in \langle 0, 2\pi \rangle, \tag{7.4}$$

is valid for $\varphi \in \langle 0, 2\pi \rangle$, where $P''^{(-)}$ is an intersection point of the line PP'' with a surface of the solid continuum for $x_k < 0$. Strictly speaking, the crack formation plane $x_i x_j$ is required to represent a plane of symmetry of $W_c^{(ij)} = W_c^{(ij)}(x_{ij}, \varphi)$ regarding $x_k > 0$ and $x_k < 0$.

The energy dW (see Eqs. (7.1)–(7.3)) is in an equilibrium state with the energy

$$dW_{cs}^{(ij)} = \gamma^{(ij)} dS^{(ij)} \tag{7.5}$$

for creation of the infinitesimal crack surface $dS^{(ij)}$ (see Figure 7.2). The crack surface energy $\gamma^{(ij)}$ per unit surface in the crack formation plane $x_i x_j$ has the form [71]–[74]

$$\gamma^{(ij)} = s_{kk} \left[K_{IC}^{(ij)}\right]^2, \quad s_{kk} = \frac{1}{E_k}, \tag{7.6}$$

where $K_{IC}^{(ij)}$ is fracture toughness along the axis x_{ij}, and s_{kk} is an elastic modulus along the axis x_k which is perpendicular to the crack formation plane $x_i x_j$ of the crack formation. E_k is the Young's modulus along the axis x_k.

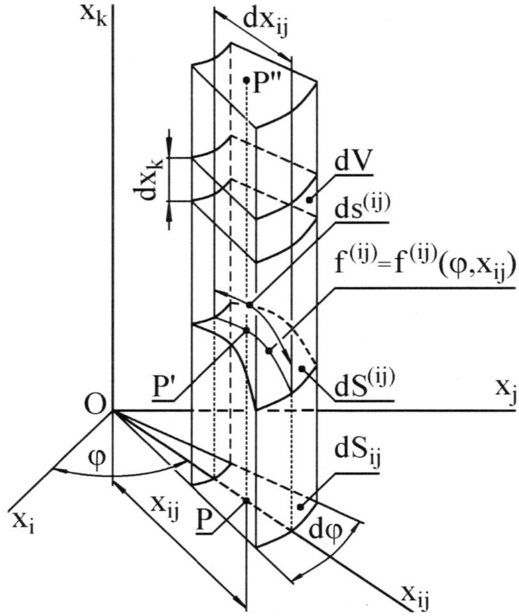

Figure 7.2. The infinitesimal prism with the height $\left|\overline{PP''}\right|$ (see Figure 7.1b) and the surface area $dS_{ij} = x_{ij}\,d\varphi\,dx_{ij}$ of a basis in the the crack formation plane x_ix_j; $dV = dS_{ij}\,dx_k$ is infinitesimal volume of the infinitesimal prism; $ds^{(ij)}$ is length of an infinitesimal part of the crack at the arbitrary point P'; $dS^{(ij)}$ is an infinitesimal surface of the crack. Strictly speaking, $dS^{(ij)}$ represents such crack surface which is related to the infinitesimal length $ds^{(ij)}$ at the arbitrary point P' of the curve *6* on the crack surface (see Figure 7.1b). The function $f^{(ij)} = f^{(ij)}\left(x_{ij}, \varphi\right)$ describes the crack shape in the plane $x_{ij}x_k$ which is perpendicular to the crack formation plane x_ix_j.

This infinitesimal crack surface with the surface area

$$dS^{(ij)} = ds^{(ij)}\,x_{ij}\,d\varphi \tag{7.7}$$

represents such crack surface which is related to the infinitesimal length $ds^{(ij)}$ (see Figure 7.2) at the arbitrary point P' (see Figure 7.1b) of the curve *6* on the crack surface.

The infinitesimal crack length $ds^{(ij)}$ is derived as [113]

$$ds^{(ij)} = dx_{ij} \times \sqrt{1 + \left[\frac{\partial f^{(ij)}}{\partial x_{ij}}\right]^2}, \tag{7.8}$$

where the function $f^{(ij)} = f^{(ij)}\left(x_{ij}, \varphi\right)$ describes the crack shape in the plane $x_{ij}x_k$ which is perpendicular to the crack formation plane x_ix_j.

7.1.2. Cell Model

Let the multi-particle-matrix system (see Figure 2.1a) be considered instead of the solid continuum in Figure 7.1. Due to the symmetry of the multi-particle-matrix system (see

Section 3.1.2.), the following analysis is thus sufficient to be considered within one eighth of the cubic cell (see Figure 7.3), i.e., $\varphi \in \langle 0, \pi/2 \rangle$ and $\nu \in \langle 0, \pi/2 \rangle$.

Let the equilibrium condition $dW = dW_{cs}^{(ij)}$ (see Eqs. (7.2), (7.3), (7.5)–(7.8)) be considered. Accordingly, we get

$$\frac{\partial f_q^{(ij)}}{\partial x_{ij}} = \pm \frac{1}{s_{kkq}\left[K_{ICq}^{(ij)}\right]^2} \sqrt{\left[W_{cq}^{(ij)}\right]^2 - \left\{s_{kkq}\left[K_{ICq}^{(ij)}\right]^2\right\}^2}, \quad q = p, m, \tag{7.9}$$

where the subscript $q = p$ or $q = m$ is considered for the crack formation in the spherical particle or cell matrix, respectively.

In case of the thermal stresses, $W_{cp}^{(ij)} = W_{cp}^{(ij)}(x_{ij}, \varphi, R_1, v)$ and $W_{cm}^{(ij)} = W_{cm}^{(ij)}(x_{ij}, R_1, v\varphi)$ represent increasing and decreasing functions of the variable x_{ij}, respectively, where $\varphi \in \langle 0, \pi/2 \rangle$ and $v \in (0, \pi/6\rangle$ (see Eqs. (2.1), (2.4)). Accordingly, $f_p^{(ij)} = f_p^{(ij)}(x_{ij}, \varphi, R_1, v)$ and $f_m^{(ij)} = f_m^{(ij)}(x_{ij}, \varphi, R_1, v)$ are assumed to be also increasing and decreasing functions of x_{ij}, where the sign "+" and "-" in Eq. (7.9) is considered, respectively. Additionally, the formula (7.9) is considered for $W_{cq}^{(ij)} - s_{kkq}\left[K_{ICq}^{(ij)}\right]^2 \geq 0$.

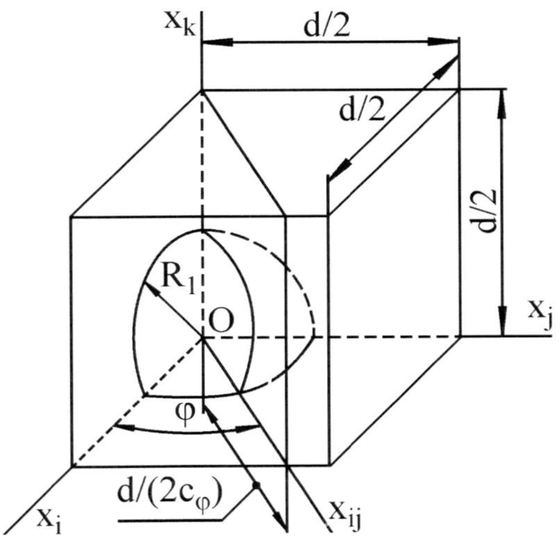

Figure 7.3. One eighth of the cubic cell with the central spherical particle with the radius R_1, where d is the cubic cell dimension. The general position of the axis x_{ij} of the plane $x_{ij}x_k$ is determined by the angle $\varphi = \angle(x_i, x_{ij}) \in \langle 0, 2\pi \rangle$, where $x_{ij} \subset x_i x_j$. The coefficient $c_\varphi = c_\varphi(\varphi)$ is given by Eqs. (3.2), (3.3).

Consequently, the condition

$$\left[W_{cq}^{(ij)}(x_{ij}, \varphi, R_1, v)\right]_{x_{ij}=R_1} - s_{kkq}\left[K_{ICq}^{(ij)}\right]^2 = 0, \quad q = p, m, \tag{7.10}$$

represents an equation with the variable R_1 and the parameters φ, v. The root $R_{1cq}^{(ij)} = R_{1cq}^{(ij)}(\varphi, v)$ of Eq. (7.10) represents a critical particle radius which is a reason of the crack initiation in the crack formation plane $x_i x_j$ in the spherical particle ($q = p$) or cell matrix ($q = m$) of the multi-particle-matrix system. The determination of $R_{1cp}^{(ij)}$ or $R_{1cm}^{(ij)}$ is related to such position $x_{ij} \in \langle 0, R_1 \rangle$ or $x_{ij} \in \left\langle R_1, d/\left(2c_\varphi\right)\right\rangle$ (see Eqs. (3.2), (3.3)) at which the dependence $W_{cp}^{(ij)} - x_{ij}$ or $W_{cm}^{(ij)} - x_{ij}$ is maximal, respectively. As mentioned above, in case of the thermal stresses, we get $x_{ij} = R_1$ for $W_{cp}^{(ij)} - x_{ij}$ and $W_{cm}^{(ij)} - x_{ij}$.

The critical particle radius $R_{1cq}^{(ij)} = R_{1cq}^{(ij)}(\varphi, v)$ thus defines a *limit state* with respect to the crack initiation in the crack formation plane $x_i x_j$ in the spherical particle ($q = p$) or cell matrix ($q = m$). This limit state *is related to the plane* $x_{ij}x_k$, where $x_{ij} \subset x_i x_j$, $x_{ij}x_k \perp x_i x_j$. Strictly speaking, this limit state is related to one value of the variable $\varphi = \angle(x_i, x_{ij}) \in \langle 0, \pi/2 \rangle$ as well as to one value of the interval $v \in \langle 0, \pi/6 \rangle$ (see Eqs. (2.1), (2.4)).

The crack initiation in the plane $x_{ij}x_k$ is then followed by the crack propagation for $R_1 > R_{1cq}^{(ij)}$.

Let the condition $R_1 > R_{1cq}^{(ij)}(\varphi, v)$ be valid. The condition

$$W_{cq}^{(ij)}(x_{ij}, \varphi, R_1, v) - s_{kkq}\left[K_{ICq}^{(ij)}\right]^2 = 0, \quad R_1 > R_{1cq}^{(ij)}, \quad q = p, m, \tag{7.11}$$

represents an equation with the variable x_{ij} and the parameters φ, R_1, v. The root $x_{0q} = x_{0q}(\varphi, R_1, v)$ which is a function of the variable φ defines a position of the crack tip point P_0 on the axis x_{ij} (see Figure 7.1b), where R_1, v represent parameters of this function, and $x_{ij} \subset x_i x_j$, $\varphi = \angle(x_i, x_{ij}) \in \langle 0, 2\pi \rangle$. The point $P_0 \in x_i x_j$ with the coordinate $x_{ij} = x_{0q}(\varphi, R_1, v)$ represents the crack tip related to the plane $x_{ij}x_k$. The dependence $x_{0q} - \varphi$ thus defines the crack shape in the crack formation plane $x_i x_j$ in the spherical particle ($q = p$) or cell matrix ($q = m$)[2].

The increasing and decreasing functions $f_p^{(ij)} = f_p^{(ij)}(x_{ij}, \varphi, R_1, v)$ and $f_m^{(ij)} = f_m^{(ij)}(x_{ij}, \varphi, R_1, v)$ of the variable $x_{ij} \in \langle x_{0p}, R_1 \rangle$ and $x_{ij} \in \langle R_1, x_{0m} \rangle$ which describe the crack shape in the plane $x_{ij}x_k$ in the spherical particle and the cell matrix, respectively, are derived as (see Figure 7.4)

$$f_p^{(ij)} = \frac{1}{s_{kkp}\left[K_{ICp}^{(ij)}\right]^2}\left[\int \sqrt{\left[W_{cp}^{(ij)}\right]^2 - \left\{s_{kkp}\left[K_{ICp}^{(ij)}\right]^2\right\}^2}\, dx_{ij} - C_p^{(ij)}\right],$$
$$x_{ij} \in \langle x_{0p}, R_1 \rangle, \quad R_1 > R_{1cp}^{(ij)}, \tag{7.12}$$

$$f_m^{(ij)} = \frac{1}{s_{kkm}\left[K_{ICm}^{(ij)}\right]^2}\left[C_m^{(ij)} - \int \sqrt{\left[W_{cm}^{(ij)}\right]^2 - \left\{s_{kkm}\left[K_{ICm}^{(ij)}\right]^2\right\}^2}\, dx_{ij}\right],$$
$$x_{ij} \in \langle R_1, x_{0m} \rangle, \quad R_1 > R_{1cm}^{(ij)}, \tag{7.13}$$

[2]The dependence $x_{0q} - \varphi$ thus defines the curve *7* in Figure 7.1b.

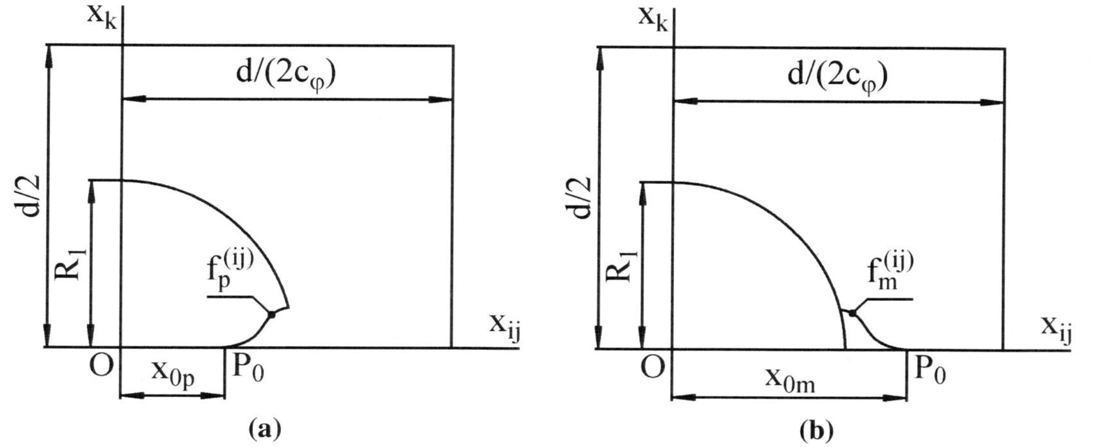

Figure 7.4. A schematic illustration of a crack shape of cracks in the plane $x_{ij}x_k$ (see Figure 7.3). The crack with a tip in the point P_0 is formed in the plane x_ix_j in (a) the spherical particle and (b) the cell matrix. The crack shape in the plane $x_{ij}x_k$ is described by the increasing and decreasing functions $f_p^{(ij)} = f_p^{(ij)}(x_{ij}, \varphi, R_1, v)$ and $f_m^{(ij)} = f_m^{(ij)}(x_{ij}, \varphi, R_1, v)$ of the variable $x_{ij} \in \langle x_{0p}, R_1\rangle$ and $x_{ij} \in \langle R_1, x_{0m}\rangle$ for $R_1 > R_{1cp}^{(ij)}$ and $R_1 > R_{1cm}^{(ij)}$ (see Eqs. (7.12)–(7.14)), respectively. The coefficient $c_\varphi = c_\varphi(\varphi)$ is given by Eqs. (3.2), (3.3).

where φ, R_1 and $v \in (0, \pi/6\rangle$ (see Eqs. (2.1), (2.4)) represent parameters of these functions. The integration constant $C_q^{(ij)}$ $(q = p,m)$ which is determined by the boundary condition $\left[f_q^{(ij)}\right]_{x_{ij}=x_{0q}} = 0$ has the form

$$C_q^{(ij)} = \left[\int \sqrt{\left[W_{cq}^{(ij)}\right]^2 - \left\{s_{kkq}\left[K_{ICq}^{(ij)}\right]^2\right\}^2}\, dx_{ij}\right]_{x_{ij}=x_{0q}}, \quad q = p,m. \tag{7.14}$$

As mentioned above, the critical particle radius $R_{1cq}^{(ij)} = R_{1cq}^{(ij)}(\varphi, v)$ defines such limit state which is related to one value of the variable $\varphi \in \langle 0, \pi/2\rangle$. Consequently, the minimal value $R_{1cq\,min}^{(ij)}$ of the dependence $R_{1cq}^{(ij)} - \varphi$ defines a limit state related to such plane $x_{ij}x_k$ which corresponds to maximal value of the dependence $W_{cq}^{(ij)}$. The maximal value $R_{1cq\,max}^{(ij)}$ of the dependence $R_{1cq}^{(ij)} - \varphi$ then defines a 'total' limit state of the crack initiation in the crack formation plane x_ix_j in the spherical particle $(q = p)$ or cell matrix $(q = m)$.

Finally, the curve integral $W_{cq}^{(ij)}$ $(q = p,m)$ is determined in Section 7.1.3..

7.1.3. Determination of $W_{cp}^{(ij)}$ and $W_{cm}^{(ij)}$

Section 7.1.3. deals with the determination of the curve integrals $W_{cp}^{(ij)}$ for $x_{ij} \in \langle 0, R_1\rangle$ and $W_{cm}^{(ij)}$ for $x_{ij} \in \langle R_1, d/(2c_\varphi)\rangle$ which are considered for the analytical model of the crack formation in the spherical particle and cell matrix of the multi-particle-matrix system (see Figure 2.1a), respectively, where the coefficient $c_\varphi = c_\varphi(\varphi)$ is given by Eqs. (3.2), (3.3).

A crack is formed either in the spherical particle or cell matrix on the condition $\beta_p > \beta_m$ or $\beta_p < \beta_m$ (see Section 3.5.1.2), respectively. The determination of $W_{cq}^{(ij)}$ $(q = p,m)$ considers the conditions $\beta_p(\varphi,\nu) \neq \beta_m(\varphi,\nu)$ for $\varphi,\nu \in \langle 0,\pi/2\rangle$ and $\beta_p(\varphi,\nu) = \beta_m(\varphi,\nu)$ for $\nu = \nu_{pm}$.

The function $\nu_{pm} = \nu_{pm}(\varphi)$ and definition intervals for the variable φ are determined in Section 7.1.3.2. As presented in Section 7.1.3.2, this function results from the condition $\alpha_p(\varphi,\nu) = \alpha_m(\varphi,\nu)$ (see Eq. (3.60)), i.e., from the condition $\beta_p(\varphi,\nu) = \beta_m(\varphi,\nu)$ for $\varepsilon'_{11tp}(\varphi,\nu) = \varepsilon'_{11tm}(\varphi,\nu) = 0$ at $T \in \langle T_f, T_r\rangle$ (see Eqs. (3.56)–(3.60)), where T_f is final temperature of a cooling process, and T_r is relaxation temperature (see Section 3.5.1..1).

If $\varepsilon'_{11tp}(\varphi,\nu) \neq \varepsilon'_{11tm}(\varphi,\nu)$ at $T \in \langle T_f, T_r\rangle$, then the condition $\beta_p(\varphi,\nu) - \beta_m(\varphi,\nu) = 0$ represents a transcendental equation regarding the variables $\varphi,\nu \in \langle 0,\pi/2\rangle$. The function $\nu_{pm} = \nu_{pm}(\varphi)$ and the definition intervals for φ are then determined by a numerical method for a real two-component material (see Section 2.1.).

7.1.3.1. Condition $\beta_p(\varphi,\nu) \neq \beta_m(\varphi,\nu)$

Let the condition $\beta_p(\varphi,\nu) \neq \beta_m(\varphi,\nu)$ be valid for $\varphi \in \langle 0,\pi/2\rangle$ and $\nu \in \langle 0,\pi/2\rangle$. If $\beta_p > \beta_m$, then a crack is formed in the spherical particle, and the curve integral $W_{cp}^{(ij)}$ for $x_{ij} \in \langle 0,R_1\rangle$ is derived as (see Figure 7.5a)

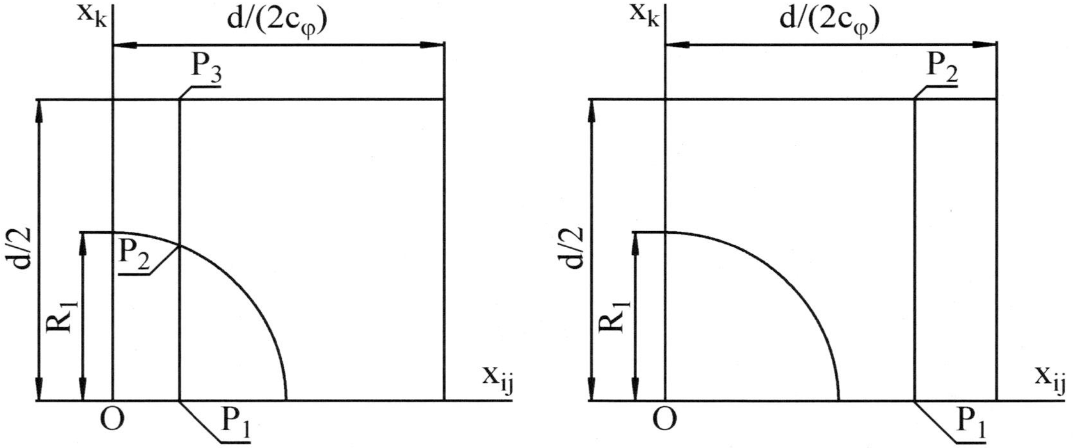

Figure 7.5. The abscissae (a) $\overline{P_1P_2P_3}$ and (b) $\overline{P_1P_2}$ along which the curve integrals $W_{cp}^{(ij)}$ for $x_{ij} \in \langle 0,R_1\rangle$ and $W_{cm}^{(ij)}$ for $x_{ij} \in \langle R_1, d/(2c_\varphi)\rangle$ are determined (see Eqs. (7.15), (7.16)), respectively. The abscissae $\overline{P_1P_2P_3}$, $\overline{P_1P_2}$ are related to the plane $x_{ij}x_k$ (see Figure 7.5). The coefficient $c_\varphi = c_\varphi(\varphi)$ is given by Eqs. (3.2), (3.3).

$$W_{cp}^{(ij)} = \int_{\overline{P_1P_2}} w_p^{(ij)}\,dx_k + \int_{\overline{P_2P_3}} w_m^{(ij)}\,dx_k = \int_0^{\sqrt{R_1^2-x_{ij}^2}} w_p^{(ij)}\,dx_k + \int_{\sqrt{R_1^2-x_{ij}^2}}^{d/2} w_m^{(ij)}\,dx_k,$$

$$x_{ij} \in \langle 0, R_1 \rangle . \tag{7.15}$$

If $\beta_p < \beta_m$, then a crack is formed in the cell matrix, and the curve integral $W_{cm}^{(ij)}$ for $x_{ij} \in \langle R_1, d/(2c_\varphi)\rangle$ is derived as (see Figure 7.5b)

$$W_{cm}^{(ij)} = \int_{\overline{P_1P_2}} w_m^{(ij)}\,dx_k = \int_0^{d/2} w_m^{(ij)}\,dx_k, \quad x_{ij} \in \left\langle R_1, \frac{d}{2c_\varphi}\right\rangle . \tag{7.16}$$

7.1.3.2. Condition $\beta_p(\varphi,\nu) = \beta_m(\varphi,\nu)$ for $\nu = \nu_{pm}(\varphi)$

Section 7.1.3.2 deals with the determination of intervals of the angles φ, ν which result in the validity of the conditions $\beta_p = \beta_m$. A reason of this determination is a dependence of the coefficient $\beta_q = \beta_q(\varphi,\nu)$ on the angles $\varphi, \nu \in \langle 0, \pi/2\rangle$ $(q = p,m)$ (see Eqs. (3.42), (3.54)–(3.60)).

If the model system (see Section 2.2.) does not exhibit a phase transformation at the temperature $T \in \langle T_f, T_r\rangle$, then the condition $\beta_p = \beta_m$ is transformed to $\alpha_p(\varphi,\nu) = \alpha_m(\varphi,\nu)$. Consequently, with regard to Eqs. (3.42), (3.60), the condition $\alpha_p(\varphi,\nu) - \alpha_m(\varphi,\nu) = 0$ is transformed to the form

$$(\alpha_{1p} - \alpha_{1m})\cos^2\varphi\sin^2\nu + (\alpha_{2p} - \alpha_{2m})\sin^2\varphi\sin^2\nu + (\alpha_{3p} - \alpha_{3m})\cos^2\nu = 0. \tag{7.17}$$

Consequently, an analysis of $\alpha_p(\varphi,\nu) = \alpha_m(\varphi,\nu)$ is as follows.

1. If the differences $\alpha_{1p} - \alpha_{1m}$, $\alpha_{2p} - \alpha_{2m}$, $\alpha_{3p} - \alpha_{3m}$ included in Eq. (7.17) exhibits identical signs, i.e., $\alpha_{1p} \gtrless \alpha_{1m}$, $\alpha_{2p} \gtrless \alpha_{2m}$, $\alpha_{3p} \gtrless \alpha_{3m}$, then the condition $\alpha_p(\varphi,\nu) \neq \alpha_m(\varphi,\nu)$ is valid for $\varphi, \nu \in \langle 0, \pi/2\rangle$. Otherwise, the following analysis is considered.

2. If $\alpha_{1p} \gtrless \alpha_{1m}$, $\alpha_{2p} \gtrless \alpha_{2m}$, $\alpha_{3p} \lessgtr \alpha_{3m}$, then the condition $\alpha_p(\varphi,\nu) = \alpha_m(\varphi,\nu)$ is valid for $\varphi \in \langle 0, \pi/2\rangle$ and for $\nu = \nu_{pm}$, where the function $\nu_{pm} = \nu_{pm}(\varphi)$ is given by Eq. (7.18).

3. If $\alpha_{1p} \gtrless \alpha_{1m}$, $\alpha_{2p} \lessgtr \alpha_{2m}$, $\alpha_{3p} \lessgtr \alpha_{3m}$, then the condition $\alpha_p(\varphi,\nu) = \alpha_m(\varphi,\nu)$ is valid for $\varphi \in \langle 0, \varphi_{pm}\rangle$ and for $\nu = \nu_{pm}$, where the angle φ_{pm} is given by Eq. (7.19).

4. If $\alpha_{1p} \gtrless \alpha_{1m}$, $\alpha_{2p} \lessgtr \alpha_{2m}$, $\alpha_{3p} \gtrless \alpha_{3m}$, then the condition $\alpha_p(\varphi,\nu) = \alpha_m(\varphi,\nu)$ is valid for $\varphi \in \langle \varphi_{pm}, \pi/2\rangle$ and for $\nu = \nu_{pm}$.

5. If $\alpha_{3p} = \alpha_{3m}$, then the following analysis is considered.

   (a) If $\alpha_{1p} \gtrless \alpha_{1m}$, $\alpha_{2p} \gtrless \alpha_{2m}$, then the condition $\alpha_p(\varphi,\nu) = \alpha_m(\varphi,\nu)$ is valid for $\nu = 0$.

(b) If $\alpha_{1p} \gtrless \alpha_{1m}$, $\alpha_{2p} \lessgtr \alpha_{2m}$, then the condition $\alpha_p(\varphi,\nu) = \alpha_m(\varphi,\nu)$ is valid for $\varphi = \varphi_{pm}$ and for $\nu \in \langle 0, \pi/2 \rangle$.

The function $\nu_{pm} = \nu_{pm}(\varphi)$ which represents a solution of Eq. (7.17) regarding the variable ν has the form

$$\nu_{pm} = \arcsin\left(\sqrt{\frac{\alpha_{3m} - \alpha_{3p}}{(\alpha_{1p} - \alpha_{1m})\cos^2\varphi + (\alpha_{2p} - \alpha_{2m})\sin^2\varphi + \alpha_{3m} - \alpha_{3p}}}\right). \tag{7.18}$$

The angle φ_{pm} which represents a solution of Eq. (7.17) regarding the variable φ for $\alpha_{3p} = \alpha_{3m}$ is derived as

$$\varphi_{pm} = \arcsin\left(\sqrt{\frac{\alpha_{1m} - \alpha_{1p}}{\alpha_{2p} - \alpha_{2m}}}\right). \tag{7.19}$$

Let the plane x_1x_2 (see Figure 3.1) in the component which is related to the subscript q_1 be isotropic. In case of this uni-axial anisotropy, we get $\alpha_{1q_1} = \alpha_{2q_1}$.

If $\varepsilon'_{11tp}(\varphi,\nu) \neq \varepsilon'_{11tm}(\varphi,\nu)$ at $T \in \langle T_f, T_r \rangle$, then the function $\nu_{pm} = \nu_{pm}(\varphi)$ and the angle $\varphi_{pm} \in \langle 0, \pi/2 \rangle$ which both result from the condition $\beta_p(\varphi,\nu) - \beta_m(\varphi,\nu) = 0$ are determined by numerical methods for a real two-component material (see Section 2.1.). In case of $\varepsilon'_{11tp}(\varphi,\nu) \neq \varepsilon'_{11tm}(\varphi,\nu)$ at $T \in \langle T_f, T_r \rangle$, the condition $\beta_p(\varphi,\nu) - \beta_m(\varphi,\nu) = 0$ thus represents a transcendental equation regarding the variables $\varphi, \nu \in \langle 0, \pi/2 \rangle$.

Let the multi-particle-matrix system (see Figure 2.1a) consist of anisotropic and isotropic components. As an example, let the matrix of the multi-particle-matrix system be isotropic. In this case, the coefficient α_m is not a function of the variables $\varphi, \nu \in \langle 0, \pi/2 \rangle$. The analysis concerning the condition $\alpha_p(\varphi,\nu) - \alpha_m(\varphi,\nu) = 0$ is also valid for the condition $\alpha_p(\varphi,\nu) - \alpha_m = 0$. Each of the thermal expansion coefficients α_{1m}, α_{2m}, α_{3m} which are considered within this analysis is only replaced by the thermal expansion coefficient α_m.

With regard to the plane $x_{ij}x_k$ (see Figure 7.6), the angle ν_{pm} thus defines the intervals $\nu \in \langle 0, \nu_{pm})$ and $\nu \in (\nu_{pm}, \pi/2\rangle$ for which the conditions $\alpha_p \lessgtr \alpha_m$ and $\alpha_p \gtrless \alpha_m$ are valid, respectively. The determination of $W_{cp}^{(ij)}$ and $W_{cm}^{(ij)}$ then considers the intervals $\nu_{pm} \in \langle 0, \nu_R \rangle$ (see Figure 7.6a), $\nu_{pm} \in \langle \nu_R, \nu^* \rangle$ (see Figure 7.6b), $\nu_{pm} \in \langle \nu^*, \pi/2 \rangle$ (see Figure 7.6c), where the angle $\nu^* = \nu^*(\varphi)$ which is shown in Fig. 7.6c is given by Eqs. (3.1)–(3.3). The angle ν_R which is shown in Fig. 7.6c has the form

$$\nu_R = \arctan\left(\frac{2R_1}{d}\right). \tag{7.20}$$

A. Interval $\nu_{pm} \in \langle 0, \nu_R \rangle$. If $\alpha_p > \alpha_m$ for $\nu \in (\nu_{pm}, \pi/2\rangle$, then a crack is formed in the spherical particle, and the curve integral $W_{cp}^{(ij)}$ for $x_{ij} \in \langle 0, x_{ijp} \rangle$, $x_{ij} \in \langle x_{ijp}, x_{ijm} \rangle$ and $x_{ij} \in \langle x_{ijm}, R_1 \rangle$ is derived as (see Figure 7.6a)

$$W_{cp}^{(ij)} = \int_{P_1P_2} w_p^{(ij)}\, dx_k = \int_0^{x_{k0}} w_p^{(ij)}\, dx_k, \quad x_{ij} \in \langle 0, x_{ijp} \rangle, \tag{7.21}$$

$$W_{cp}^{(ij)} = \int\limits_{\overline{P_3P_4}} w_p^{(ij)}\,dx_k + \int\limits_{\overline{P_4P_5}} w_m^{(ij)}\,dx_k = \int\limits_{0}^{\sqrt{R_1^2-x_{ij}^2}} w_p^{(ij)}\,dx_k + \int\limits_{\sqrt{R_1^2-x_{ij}^2}}^{x_{k0}} w_m^{(ij)}\,dx_k,$$

$$x_{ij} \in \left\langle x_{ijp}, x_{ijm} \right\rangle, \tag{7.22}$$

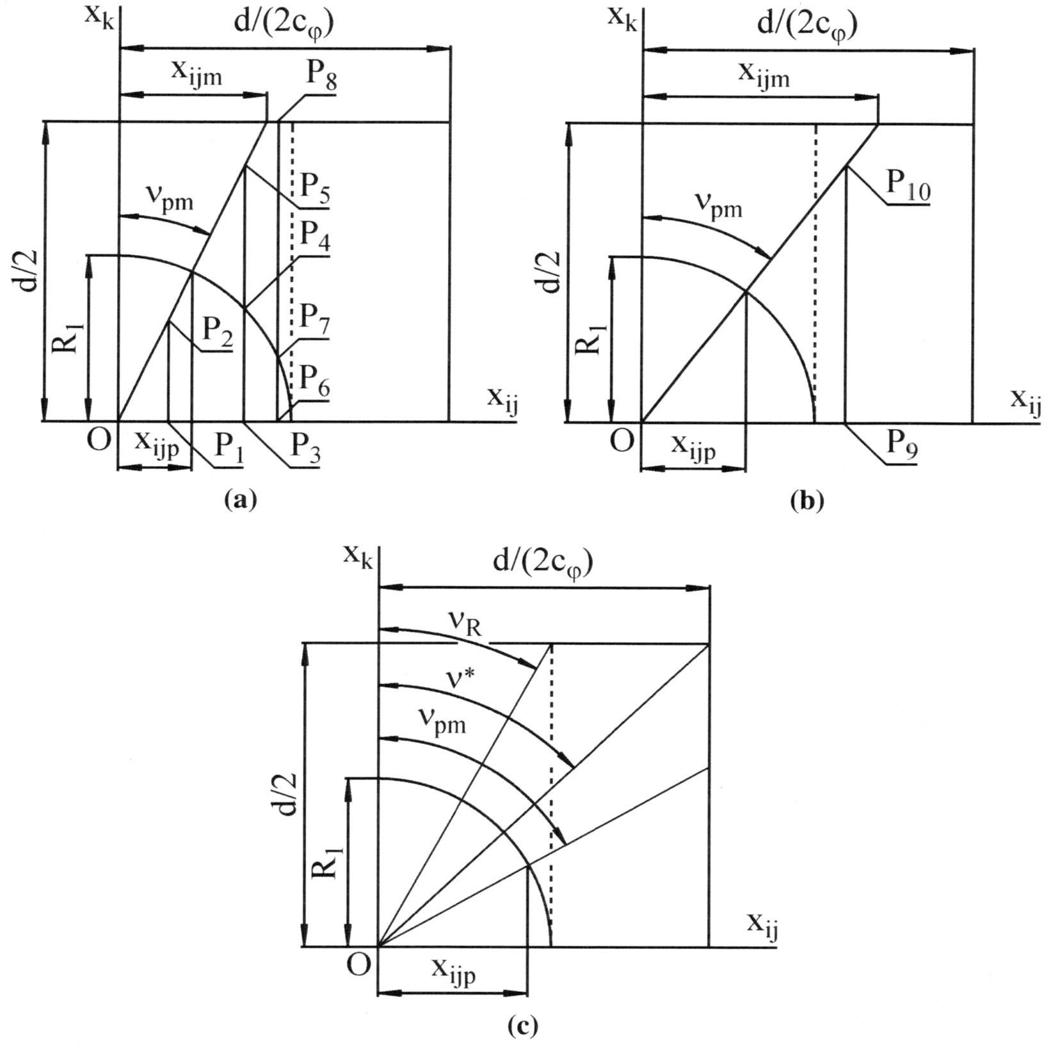

Figure 7.6. The intervals (a) $\nu_{pm} \in \langle 0, \nu_R \rangle$, (b) $\nu_{pm} \in \langle \nu_R, \nu^* \rangle$, (c) $\nu_{pm} \in \langle \nu^*, \pi/2 \rangle$ of the angle $\nu_{pm} = \nu_{pm}(\varphi)$ (see Eqs. (7.18), (7.20)). The abscissae $\overline{P_1P_2}$, $\overline{P_3P_4P_5}$, $\overline{P_6P_7P_8}$ and $\overline{P_9P_{10}}$ along which the curve integrals $W_{cp}^{(ij)}$ and $W_{cm}^{(ij)}$ are determined (see Eqs. (7.21)–(7.25)), respectively. The abscissae $\overline{P_1P_2}$, $\overline{P_3P_4P_5}$, $\overline{P_6P_7P_8}$, $\overline{P_9P_{10}}$ are related to the plane $x_{ij}x_k$ (see Figure 7.5). The coefficient $c_\varphi = c_\varphi(\varphi)$ is given by Eqs. (3.2), (3.3).

$$W_{cp}^{(ij)} = \int\limits_{\overline{P_6P_7}} w_p^{(ij)}\,dx_k + \int\limits_{\overline{P_7P_8}} w_m^{(ij)}\,dx_k = \int\limits_{0}^{\sqrt{R_1^2-x_{ij}^2}} w_p^{(ij)}\,dx_k + \int\limits_{\sqrt{R_1^2-x_{ij}^2}}^{d/2} w_m^{(ij)}\,dx_k,$$

$$x_{ij} \in \langle x_{ijm}, R_1 \rangle, \tag{7.23}$$

where x_{ijp}, x_{ijm}, x_{k0} have the forms

$$x_{ijp} = R_1 \sin\nu_{pm}, \quad x_{ijm} = R_1 \tan\nu_{pm}, \quad x_{k0} = x_{ij}\cot\nu_{pm}. \tag{7.24}$$

If $\alpha_p < \alpha_m$ for $\nu \in (\nu_{pm}, \pi/2\rangle$, then a crack is formed in the cell matrix, and the curve integral $W_{cm}^{(ij)}$ for $x_{ij} \in \langle R_1, d/(2c_\varphi)\rangle$ (see Figure 7.6a) is given by Eq. (7.16), where the coefficient c_φ is given by Eqs. (3.2), (3.2).

B. Interval $\nu_{pm} \in \langle \nu_R, \nu^* \rangle$**.** If $\alpha_p > \alpha_m$ for $\nu \in (\nu_{pm}, \pi/2\rangle$, then the curve integral $W_{cp}^{(ij)}$ for $x_{ij} \in \langle 0, x_{ijp}\rangle$ and $x_{ij} \in \langle x_{ijp}, R_1\rangle$ (see Figure 7.6b) is given by Eqs. (7.21) and (7.22), respectively.

If $\alpha_p < \alpha_m$ for $\nu \in (\nu_{pm}, \pi/2\rangle$, then the curve integral $W_{cm}^{(ij)}$ for $x_{ij} \in \langle x_{ijm}, d/(2c_\varphi)\rangle$ is given by Eq. (7.16). The curve integral $W_{cm}^{(ij)}$ for $x_{ij} \in \langle R_1, x_{ijm}\rangle$ is derived as (see Figure 7.6b)

$$W_{cm}^{(ij)} = \int\limits_{\overline{P_9P_{10}}} w_m^{(ij)}\,dx_k = \int\limits_{0}^{x_{k0}} w_m^{(ij)}\,dx_k, \quad x_{ij} \in \langle R_1, x_{ijm}\rangle. \tag{7.25}$$

C. Interval $\nu_{pm} \in \langle \nu^*, \pi/2 \rangle$**.**

Method 1. If $\alpha_p > \alpha_m$ for $\nu \in (\nu_{pm}, \pi/2\rangle$, then the curve integral $W_{cp}^{(ij),1}$ for $x_{ij} \in \langle 0, x_{ijp}\rangle$ and $x_{ij} \in \langle x_{ijp}, R_1\rangle$ (see Figure 7.6c) is given by Eqs. (7.21) and (7.22), respectively. If $\alpha_p < \alpha_m$ for $\nu \in (\nu_{pm}, \pi/2\rangle$, then the curve integral $W_{cm}^{(ij),1}$ for $x_{ij} \in \langle R_1, d/(2c_\varphi)\rangle$ is given by Eq. (7.25) [3].

Method 2. As mentioned above, the angle $\nu_{pm} = \nu_{pm}(\varphi)$ thus defines two volumes in the cubic cell which are determined by the intervals $\nu \in \langle 0, \nu_{pm})$ and $\nu \in (\nu_{pm}, \pi/2\rangle$, where definition intervals for the variable φ are presented in Items 1–5, p. 113. Additionally, the conditions $\alpha_p \lesseqgtr \alpha_m$ and $\alpha_p \gtreqless \alpha_m$ are valid for $\nu \in \langle 0, \nu_{pm})$ and $\nu \in (\nu_{pm}, \pi/2\rangle$, respectively.

The method 1 for the determination of $W_{cp}^{(ij),1}$ and $W_{cm}^{(ij),1}$ is based on an assumption that the crack is formed in the spherical particle or cell matrix in the plane x_ix_j provided that $\alpha_p > \alpha_m$ or $\alpha_p < \alpha_m$ for $\nu \in (\nu_{pm}, \pi/2\rangle$ (see Figure 7.6c), respectively.

On the contrary, the method 2 is based on an assumption that the crack can be formed in the spherical particle or cell matrix in the plane x_ix_j in spite of the condition $\beta_p < \beta_m$

[3] The superscript *1* in $W_{cp}^{(ij),1}$ and $W_{cm}^{(ij),1}$ is related to the method 1.

or $\beta_p > \beta_m$, respectively. The following analysis can then explain such paradoxical behaviour of the cracking which is experimentally observed in a real two-component material [71]–[74]. This analysis results from an assumption that the release of energy of a system considers 'minimal resistance' of the system, i.e., the energy of a system is released through 'minimal resistance' of the system. This analysis is determined for the spherical particle and cell matrix.

SPHERICAL PARTICLE. Let the condition $\alpha_p < \alpha_m$ be valid for $\nu \in (\nu_{pm}, \pi/2\rangle$.

Let $R_{1cp}^{(jk),1} = R_{1cp}^{(jk),1}(\varphi, v)$ represent a critical particle radius for the crack initiation in the spherical particle in the plane x_jx_k. Let $R_{1cp}^{(jk),1} = R_{1cp}^{(jk),1}(\varphi, v)$ be determined by the curve integral $W_{cp}^{(jk),1} = W_{cp}^{(jk),1}\left(x_{jk}, \varphi, R_1, v\right)$ for $\varphi = \angle\left(x_j, x_{jk}\right)$, $x_{jk} \subset x_jx_k$ (see Eq. (7.10)). Let $W_{cp}^{(jk),1}$ be determined by the *method 1*, where the transformations in Section 7.1.4. are considered within this determination.

Let $R_{1cp}^{(ki),1} = R_{1cp}^{(ki),1}(\varphi, v)$ represent a critical particle radius for the crack initiation in the spherical particle in the plane x_kx_i. Let $R_{1cp}^{(ki),1} = R_{1cp}^{(ki),1}(\varphi, v)$ be determined by the curve integral $W_{cp}^{(ki),1} = W_{cp}^{(ki),1}(x_{ki}, \varphi, R_1, v)$ for $\varphi = \angle(x_k, x_{ki})$, $x_{ki} \subset x_kx_i$ (see Eq. (7.10)). Let $W_{cp}^{(ki),1}$ be also determined by the *method 1*, where the transformations in Section 7.1.4. are considered within this determination.

Let $R_{1cm}^{(ij),1} = R_{1cm}^{(ij),1}(\varphi, v)$ represent a critical particle radius for the crack initiation in the cell matrix in the plane x_kx_i. Let $R_{1cm}^{(ij),1} = R_{1cm}^{(ij),1}(\varphi, v)$ be determined by the curve integral $W_{cm}^{(ij),1} = W_{cm}^{(ij),1}(x_{ij}, \varphi, R_1, v)$ for $\varphi = \angle(x_i, x_{ij})$, $x_{ij} \subset x_ix_j$ (see Eq. (7.10)). Let $W_{cm}^{(ij),1}$ be also determined by the *method 1*, where the transformations in Section 7.1.4. are considered within this determination.

Finally, let $R_{1cp}^{(ij),2} = R_{1cp}^{(ij),2}(\varphi, v)$ represent a critical particle radius for the crack initiation in the spherical particle in the plane x_kx_i. Let $R_{1cp}^{(ij),2} = R_{1cp}^{(ij),2}(\varphi, v)$ be determined by the curve integral $W_{cp}^{(ij),2} = W_{cp}^{(ij),2}(x_{ij}, \varphi, R_1, v)$ for $\varphi = \angle(x_i, x_{ij})$, $x_{ij} \subset x_ix_j$ (see Eq. (7.10)). With regard to Figure 7.7, the curve integral $W_{cp}^{(ij),2}$ which is determined by the *method 2* for $x_{ij} \in \langle 0, x_{ijp}\rangle$ and $x_{ij} \in \langle x_{ijp}, R_1\rangle$ have the form [4]

$$W_{cp}^{(ij),2} = \int\limits_{P_2P_3} w_p^{(ij)}\, dx_k + \int\limits_{P_3P_4} w_m^{(ij)}\, dx_k - \int\limits_{P_1P_2} w_p^{(ij)}\, dx_k$$

$$= \int\limits_{x_{k0}}^{\sqrt{R_1^2 - x_{ij}^2}} w_p^{(ij)}\, dx_k + \int\limits_{\sqrt{R_1^2 - x_{ij}^2}}^{d/2} w_m^{(ij)}\, dx_k - \int\limits_{0}^{x_{k0}} w_p^{(ij)}\, dx_k \geq 0, \quad x_{ij} \in \langle 0, x_{ijp}\rangle, \tag{7.26}$$

[4]The superscript 2 in $W_{cp}^{(ij),2}$ and $W_{cm}^{(ij),2}$ (see Eqs. (7.26), (7.27), (7.29)) is related to the method 2.

$$W_{cp}^{(ij),2} = \int\limits_{\overline{P_7P_8}} w_m^{(ij)}\,dx_k - \left[\int\limits_{\overline{P_5P_6}} w_p^{(ij)}\,dx_k + \int\limits_{\overline{P_6P_7}} w_m^{(ij)}\,dx_k\right]$$

$$= \int_{x_{k0}}^{d/2} w_m^{(ij)}\,dx_k - \left[\int_0^{\sqrt{R_1^2-x_{ij}^2}} w_p^{(ij)}\,dx_k + \int_{\sqrt{R_1^2-x_{ij}^2}}^{x_{k0}} w_m^{(ij)}\,dx_k\right] \geq 0, \quad x_{ij} \in \left\langle x_{ijp}, R_1\right\rangle. \tag{7.27}$$

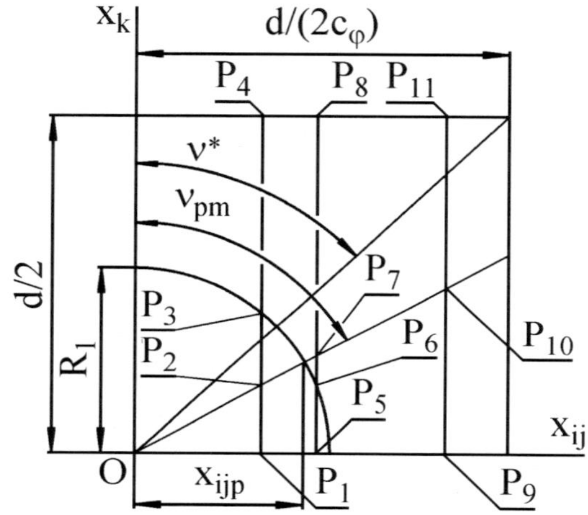

Figure 7.7. The abscissae $\overline{P_1P_2P_3P_4}$, $\overline{P_5P_6P_7P_8}$ and $\overline{P_9P_{10}P_{11}}$ along which the curve integrals $W_{cp}^{(ij),2}$ and $W_{cm}^{(ij),2}$ are determined for $\nu_{pm} \in \langle \nu^*, \pi/2\rangle$ (see Eqs. (7.26), (7.27), (7.29)), respectively. The abscissae $\overline{P_1P_2P_3P_4}$, $\overline{P_5P_6P_7P_8}$ and $\overline{P_9P_{10}P_{11}}$ are related to the plane $x_{ij}x_k$ (see Figure 7.5). The coefficient $c_\varphi = c_\varphi(\varphi)$ is given by Eqs. (3.2), (3.3).

Additionally, $W_{cp}^{(ij),2} = W_{cp}^{(ij),2}(x_{ij}, \varphi, R_1, v)$ is defined for such interval $x_{ij} \in \left\langle x_{ijp0}, R_1\right\rangle \subset \langle 0, R_1\rangle$ for which the condition $W_{cp}^{(ij),2} \geq 0$ is valid, where x_{ijp0} is determined by the condition $W_{cp}^{(ij),2} = 0$. In case of the thermal stresses, $W_{cp}^{(ij),2}$ represents an increasing function of $x_{ij} \in \left\langle x_{ijp0}, R_1\right\rangle$.

If the conditions

$$R_{1cp}^{(ij),2} < R_{1cp}^{(jk),1}, \quad R_{1cp}^{(ij),2} < R_{1cp}^{(ki),1}, \quad R_{1cp}^{(ij),2} < R_{1cm}^{(ij),1} \tag{7.28}$$

are simultaneously valid, then the crack is formed in the plane x_ix_j in the spherical particle

- in spite of the fact that the condition $\alpha_p < \alpha_m$ for $\nu \in (\nu_{pm}, \pi/2\rangle$ is valid,
- and in spite of the fact that the crack would be expected to be formed in the cell matrix regarding $\alpha_p < \alpha_m$ for $\nu \in (\nu_{pm}, \pi/2\rangle$.

The critical particle radius $R_{1cp}^{(ij),2}$ which is a minimal value of the set $\left\{R_{1cp}^{(ij),2}, R_{1cp}^{(jk),1}, R_{1cp}^{(ki),1}, R_{1cm}^{(ij),1}\right\}$ thus represents the 'minimal resistance' through which energy of the cubic cell is released.

<u>CELL MATRIX.</u> Let the condition $\alpha_p > \alpha_m$ be valid for $\nu \in (\nu_{pm}, \pi/2\rangle$.

Let $R_{1cm}^{(jk),1} = R_{1cm}^{(jk),1}(\varphi, v)$ represent a critical particle radius for the crack initiation in the cell matrix in the plane x_jx_k. Let $R_{1cm}^{(jk),1} = R_{1cm}^{(jk),1}(\varphi, v)$ be determined by the curve integral $W_{cm}^{(jk),1} = W_{cm}^{(jk),1}\left(x_{jk}, \varphi, R_1, v\right)$ for $\varphi = \angle\left(x_j, x_{jk}\right)$, $x_{jk} \subset x_jx_k$ (see Eq. (7.10)). Let $W_{cm}^{(jk),1}$ be determined by the *method 1*, where the transformations in Section 7.1.4. are considered within this determination.

Let $R_{1cm}^{(ki),1} = R_{1cm}^{(ki),1}(\varphi, v)$ represent a critical particle radius for the crack initiation in the cell matrix in the plane x_kx_i. Let $R_{1cm}^{(ki),1} = R_{1cm}^{(ki),1}(\varphi, v)$ be determined by the curve integral $W_{cm}^{(ki),1} = W_{cm}^{(ki),1}(x_{ki}, \varphi, R_1, v)$ for $\varphi = \angle(x_k, x_{ki})$, $x_{ki} \subset x_kx_i$ (see Eq. (7.10)). Let $W_{cm}^{(ki),1}$ be also determined by the *method 1*, where the transformations in Section 7.1.4. are considered within this determination.

Let $R_{1cp}^{(ij),1} = R_{1cp}^{(ij),1}(\varphi, v)$ represent a critical particle radius for the crack initiation in the spherical particle in the plane x_kx_i. Let $R_{1cp}^{(ij),1} = R_{1cp}^{(ij),1}(\varphi, v)$ be determined by the curve integral $W_{cp}^{(ij),1} = W_{cp}^{(ij),1}(x_{ij}, \varphi, R_1, v)$ for $\varphi = \angle(x_i, x_{ij})$, $x_{ij} \subset x_ix_j$ (see Eq. (7.10)). Let $W_{cp}^{(ij),1}$ be also determined by the *method 1*, where the transformations in Section 7.1.4. are considered within this determination.

Finally, let $R_{1cm}^{(ij),2} = R_{1cm}^{(ij),2}(\varphi, v)$ represent a critical particle radius for the crack initiation in the cell matrix in the plane x_kx_i. Let $R_{1cm}^{(ij),2} = R_{1cm}^{(ij),2}(\varphi, v)$ be determined by the curve integral $W_{cm}^{(ij),2} = W_{cm}^{(ij),2}(x_{ij}, \varphi, R_1, v)$ for $\varphi = \angle(x_i, x_{ij})$, $x_{ij} \subset x_ix_j$ (see Eq. (7.10)). With regard to Figure 7.7, the curve integral $W_{cm}^{(ij),2}$ which is determined by the *method 2* for $x_{ij} \in \left\langle R_1, d/\left(2c_\varphi\right)\right\rangle$ have the form

$$W_{cm}^{(ij),2} = \int\limits_{\overline{P_{10}P_{11}}} w_m^{(ij)}\, dx_k - \int\limits_{\overline{P_9P_{10}}} w_m^{(ij)}\, dx_k = \int\limits_{x_{k0}}^{d/2} w_m^{(ij)}\, dx_k - \int\limits_{0}^{x_{k0}} w_m^{(ij)}\, dx_k,$$

$$x_{ij} \in \left\langle R_1, \frac{d}{2c_\varphi}\right\rangle. \tag{7.29}$$

Additionally, $W_{cm}^{(ij),2} = W_{cm}^{(ij),2}(x_{ij}, \varphi, R_1, v)$ is defined for such interval $x_{ij} \in \left\langle R_1, x_{ijm0}\right\rangle \subset \left\langle R_1, d/\left(2c_\varphi\right)\right\rangle$ for which the condition $W_{cm}^{(ij),2} \geq 0$ is valid, where x_{ijm0} is determined by the condition $W_{cm}^{(ij),2} = 0$. In case of the thermal stresses, $W_{cm}^{(ij),2}$ represents a decreasing function of $x_{ij} \in \left\langle R_1, x_{ijm0}\right\rangle$.

If the conditions

$$R_{1cm}^{(ij),2} < R_{1cm}^{(jk),1}, \quad R_{1cm}^{(ij),2} < R_{1cm}^{(ki),1}, \quad R_{1cm}^{(ij),2} < R_{1cp}^{(ij),1} \tag{7.30}$$

are simultaneously valid, then the crack is formed in the plane x_ix_j in the cell matrix

- in spite of the fact that the condition $\alpha_p > \alpha_m$ for $\nu \in (\nu_{pm}, \pi/2\rangle$ is valid,

- and in spite of the fact that the crack would be expected to be formed in the spherical particle regarding $\alpha_p > \alpha_m$ for $\nu \in (\nu_{pm}, \pi/2\rangle$.

The critical particle radius $R_{1cm}^{(ij),2}$ which is a minimal value of the set $\left\{R_{1cm}^{(ij),2}, R_{1cm}^{(jk),1}, R_{1cm}^{(ki),1}, R_{1cp}^{(ij),1}\right\}$ thus represents the 'minimal resistance' through which energy of the cubic cell is released.

7.1.4. Transformations Concerning Cracking in Planes x_ix_j, x_jx_k, x_kx_i

As mentioned above, the method 2 in Section 7.1.3.2 considers the curve integrals $W_{cq}^{(ij),2} = W_{cq}^{(ij),2}(x_{ij}, \varphi, R_1, v)$, $W_{cq}^{(jk),1} = W_{cq}^{(jk),1}\left(x_{jk}, \varphi, R_1, v\right)$, $W_{cq}^{(ki),1} = W_{cq}^{(ki),1}(x_{ki}, \varphi, R_1, v)$ $(q = p,m)$ in the planes $x_{ij}x_k$, $x_{jk}x_i$, $x_{ki}x_j$, respectively, where $\varphi = \angle(x_i, x_{ij})$, $x_{ij} \subset x_ix_j$; $\varphi = \angle\left(x_j, x_{jk}\right)$, $x_{jk} \subset x_jx_k$; $\varphi = \angle(x_k, x_{ki})$, $x_{ki} \subset x_kx_i$.

The determination of $W_{cq}^{(ij),2}$, $W_{cq}^{(jk),1}$, $W_{cq}^{(ki),1}$ is required to consider the following transformations which concerns the crack formation in the planes x_ix_j, x_jx_k, x_kx_i.

The analysis in Sections 7.1.1.–7.1.3. which deals with the crack formation in the plane x_ix_j (see Figure 7.1) is also valid for the crack formation

- in the plane x_jx_k, and then $w^{(ij)} = w^{(ij)}(x_{ij}, \varphi, x_k)$, $W_c^{(ij)} = W_c^{(ij)}(x_{ij}, \varphi)$, $f^{(ij)} = f^{(ij)}(x_{ij}, \varphi)$, s_{kk} are replaced by $w^{(jk)} = w^{(jk)}\left(x_{jk}, \varphi, x_i\right)$, $W_c^{(jk)} = W_c^{(jk)}\left(x_{jk}, \varphi\right)$, $f^{(jk)} = f^{(jk)}\left(x_{jk}, \varphi\right)$, s_{ii}, respectively, where $\varphi = \angle\left(x_j, x_{jk}\right) \in \langle 0, 2\pi\rangle$.

- in the plane x_jx_k, and then $w^{(ij)} = w^{(ij)}(x_{ij}, \varphi, x_k)$, $W_c^{(ij)} = W_c^{(ij)}(x_{ij}, \varphi)$, $f^{(ij)} = f^{(ij)}(x_{ij}, \varphi)$, s_{kk} are replaced by $w^{(ki)} = w^{(ki)}(x_{ki}, \varphi, x_j)$, $W_c^{(ki)} = W_c^{(ki)}(x_{ki}, \varphi)$, $f^{(ki)} = f^{(ki)}(x_{ki}, \varphi)$, s_{jj}, respectively, where $\varphi = \angle(x_k, x_{ki}) \in \langle 0, 2\pi\rangle$.

With regard to the cracking in the planes x_ix_j, x_jx_k, x_kx_i, i.e., in the planes x_1x_2, x_2x_3, x_3x_1, the following analyses and transformations concerning

- the coefficients a_{11}, a_{12}, a_{13}, a_{31}, a_{32}, a_{33} (see Eq. (3.42)),
- the elastic moduli s_{11q}, $s_{12q}, \ldots, s_{56q}$, s_{66q} (see Eq. (3.41)),
- the thermal expansion coefficient α_{iq} $(i = 1,2,3)$ (see Eqs. (3.42), (3.60))

are required to be considered.

With regard to the coefficients a_{11}, a_{12}, a_{13}, a_{31}, a_{32}, a_{33} included in Eqs. (3.41), (3.60), we get:

- In case of the cracking in the plane x_ix_j, the coefficients a_{11}, a_{12}, a_{13}, a_{31}, a_{32}, a_{33} have the forms

$$a_{11} = \frac{x_{ij}\cos\varphi}{\sqrt{x_{ij}^2 + x_k^2}}, \quad a_{12} = \frac{x_{ij}\sin\varphi}{\sqrt{x_{ij}^2 + x_k^2}}, \quad a_{13} = \frac{x_k}{\sqrt{x_{ij}^2 + x_k^2}},$$

$$a_{31} = -\frac{x_k\cos\varphi}{\sqrt{x_{ij}^2 + x_k^2}}, \quad a_{32} = -\frac{x_k\sin\varphi}{\sqrt{x_{ij}^2 + x_k^2}}, \quad a_{33} = -\frac{x_{ij}}{\sqrt{x_{ij}^2 + x_k^2}}. \tag{7.31}$$

- In case of the cracking in the plane $x_j x_k$, the variables x_{ij} and x_k in Eq. (7.31) are replaced by x_{jk} and x_i, respectively.
- In case of the cracking in the plane $x_k x_i$, the variables x_{ij} and x_k in Eq. (7.31) are replaced by x_{ki} and x_j, respectively.

Consequently, the elastic moduli $s_{11q}, s_{12q}, \ldots, s_{56q}, s_{66q}$ included in Eq. (3.41) and the thermal expansion coefficient α_{iq} $(i=1,2,3)$ included in Eq. (3.60) are replaced by $s_{11q}^{(y)}$, $s_{12q}^{(y)}, \ldots, s_{56q}^{(y)}, s_{66q}^{(y)}$ and $\alpha_{iq}^{(y)}$, respectively, where $y = ij$, jk, ki. In case of the cracking in the planes $x_i x_j$, $x_j x_k$ and $x_k x_i$, we get $y = ij$, $y = jk$ and $y = ki$, respectively.

The transformations of $s_{11q}^{(y)}, s_{12q}^{(y)}, \ldots, s_{56q}^{(y)}, s_{66q}^{(y)}$ and $\alpha_{iq}^{(y)}$ due to the cracking in the planes x_1x_2, x_2x_3, x_3x_1 are as follows. Let the cracking in the plane $x_i x_j$ be related to the Cartesian system $\left(Ox_1^{(y)} x_2^{(y)} x_3^{(y)}\right)$, where $y \equiv ij = 12,\ 23,\ 31$. Relationships between $x_i^{(y_1)}$, $x_j^{(y_2)}$ $(i, j = 1,2,3;\ i \neq j;\ y_1, y_2 = 12,\ 23,\ 31;\ y_1 \neq y_2)$ are derived as

$$x_1^{(12)} = x_3^{(23)} = x_2^{(31)}, \quad x_2^{(12)} = x_1^{(23)} = x_3^{(31)}, \quad x_3^{(12)} = x_2^{(23)} = x_1^{(31)}. \tag{7.32}$$

The elastic modulus $s_{rstuq}^{(y)}$ $(\equiv s_{rstuq})$ and the thermal expansion coefficient $\alpha_{iq}^{(y)}$ $(\equiv \alpha_{iq})$ $(i, r, s, t, u = 1,2,3;\ y = 12,\ 23,\ 31)$ are related to the axes $x_r^{(y)}$, $x_s^{(y)}$, $x_t^{(y)}$, $x_u^{(y)}$ and $x_i^{(y)}$, respectively.

The transformations $s_{ijklq}^{(y_1)} \rightarrow s_{stuvq}^{(y_2)}$ and $\alpha_{iq}^{(y_1)} \rightarrow \alpha_{jq}^{(y_2)}$ $(i, j, k, l, s, t, u, v = 1,2,3;\ y_1, y_2 = 12, 23, 31;\ y_1 \neq y_2)$ can be thus determined using the relationships given by Eq. (7.32). Strictly speaking, if the numerical values $s_{ijklq}^{(y_1)}$, $\alpha_{iq}^{(y_1)}$ are known, then the numerical values $s_{stuvq}^{(y_2)}$, $\alpha_{jq}^{(y_2)}$ can be determined by Eq. (7.32). Finally, with regard to the subscripts *ijkl*, *stuv*, the subscript transformations given by Eq. (3.43) are considered. As an example, considering Eqs. (7.32), (3.43), the elastic modulus $s_{45q}^{(23)}$ in the Cartesian system $\left(Ox_1^{(23)} x_2^{(23)} x_3^{(23)}\right)$ which is derived regarding the Cartesian system $\left(Ox_1^{(31)} x_2^{(31)} x_3^{(31)}\right)$ has the form

$$s_{45q}^{(23)} = s_{2313q}^{(23)} = s_{1232q}^{(31)} = s_{1223q}^{(31)} = s_{64q}^{(31)} = s_{46q}^{(31)}. \tag{7.33}$$

Accordingly, the transformations $s_{ijklq}^{(y_1)} \rightarrow s_{stuvq}^{(y_2)}$ and $\alpha_{iq}^{(y_1)} \rightarrow \alpha_{jq}^{(y_2)}$ are derived as

$$\begin{aligned}
&s_{11q}^{(12)} = s_{33q}^{(23)} = s_{22q}^{(31)} \quad s_{12q}^{(12)} = s_{13q}^{(23)} = s_{23q}^{(31)}, \quad s_{13q}^{(12)} = s_{23q}^{(23)} = s_{12q}^{(31)}, \\
&s_{14q}^{(12)} = s_{36q}^{(23)} = s_{25q}^{(31)}, \quad s_{15q}^{(12)} = s_{34q}^{(23)} = s_{26q}^{(31)}, \quad s_{16q}^{(12)} = s_{35q}^{(23)} = s_{24q}^{(31)}, \\
&s_{22q}^{(12)} = s_{11q}^{(23)} = s_{33q}^{(31)}, \quad s_{23q}^{(12)} = s_{12q}^{(23)} = s_{13q}^{(31)}, \quad s_{24q}^{(12)} = s_{16q}^{(23)} = s_{35q}^{(31)}, \\
&s_{25q}^{(12)} = s_{14q}^{(23)} = s_{36q}^{(31)}, \quad s_{26q}^{(12)} = s_{15q}^{(23)} = s_{34q}^{(31)}, \quad s_{33q}^{(12)} = s_{22q}^{(23)} = s_{11q}^{(31)}, \\
&s_{34q}^{(12)} = s_{26q}^{(23)} = s_{15q}^{(31)}, \quad s_{35q}^{(12)} = s_{24q}^{(23)} = s_{16q}^{(31)}, \quad s_{36q}^{(12)} = s_{25q}^{(23)} = s_{14q}^{(31)}, \\
&s_{44q}^{(12)} = s_{66q}^{(23)} = s_{55q}^{(31)}, \quad s_{45q}^{(12)} = s_{46q}^{(23)} = s_{56q}^{(31)}, \quad s_{46q}^{(12)} = s_{56q}^{(23)} = s_{45q}^{(31)}, \\
&s_{55q}^{(12)} = s_{44q}^{(23)} = s_{66q}^{(31)}, \quad s_{56q}^{(12)} = s_{45q}^{(23)} = s_{46q}^{(31)}, \quad s_{66q}^{(12)} = s_{55q}^{(23)} = s_{44q}^{(31)},
\end{aligned} \tag{7.34}$$

$$\alpha_{1q}^{(12)} = \alpha_{3q}^{(23)} = \alpha_{2q}^{(31)}, \quad \alpha_{2q}^{(12)} = \alpha_{1q}^{(23)} = \alpha_{3q}^{(31)}, \quad \alpha_{3q}^{(12)} = \alpha_{2q}^{(23)} = \alpha_{1q}^{(31)}. \tag{7.35}$$

7.2. Analytical Model of Energy Barrier

7.2.1. General Analysis

The analytical model of an energy barrier which also represents an energy gradient includes

- the definition of the energy barriers $W_{bi} = W_{bi}(x_i)$, $W_{bj} = W_{bj}(x_j)$, $W_{bk} = W_{bk}(x_k)$ (see Eqs. (7.37)–(7.39)) along the axes x_i, x_j, x_k ($i, j, k = 1,2,3$; $i \neq j \neq k$) (see Figure 7.8), respectively,
- the determination of the integration intervals $x_i \in \langle x_{i1q}, x_{i2q} \rangle$, $x_j \in \langle x_{j1q}, x_{j2q} \rangle$, $x_k \in \langle x_{k1q}, x_{k2q} \rangle$ for the spherical particle ($q = p$), the spherical envelope ($q = e$) and the cell matrix ($q = m$) which represent components of the multi-particle-matrix and multi-particle-envelope-matrix systems (see Figure 2.1).

The definition of the energy barriers $W_{bi} = W_{bi}(x_i)$, $W_{bj} = W_{bj}(x_j)$, $W_{bk} = W_{bk}(x_k)$ is as follows.

Let a solid continuum with a general shape shown in Figure 7.8 be considered. Let W be energy accumulated in the volume V of this solid continuum, and then we get

$$W = \int_V w\,dV = \int_{x_{i1}}^{x_{i2}} \int_{x_{j1}}^{x_{j2}} \int_{x_{k1}}^{x_{k2}} w\,dx_i\,dx_j\,dx_k$$

$$= \int_{x_{i1}}^{x_{i2}} \int_{S_i} w\,dx_i\,dS_i = \int_{x_{j1}}^{x_{j2}} \int_{S_j} w\,dx_j\,dS_j = \int_{x_{k1}}^{x_{k2}} \int_{S_k} w\,dx_k\,dS_k, \tag{7.36}$$

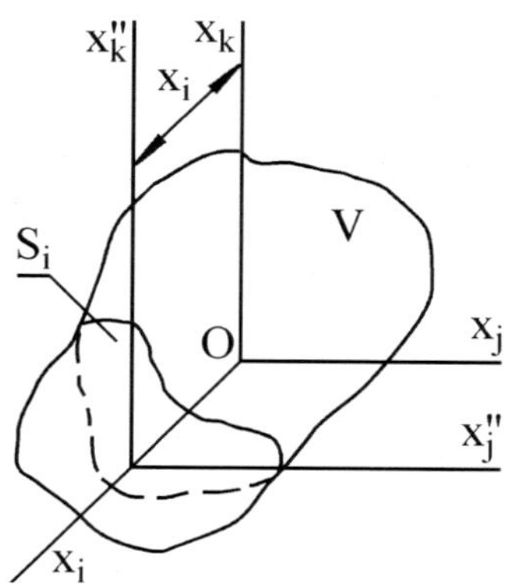

Figure 7.8. The solid continuum with a general shape with the volume V, and with the surface S_i in a position given by $x_i \in \langle x_{i1}, x_{i2} \rangle$, where the difference $x_{i2} - x_{i1}$ represents length of the solid continuum along the axis x_i. The surface S_i with the normal x_i is a cross-section of the solid continuum. The cross-section is given by the plane $x_j'' x_k''$ which is parallel to the plane $x_j x_k$.

where $w = w(x_i, x_j, x_k)$ is thermal-stress induced elastic energy density. x_{i1}, x_{i2}; x_{j1}, x_{j2}; x_{k1}, x_{k2} are integration boundaries related to the variables x_i; x_j; x_k, respectively. The surfaces

S_i, S_j, S_k in positions given by $x_i \in \langle x_{1i}, x_{2i} \rangle$, $x_j \in \left\langle x_{1j}, x_{2j} \right\rangle$, $x_k \in \langle x_{1k}, x_{2k} \rangle$ represent cross-sections of the solid continuum, where these surfaces are perpendicular to the axes x_i, x_j, x_k, respectively. As an example (see Figure 7.8), the surface S_i with the normal x_i is related to the plane $x_j'' x_k''$ which is parallel to $x_j x_k$.

Consequently, $W_{bi} = W_{bi}(x_i)$, $W_{bj} = W_{bj}(x_j)$, $W_{bk} = W_{bk}(x_k)$ which represent surface integrals related to S_i, S_j, S_k, respectively, have the forms

$$W_{bi} = W_{bi}(x_i) = \frac{\partial W}{\partial x_i} = \int_{S_i} w \, dS_i = \int_{x_{j1}}^{x_{j2}} \int_{x_{k1}}^{x_{k2}} w \, dx_j \, dx_k, \tag{7.37}$$

$$W_{bj} = W_{bj}(x_j) = \frac{\partial W}{\partial x_j} = \int_{S_j} w \, dS_j = \int_{x_{i1}}^{x_{i2}} \int_{x_{k1}}^{x_{k2}} w \, dx_i \, dx_k, \tag{7.38}$$

$$W_{bk} = W_{bk}(x_k) = \frac{\partial W}{\partial x_k} = \int_{S_k} w \, dS_k = \int_{x_{i1}}^{x_{i2}} \int_{x_{j1}}^{x_{j2}} w \, dx_i \, dx_j. \tag{7.39}$$

This definition of an energy barrier, i.e., $W_{bi} = \partial W / \partial x_i$, $W_{bj} = \partial W / \partial x_j$, $W_{bk} = \partial W / \partial x_k$, is also presented in [38].

If $w = w(x_i, x_j, x_k) = w(r, \varphi, \nu)$ is determined by the spherical coordinates (r, φ, ν), then the following relationships between the coordinates (x_i, x_j, x_k) and (r, φ, ν) are considered:

1. If $\varphi = \angle(x_i, x_{ij})$, $\nu = \angle(x_1', x_k)$ (see Figure 3.1), $x_{ij} \subset x_i x_j$ (see Figure 7.1), then we get

$$x_i = r \cos\varphi \sin\nu, \quad x_j = r \sin\varphi \sin\nu, \quad x_k = r \cos\nu. \tag{7.40}$$

2. If $\varphi = \angle\left(x_j, x_{jk}\right)$, $\nu = \angle(x_1', x_i)$, $x_{jk} \subset x_j x_k$, then we get

$$x_i = r \cos\nu, \quad x_j = r \cos\varphi \sin\nu, \quad x_k = r \sin\varphi \sin\nu. \tag{7.41}$$

3. If $\varphi = \angle(x_k, x_{ki})$, $\nu = \angle(x_1', x_j)$, $x_{ki} \subset x_k x_i$, then we get

$$x_i = r \sin\varphi \sin\nu, \quad x_j = r \cos\nu, \quad x_k = r \cos\varphi \sin\nu. \tag{7.42}$$

In general, energy barriers influence motion of dislocations, magnetic domain walls, etc., where a magnetic domain wall represents a front between magnetic domains with and without ordered magnetic moments [112].

As an example, a planar magnetic domain wall in a magnetic material is shifted along the axis x from the position $x = a$ to the position $x = b$ due to the change $\Delta H_{ba} = H_b - H_a$ of the magnetic field intensity H (see Fig 7.9). As presented in Fig 7.9, the point B with the position $x = b$ represents a maximum of the energy barrier *2*. Consequently, this magnetic domain wall moves along the axis x to the position $x = c$ at the constant intensity $H_b = H_c$. The wall is then stopped by the energy barrier *3* which is higher than the energy barrier *2*. The position change $\Delta x_{cb} = c - b$ of the wall results in the change ΔM_{cb} of the magnetic moment M of the magnetic material. Accordingly, ΔM_{ba} related to $\Delta x_{cb} = c - b$ results

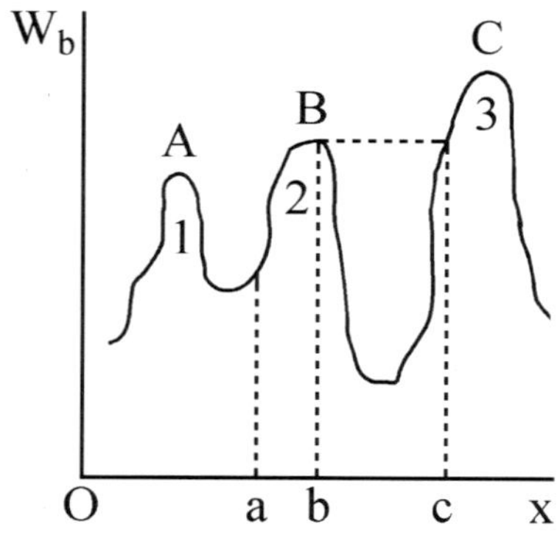

Figure 7.9. The energy barriers *1*, *2*, *3* with maximal values in the points *A*, *B*, *C*, respectively. The energy barrier *2* exhibits the maximum in the position $x = b$.

in a discrete change of the magnetization hysteresis loop at zero change of H, i.e., for $\Delta H_{cb} = H_c - H_b = 0$. This discrete change at the constant intensity $H_b = H_c$ induces a voltage impulse in a pickup coil. The voltage impulse is known as the Barkhausen jump. Additionally, maximal values of the energy barriers *1*, *2*, *3* are reasons of the coercivity of magnetic materials including magnetic multi-component materials [112].

Sections 7.2.2., 7.2.3. presents the determination of $W_{bi} = W_{bi}(x_i)$, $W_{bj} = W_{bj}(x_j)$, $W_{bk} = W_{bk}(x_k)$ in multi-particle-matrix and multi-particle-envelope-matrix systems (see Figure 2.1). This determination is thus sufficient to be performed within one eighth of the cubic cell, i.e., for $x_i \in \langle 0, d/2 \rangle$, $x_j \in \langle 0, d/2 \rangle$ and $x_k \in \langle 0, d/2 \rangle$. This is a consequence of symmetry of these systems due to the matrix infinity and the periodical distribution of the spherical particles and the spherical envelopes (see Section 3.1.2.).

7.2.2. Multi-Particle-Matrix System

The surface integral $W_{bi} = W_{bi}(x_i)$ for $x_i \in \langle 0, R_1 \rangle$ and $x_i \in \langle R_1, d/2 \rangle$ has the forms

$$W_{bi} = W_{bi}(x_i) = 4\left(\int_{x_{j1p}}^{x_{j2p}} \int_{x_{k1p}}^{x_{k2p}} w_p \, dx_j \, dx_k + \int_{x_{j1m}}^{x_{j2m}} \int_{x_{k1m}}^{x_{k2m}} w_m \, dx_j \, dx_k\right),$$

$$x_i \in \langle 0, R_1 \rangle;$$

$$\langle x_{j1p}, x_{j2p} \rangle = \left\langle 0, \sqrt{R_1^2 - x_i^2} \right\rangle, \quad \langle x_{k1p}, x_{k2p} \rangle = \left\langle 0, \sqrt{R_1^2 - \left(x_i^2 + x_j^2\right)} \right\rangle;$$

$$\langle x_{j1m}, x_{j2m} \rangle = \left\langle 0, \sqrt{R_1^2 - x_i^2} \right\rangle, \quad \langle x_{k1m}, x_{k2m} \rangle = \left\langle \sqrt{R_1^2 - \left(x_i^2 + x_j^2\right)}, \frac{d}{2} \right\rangle;$$

$$\langle x_{j1m}, x_{j2m} \rangle = \left\langle \sqrt{R_1^2 - x_i^2}, \frac{d}{2} \right\rangle, \quad \langle x_{k1m}, x_{k2m} \rangle = \left\langle 0, \frac{d}{2} \right\rangle, \tag{7.43}$$

$$W_{bi} = W_{bi}(x_i) = 4 \int\limits_{x_{j1m}}^{x_{j2m}} \int\limits_{x_{k1m}}^{x_{k2m}} w_m \, dx_j \, dx_k, \quad x_i \in \left\langle R_1, \frac{d}{2} \right\rangle;$$

$$\left\langle x_{j1m}, x_{j2m} \right\rangle = \left\langle x_{k1m}, x_{k2m} \right\rangle = \left\langle 0, \frac{d}{2} \right\rangle, \tag{7.44}$$

where the integration intervals $x_j \in \left\langle x_{j1q}, x_{j2q} \right\rangle$, $x_k \in \left\langle x_{k1q}, x_{k2q} \right\rangle$ are related to the surface S_{iq} which represents a cross-section within the spherical particle ($q = p$) and the cell matrix ($q = m$). The cross-section S_{iq} with the normal x_i is related to the plane $x_j'' x_k''$ (see Figure 7.8) in a position for $x_i \in \langle 0, d/2 \rangle$, where $x_j'' x_k'' \parallel x_j x_k$.

The surfaces integrals $W_{bj} = W_{bj}(x_j)$ and $W_{bk} = W_{bk}(x_k)$ (see Eqs. (7.38), (7.39)) are given by Eqs. (7.43), (7.44), where the transformations $i \to j$, $j \to i$, $k \to k$ and $i \to k$, $j \to j$, $i \to k$ of the subscripts i, j, k are considered in Eqs. (7.43), (7.44), respectively.

7.2.3. Multi-Particle-Envelope-Matrix System

The surface integral $W_{bi} = W_{bi}(x_i)$ for $x_i \in \langle R_2, d/2 \rangle$ is given by Eq. (7.44). Consequently, $W_{bi} = W_{bi}(x_i)$ for $x_i \in \langle 0, R_1 \rangle$ and $x_i \in \langle R_1, R_2 \rangle$ has the forms

$$W_{bi} = W_{bi}(x_i) = 4 \left(\int\limits_{x_{j1p}}^{x_{j2p}} \int\limits_{x_{k1p}}^{x_{k2p}} w_p \, dx_j \, dx_k + \int\limits_{x_{j1e}}^{x_{j2e}} \int\limits_{x_{k1e}}^{x_{k2e}} w_e \, dx_j \, dx_k \right.$$
$$\left. + \int\limits_{x_{j1m}}^{x_{j2m}} \int\limits_{x_{k1m}}^{x_{k2m}} w_m \, dx_j \, dx_k \right), \quad x_i \in \langle 0, R_1 \rangle;$$

$$\left\langle x_{j1p}, x_{j2p} \right\rangle = \left\langle 0, \sqrt{R_1^2 - x_i^2} \right\rangle, \quad \left\langle x_{k1p}, x_{k2p} \right\rangle = \left\langle 0, \sqrt{R_1^2 - \left(x_i^2 + x_j^2\right)} \right\rangle;$$

$$\left\langle x_{j1e}, x_{j2e} \right\rangle = \left\langle 0, \sqrt{R_1^2 - x_i^2} \right\rangle,$$

$$\left\langle x_{k1e}, x_{k2e} \right\rangle = \left\langle \sqrt{R_1^2 - \left(x_i^2 + x_j^2\right)}, \sqrt{R_2^2 - \left(x_i^2 + x_j^2\right)} \right\rangle;$$

$$\left\langle x_{j1e}, x_{j2e} \right\rangle = \left\langle \sqrt{R_1^2 - x_i^2}, \sqrt{R_2^2 - x_i^2} \right\rangle,$$

$$\left\langle x_{k1e}, x_{k2e} \right\rangle = \left\langle 0, \sqrt{R_2^2 - \left(x_i^2 + x_j^2\right)} \right\rangle;$$

$$\left\langle x_{j1m}, x_{j2m} \right\rangle = \left\langle 0, \sqrt{R_2^2 - x_i^2} \right\rangle, \quad \left\langle x_{k1m}, x_{k2m} \right\rangle = \left\langle \sqrt{R_2^2 - \left(x_i^2 + x_j^2\right)}, \frac{d}{2} \right\rangle;$$

$$\left\langle x_{j1m}, x_{j2m} \right\rangle = \left\langle \sqrt{R_2^2 - x_i^2}, \frac{d}{2} \right\rangle, \quad \left\langle x_{k1m}, x_{k2m} \right\rangle = \left\langle 0, \frac{d}{2} \right\rangle, \tag{7.45}$$

$$W_{bi} = W_{bi}(x_i) = 4 \left(\int\limits_{x_{j1e}}^{x_{j2e}} \int\limits_{x_{k1e}}^{x_{k2e}} w_e \, dx_j \, dx_k + \int\limits_{x_{j1m}}^{x_{j2m}} \int\limits_{x_{k1m}}^{x_{k2m}} w_m \, dx_j \, dx_k \right),$$

$$x_i \in \langle R_1, R_2 \rangle ;$$

$$\langle x_{j1e}, x_{j2e} \rangle = \left\langle 0, \sqrt{R_2^2 - x_i^2} \right\rangle, \quad \langle x_{k1e}, x_{k2e} \rangle = \left\langle 0, \sqrt{R_2^2 - \left(x_i^2 + x_j^2\right)} \right\rangle ;$$

$$\langle x_{j1m}, x_{j2m} \rangle = \left\langle 0, \sqrt{R_2^2 - x_i^2} \right\rangle, \quad \langle x_{k1m}, x_{k2m} \rangle = \left\langle \sqrt{R_2^2 - \left(x_i^2 + x_j^2\right)}, \frac{d}{2} \right\rangle ;$$

$$\langle x_{j1m}, x_{j2m} \rangle = \left\langle \sqrt{R_2^2 - x_i^2}, \frac{d}{2} \right\rangle, \quad \langle x_{k1m}, x_{k2m} \rangle = \left\langle 0, \frac{d}{2} \right\rangle, \tag{7.46}$$

where the integration intervals $x_j \in \langle x_{j1q}, x_{j2q} \rangle$, $x_k \in \langle x_{k1q}, x_{k2q} \rangle$ are related to the surface S_{iq} which represents a cross-section within the spherical particle ($q = p$), the spherical envelope ($q = e$) and the cell matrix ($q = m$). The cross-section S_{iq} with the normal x_i is related to the plane $x_j'' x_k''$ (see Figure 7.8) in a position for $x_i \in \langle 0, d/2 \rangle$, where $x_j'' x_k'' \parallel x_j x_k$.

The surfaces integrals $W_{bj} = W_{bj}(x_j)$ and $W_{bk} = W_{bk}(x_k)$ (see Eqs. (7.38), (7.39)) are given by Eqs. (7.45), (7.46), where the transformations $i \to j$, $j \to i$, $k \to k$ and $i \to k$, $j \to j$, $i \to k$ of the subscripts i, j, k are considered in Eqs. (7.45), (7.46), respectively.

7.3. Analytical Model of Strengthening

7.3.1. General Analysis

The analytical model of the micro-strengthening $\sigma_{si} = \sigma_{si}(x_i)$, $\sigma_{sj} = \sigma_{si}(x_j)$, $\sigma_{sk} = \sigma_{si}(x_k)$ and the macro-strengthening $\overline{\sigma}_{si}$, $\overline{\sigma}_{sj}$, $\overline{\sigma}_{sk}$ along the axes x_i, x_j, x_k ($i, j, k = 1,2,3$; $i \neq j \neq k$) (see Figure 7.8) is based on the following energy analysis.

Let a solid continuum with a general shape and with the volume V shown in Figure 7.8 be considered. Let the surface S_i in a position given by $x_i \in \langle x_{1i}, x_{2i} \rangle$ be a cross-section of this solid continuum, where $S_i(x_i)$ is an area of S_i. The cross-section S_i with the normal x_i is related to the plane $x_j'' x_k''$ (see Figure 7.8), where $x_j'' x_k'' \parallel x_j x_k$. The surface S_i is described by the coordinates $x_j \in \langle x_{j1}, x_{j2} \rangle$ and $x_k \in \langle x_{k1}, x_{k2} \rangle$.

Let the micro-strengthening $\sigma_{si} = \sigma_{si}(x_i)$ along the axis x_i which is required to be determined within this analysis represent a normal stress acting on the surface S_i. Additionally, let $\sigma_{si} = \sigma_{si}(x_i)$ be constant regarding each point of the surface S_i, i.e., $\sigma_{si} = \sigma_{si}(x_i) \neq f(x_j, x_k)$ is not a function of $x_j \in \langle x_{j1}, x_{j2} \rangle$ and $x_k \in \langle x_{k1}, x_{k2} \rangle$.

Consequently, let the stress $\sigma_{si} = \sigma_{si}(x_i)$ induce the elastic strain $\varepsilon_{sii} = \varepsilon_{sii}(x_i) = s_{ii}\sigma_{si}(x_i)$ along the axis x_i. Accordingly, the elastic energy density $W_{\sigma_{si}} = W_{\sigma_{si}}(x_i)$ accumulated on the surface area $S_i = S_i(x_i)$ has the form

$$W_{\sigma_{si}} = W_{\sigma_{si}}(x_i) = \frac{\sigma_{si}\,\varepsilon_{sii}\,S_i}{2} = \frac{s_{ii}\,S_i\,\sigma_{si}^2}{2}, \quad s_{ii} = \frac{1}{E_i}, \tag{7.47}$$

where E_i is the Young's modulus along the axis x_i.

Let $\sigma_i = \sigma_i(x_i, x_j, x_k)$ be a thermal stress which acts along the axis x_i in the solid continuum. Let $w_i = w_i(x_i, x_j, x_k)$ be elastic energy density which is induced by $\sigma_i = \sigma_i(x_i, x_j, x_k)$, where σ_i and w_i are given by Eqs. (4.76)–(4.80). Accordingly, the thermal-stress induced elastic energy density $W_i = W_i(x_i)$ accumulated on the surface area $S_i = S_i(x_i)$ is derived as

$$W_i = W_i(x_i) = \int_{S_i} w_i \, dS_i. \tag{7.48}$$

With regard to a sign of the stress $\sigma_{si} = \sigma_{si}(x_i)$, the determination of the integral in Eq. (7.48) is required to consider the following conditions:

1. If $\sigma_i(x_i, x_j, x_k) < 0$ at a point with the coordinates (x_i, x_j, x_k) on the surface S_i, then $w_i = w_i(x_i, x_j, x_k)$ at this point is considered to be $-w_i$.

2. If $\sigma_i(x_i, x_j, x_k) > 0$ at a point with the coordinates (x_i, x_j, x_k) on the surface S_i, then $w_i = w_i(x_i, x_j, x_k)$ at this point is considered to be $+w_i$.

The micro-strengthening $\sigma_{si} = \sigma_{si}(x_i)$ along the axis x_i is determined by the condition

$$W_{\sigma_{si}} = W_i, \tag{7.49}$$

and then we get

$$\sigma_{si} = \sigma_{si}(x_i) = \pm \sqrt{\frac{2|W_i|}{s_{ii} S_i}}, \quad s_{ii} = \frac{1}{E_i}, \tag{7.50}$$

where the signs '+' and '-' in Eq. (7.50) are considered for $W_i > 0$ and $W_i < 0$, respectively.

The macro-strengthening $\overline{\sigma}_{si}$ along the axis x_i represents a mean value of $\sigma_{si} = \sigma_{si}(x_i)$ regarding the interval $x_i \in \langle x_{1i}, x_{2i} \rangle$, and then we get

$$\overline{\sigma}_{si} = \frac{1}{x_{2i} - x_{1i}} \int_{x_{1i}}^{x_{2i}} \sigma_{si} \, dx_i = \frac{1}{x_{2i} - x_{1i}} \sqrt{\frac{2}{s_{ii}}} \int_{x_{1i}}^{x_{2i}} \pm \sqrt{\frac{|W_i|}{S_i}} \, dx_i, \quad s_{ii} = \frac{1}{E_i}, \tag{7.51}$$

where the following conditions are considered:

3. If $W_i(x_i) > 0$ in a position given by the coordinate x_i, then the sign '+' in Eq. (7.51) is considered.

4. If $W_i(x_i) < 0$ in a position given by the coordinate x_i, then the sign '-' in Eq. (7.51) is considered.

Finally, the following analysis concerning a sign of $\overline{\sigma}_{si}$ is considered:

5. If $\overline{\sigma}_{si} < 0$, then the macro-strengthening $\overline{\sigma}_{si}$ represents a 'resistance' against compressive mechanical loading.

6. If $\overline{\sigma}_{si} > 0$, then the macro-strengthening $\overline{\sigma}_{si}$ represents a 'resistance' against tensile mechanical loading.

With regard to the micro-/macro-strengthening along the axes x_j and x_k, the transformations $i \to j$, $j \to i$, $k \to k$ and $i \to k$, $j \to j$, $i \to k$ of the subscripts i, j, k are considered in Eqs. (7.47)–(7.51), respectively.

7.3.2. Strengthening in the Model Systems

The micro- and macro-strengthening in the model systems (see Figure 2.1) is sufficient to be determined within one eighth of the cubic cell, i.e., for $x_i \in \langle 0, d/2 \rangle$, $x_j \in \langle 0, d/2 \rangle$ and $x_k \in \langle 0, d/2 \rangle$. This is a consequence of symmetry of these systems due to the matrix infinity and the periodical distribution of the spherical particles and the spherical envelopes (see Section 3.1.2.).

7.3.2.1. Multi-Particle-Matrix System

If $x_i \in \langle 0, R_1 \rangle$, then the areas $S_{ip} = S_{ip}(x_i)$ and $S_{im1} = S_{im1}(x_i)$ of the surfaces S_{ip} and S_{im1} in the spherical particle and the cell matrix within one eighth of the cubic cell, respectively, are derived as

$$S_{ip} = \frac{\pi\left(R_1^2 - x_i^2\right)}{4}, \quad x_i \in \langle 0, R_1 \rangle, \tag{7.52}$$

$$S_{im1} = \frac{d^2}{4} - S_{ip} = \frac{d^2 - \pi\left(R_1^2 - x_i^2\right)}{4}, \quad x_i \in \langle 0, R_1 \rangle. \tag{7.53}$$

Consequently, the elastic energy density $W_{\sigma_{si}1} = W_{\sigma_{si}1}(x_i)$ for $x_i \in \langle 0, R_1 \rangle$ (see Eq. (7.47)) has the form

$$W_{\sigma_{si}1} = \frac{\sigma_{si}^2\left(s_{iip} S_{ip} + s_{iim} S_{im1}\right)}{2}, \quad s_{iiq} = \frac{1}{E_{iq}}, \quad x_i \in \langle 0, R_1 \rangle, \quad q = p, m, \tag{7.54}$$

where E_{iq} is the Young's modulus along the axis x_i in the spherical particle ($q = p$) and the cell matrix ($q = m$).

The elastic energy density $W_{i1} = W_{i1}(x_i)$ for $x_i \in \langle 0, R_1 \rangle$ (see Eq. (7.48)) accumulated on the surfaces S_{ip} and S_{im1} which are described by $x_j \in \langle x_{j1p}, x_{j2p} \rangle$, $x_k \in \langle x_{k1p}, x_{k2p} \rangle$ and $x_j \in \langle x_{j1m}, x_{j2m} \rangle$, $x_k \in \langle x_{k1m}, x_{k2m} \rangle$, respectively, is derived as

$$\begin{aligned}
W_{i1} &= \int\limits_{S_{ip}} w_{ip}\, dS_i + \int\limits_{S_{im1}} w_{im}\, dS_i \\
&= \int\limits_{x_{j1p}}^{x_{j2p}} \int\limits_{x_{k1p}}^{x_{k2p}} w_{ip}\, dx_j\, dx_k + \int\limits_{x_{j1m}}^{x_{j2m}} \int\limits_{x_{k1m}}^{x_{k2m}} w_{im}\, dx_j\, dx_k, \quad x_i \in \langle 0, R_1 \rangle; \\
&\langle x_{j1p}, x_{j2p} \rangle = \left\langle 0, \sqrt{R_1^2 - x_i^2} \right\rangle, \quad \langle x_{k1p}, x_{k2p} \rangle = \left\langle 0, \sqrt{R_1^2 - \left(x_i^2 + x_j^2\right)} \right\rangle; \\
&\langle x_{j1m}, x_{j2m} \rangle = \left\langle 0, \sqrt{R_1^2 - x_i^2} \right\rangle, \quad \langle x_{k1m}, x_{k2m} \rangle = \left\langle \sqrt{R_1^2 - \left(x_i^2 + x_j^2\right)}, \frac{d}{2} \right\rangle; \\
&\langle x_{j1m}, x_{j2m} \rangle = \left\langle \sqrt{R_1^2 - x_i^2}, \frac{d}{2} \right\rangle, \quad \langle x_{k1m}, x_{k2m} \rangle = \left\langle 0, \frac{d}{2} \right\rangle,
\end{aligned} \tag{7.55}$$

where the determination of the integrals in Eq. (7.55) is required to consider the conditions in Items 1, 2, Section 7.3.1.

The micro-strengthening $\sigma_{si1} = \sigma_{si1}(x_i)$ for $x_i \in \langle 0, R_1 \rangle$ along the axis x_i (see Eq. (7.50)) which is determined by the condition $W_{\sigma_{si}1} = W_{i1}$ has the form

$$\sigma_{si1} = \pm \sqrt{\frac{2\,|W_{i1}|}{s_{iip}\,S_{ip} + s_{iim}\,S_{im1}}}, \quad s_{iiq} = \frac{1}{E_{iq}}, \quad x_i \in \langle 0, R_1 \rangle, \quad q = p, m, \tag{7.56}$$

where the signs '+' and '-' in Eq. (7.56) are considered for $W_{i1} > 0$ and $W_{i1} < 0$, respectively.

If $x_i \in \langle R_1, d/2 \rangle$, then $S_{im2} = d^2/4$ is an surface area in the cell matrix within one eighth of the cubic cell. Consequently, the elastic energy density $W_{\sigma_{si}2} = W_{\sigma_{si}2}(x_i)$ for $x_i \in \langle R_1, d/2 \rangle$ (see Eq. (7.47)) has the form

$$W_{\sigma_{si}2} = \frac{s_{iim}\,d^2\,\sigma_{si}^2}{8}, \quad s_{iim} = \frac{1}{E_{im}}, \quad x_i \in \left\langle R_1, \frac{d}{2} \right\rangle. \tag{7.57}$$

The elastic energy density $W_{i2} = W_{i2}(x_i)$ for $x_i \in \langle R_1, d/2 \rangle$ (see Eq. (7.48)) accumulated on the surface S_{im2} which is described by $x_j \in \langle 0, d/2 \rangle$, $x_k \in \langle 0, d/2 \rangle$ is derived as

$$W_{i2} = \int\limits_{S_{im2}} w_{im}\,dS_i = \int\limits_0^{d/2}\int\limits_0^{d/2} w_{im}\,dx_j\,dx_k, \quad x_i \in \left\langle R_1, \frac{d}{2} \right\rangle, \tag{7.58}$$

where the determination of the integral in Eq. (7.58) is required to consider the conditions in Items 1, 2, Section 7.3.1.

The micro-strengthening $\sigma_{si2} = \sigma_{si2}(x_i)$ for $x_i \in \langle R_1, d/2 \rangle$ along the axis x_i (see Eq. (7.50)) which is determined by the condition $W_{\sigma_{si}2} = W_{i2}$ has the form

$$\sigma_{si2} = \pm \frac{2}{d}\sqrt{\frac{2\,|W_{i2}|}{s_{iim}}}, \quad s_{iim} = \frac{1}{E_{im}}, \quad x_i \in \left\langle R_1, \frac{d}{2} \right\rangle, \tag{7.59}$$

where the signs '+' and '-' in Eq. (7.59) are considered for $W_{i2} > 0$ and $W_{i2} < 0$, respectively. Additionally, we get $[\sigma_{si1}(x_i)]_{x_i=R_1} = [\sigma_{si2}(x_i)]_{x_i=R_1}$ due to $[S_{im1}(x_i)]_{x_i=R_1} = S_{im2}$ and $[W_{i1}(x_i)]_{x_i=R_1} = [W_{i2}(x_i)]_{x_i=R_1}$.

The macro-strengthening $\overline{\sigma}_{si}$ along the axis x_i (see Eq. (7.51)) which represents a mean value of $\sigma_{si} = \sigma_{si1}(x_i) + \sigma_{si2}(x_i)$ regarding the interval $x_i \in \langle x_{1i}, x_{2i} \rangle = \langle 0, d/2 \rangle$ is derived as

$$\begin{aligned}
\overline{\sigma}_{si} &= \frac{2}{d}\left(\int\limits_0^{R_1} \sigma_{si1}\,dx_i + \int\limits_{R_1}^{d/2} \sigma_{si2}\,dx_i \right) \\
&= \frac{2\sqrt{2}}{d}\left(\int\limits_0^{R_1} \pm \sqrt{\frac{|W_{i1}|}{s_{iip}\,S_{ip} + s_{iim}\,S_{im1}}}\,dx_i + \frac{2}{d \times \sqrt{s_{iim}}} \int\limits_{R_1}^{d/2} \pm \sqrt{|W_{i2}|}\,dx_i \right), \\
s_{iiq} &= \frac{1}{E_{iq}}, \quad q = p, m,
\end{aligned} \tag{7.60}$$

where the conditions and the analysis in Items 3–6, Section 7.3.1. are considered.

7.3.2.2. Multi-Particle-Envelope-Matrix System

If $x_i \in \langle 0, R_1 \rangle$, then the area $S_{ip} = S_{ip}(x_i)$ of the surface S_{ip} in the spherical particle is given by Eq. (7.52). The areas $S_{ie1} = S_{ie1}(x_i)$ and $S_{im3} = S_{im3}(x_i)$ of the surfaces S_{ie1} and S_{im3} for $x_i \in \langle 0, R_1 \rangle$ in the spherical envelope and the cell matrix within one eighth of the cubic cell, respectively, are derived as

$$S_{ie1} = \frac{\pi \left(R_2^2 - x_i^2\right)}{4} - S_{ip} = \frac{\pi \left(R_2^2 - R_1^2\right)}{4}, \quad x_i \in \langle 0, R_1 \rangle, \tag{7.61}$$

$$S_{im3} = \frac{d^2 - \pi \left(R_2^2 - x_i^2\right)}{4}, \quad x_i \in \langle 0, R_1 \rangle. \tag{7.62}$$

Consequently, the elastic energy density $W_{\sigma_{si}3} = W_{\sigma_{si}3}(x_i)$ for $x_i \in \langle 0, R_1 \rangle$ (see Eq. (7.47)) has the form

$$W_{\sigma_{si}3} = \frac{\sigma_{si}^2 \left(s_{iip} S_{ip} + s_{iie} S_{ie1} + s_{iim} S_{im3}\right)}{2}, \quad s_{iiq} = \frac{1}{E_{iq}},$$
$$x_i \in \langle 0, R_1 \rangle, \quad q = p, e, m, \tag{7.63}$$

where E_{iq} is the Young's modulus along the axis x_i in the spherical particle ($q = p$), the spherical envelope ($q = e$) and the cell matrix ($q = m$).

The elastic energy density $W_{i3} = W_{i3}(x_i)$ for $x_i \in \langle 0, R_1 \rangle$ (see Eq. (7.48)) accumulated on the surfaces S_{ip}; S_{ie1} and S_{im3} which are described by $x_j \in \langle x_{j1p}, x_{j2p} \rangle$, $x_k \in \langle x_{k1p}, x_{k2p} \rangle$; $x_j \in \langle x_{j1e}, x_{j2e} \rangle$, $x_k \in \langle x_{k1e}, x_{k2e} \rangle$ and $x_j \in \langle x_{j1m}, x_{j2m} \rangle$, $x_k \in \langle x_{k1m}, x_{k2m} \rangle$, respectively, is derived as

$$W_{i3} = \int\limits_{S_{ip}} w_{ip}\, dS_i + \int\limits_{S_{ie1}} w_{ie}\, dS_i + \int\limits_{S_{im3}} w_{im}\, dS_i$$
$$= \int\limits_{x_{j1p}}^{x_{j2p}} \int\limits_{x_{k1p}}^{x_{k2p}} w_{ip}\, dx_j\, dx_k + \int\limits_{x_{j1e}}^{x_{j2e}} \int\limits_{x_{k1e}}^{x_{k2e}} w_{ie}\, dx_j\, dx_k$$
$$+ \int\limits_{x_{j1m}}^{x_{j2m}} \int\limits_{x_{k1m}}^{x_{k2m}} w_{im}\, dx_j\, dx_k, \quad x_i \in \langle 0, R_1 \rangle;$$

$$\langle x_{j1p}, x_{j2p} \rangle = \left\langle 0, \sqrt{R_1^2 - x_i^2} \right\rangle, \quad \langle x_{k1p}, x_{k2p} \rangle = \left\langle 0, \sqrt{R_1^2 - \left(x_i^2 + x_j^2\right)} \right\rangle;$$

$$\langle x_{j1e}, x_{j2e} \rangle = \left\langle 0, \sqrt{R_1^2 - x_i^2} \right\rangle,$$

$$\langle x_{k1e}, x_{k2e} \rangle = \left\langle \sqrt{R_1^2 - \left(x_i^2 + x_j^2\right)}, \sqrt{R_2^2 - \left(x_i^2 + x_j^2\right)} \right\rangle;$$

$$\langle x_{j1e}, x_{j2e} \rangle = \left\langle \sqrt{R_1^2 - x_i^2}, \sqrt{R_2^2 - x_i^2} \right\rangle,$$

$$\langle x_{k1e}, x_{k2e} \rangle = \left\langle 0, \sqrt{R_2^2 - \left(x_i^2 + x_j^2\right)} \right\rangle;$$

$$\left\langle x_{j1m},x_{j2m}\right\rangle = \left\langle 0,\sqrt{R_2^2-x_i^2}\right\rangle, \quad \left\langle x_{k1m},x_{k2m}\right\rangle = \left\langle \sqrt{R_2^2-\left(x_i^2+x_j^2\right)},\frac{d}{2}\right\rangle;$$

$$\left\langle x_{j1m},x_{j2m}\right\rangle = \left\langle \sqrt{R_2^2-x_i^2},\frac{d}{2}\right\rangle, \quad \left\langle x_{k1m},x_{k2m}\right\rangle = \left\langle 0,\frac{d}{2}\right\rangle, \tag{7.64}$$

where the determination of the integrals in Eq. (7.64) is required to consider the conditions in Items 1, 2, Section 7.3.1.

The micro-strengthening $\sigma_{si3} = \sigma_{si3}(x_i)$ for $x_i \in \langle 0,R_1\rangle$ along the axis x_i (see Eq. (7.50)) which is determined by the condition $W_{\sigma_{si}3} = W_{i3}$ has the form

$$\sigma_{si3} = \pm\sqrt{\frac{2\left|W_{i3}\right|}{s_{iip}\,S_{ip}+s_{iie}\,S_{ie1}+s_{iim}\,S_{im3}}}, \quad s_{iiq} = \frac{1}{E_{iq}}, \quad x_i \in \langle 0,R_1\rangle, \quad q = p,e,m, \tag{7.65}$$

where the signs '+' and '-' in Eq. (7.65) are considered for $W_{i3} > 0$ and $W_{i3} < 0$, respectively.

If $x_i \in \langle R_1,R_2\rangle$, then the area $S_{im3} = S_{im3}(x_i)$ of the surface S_{im3} in the cell matrix, given by Eq. (7.62), is considered. The area $S_{ie2} = S_{ie2}(x_i)$ of the surface S_{ie2} for $x_i \in \langle R_1,R_2\rangle$ in the spherical envelope within one eighth of the cubic cell is derived as

$$S_{ie2} = \frac{\pi\left(R_2^2-x_i^2\right)}{4}, \quad x_i \in \langle R_1,R_2\rangle. \tag{7.66}$$

Consequently, the elastic energy density $W_{\sigma_{si}4} = W_{\sigma_{si}4}(x_i)$ for $x_i \in \langle R_1,R_2\rangle$ (see Eq. (7.47)) has the form

$$W_{\sigma_{si}4} = \frac{\sigma_{si}^2\left(s_{iie}\,S_{ie2}+s_{iim}\,S_{im3}\right)}{2}, \quad s_{iiq} = \frac{1}{E_{iq}}, \quad x_i \in \langle 0,R_1\rangle, \quad q = e,m. \tag{7.67}$$

The elastic energy density $W_{i4} = W_{i4}(x_i)$ for $x_i \in \langle R_1,R_2\rangle$ (see Eq. (7.48)) accumulated on the surfaces S_{ie2} and S_{im3} which are described by $x_j \in \left\langle x_{j1e},x_{j2e}\right\rangle$, $x_k \in \langle x_{k1e},x_{k2e}\rangle$ and $x_j \in \left\langle x_{j1m},x_{j2m}\right\rangle$, $x_k \in \langle x_{k1m},x_{k2m}\rangle$, respectively, is derived as

$$W_{i4} = \int\limits_{S_{ie2}} w_{ie}\,dS_i + \int\limits_{S_{im3}} w_{im}\,dS_i$$

$$= \int\limits_{x_{j1e}}^{x_{j2e}}\int\limits_{x_{k1e}}^{x_{k2e}} w_{ie}\,dx_j\,dx_k + \int\limits_{x_{j1m}}^{x_{j2m}}\int\limits_{x_{k1m}}^{x_{k2m}} w_{im}\,dx_j\,dx_k, \quad x_i \in \langle R_1,R_2\rangle;$$

$$\left\langle x_{j1e},x_{j2e}\right\rangle = \left\langle 0,\sqrt{R_2^2-x_i^2}\right\rangle, \quad \left\langle x_{k1e},x_{k2e}\right\rangle = \left\langle 0,\sqrt{R_2^2-\left(x_i^2+x_j^2\right)}\right\rangle;$$

$$\left\langle x_{j1m},x_{j2m}\right\rangle = \left\langle 0,\sqrt{R_2^2-x_i^2}\right\rangle, \quad \left\langle x_{k1m},x_{k2m}\right\rangle = \left\langle \sqrt{R_2^2-\left(x_i^2+x_j^2\right)},\frac{d}{2}\right\rangle;$$

$$\left\langle x_{j1m},x_{j2m}\right\rangle = \left\langle \sqrt{R_2^2-x_i^2},\frac{d}{2}\right\rangle, \quad \left\langle x_{k1m},x_{k2m}\right\rangle = \left\langle 0,\frac{d}{2}\right\rangle, \tag{7.68}$$

where the determination of the integrals in Eq. (7.68) is required to consider the conditions in Items 1, 2, Section 7.3.1.

The micro-strengthening $\sigma_{si3} = \sigma_{si4}(x_i)$ for $x_i \in \langle R_1, R_2\rangle$ along the axis x_i (see Eq. (7.50)) which is determined by the condition $W_{\sigma_{si}4} = W_{i4}$ has the form

$$\sigma_{si4} = \pm\sqrt{\frac{2|W_{i4}|}{s_{iie}S_{ie2} + s_{iim}S_{im3}}}, \quad s_{iiq} = \frac{1}{E_{iq}}, \quad x_i \in \langle R_1, R_2\rangle, \quad q = e, m, \tag{7.69}$$

where the signs '+' and '-' in Eq. (7.69) are considered for $W_{i4} > 0$ and $W_{i4} < 0$, respectively.

If $x_i \in \langle R_2, d/2\rangle$, then $S_{im2} = d^2/4$, and $W_{\sigma_{si}} = W_{\sigma_{si}}(x_i)$, $W_i = W_i(x_i)$, $\sigma_{si} = \sigma_{si}(x_i)$ (see Eqs. (7.47)–(7.50)) for $x_i \in \langle R_2, d/2\rangle$ are given by Eqs. (7.57)–(7.59).

The macro-strengthening $\overline{\sigma}_{si}$ along the axis x_i (see Eq. (7.51)) which represents a mean value of $\sigma_{si} = \sigma_{si3}(x_i) + \sigma_{si4}(x_i) + \sigma_{si2}(x_i)$ regarding the interval $x_i \in \langle x_{1i}, x_{2i}\rangle = \langle 0, d/2\rangle$ is derived as

$$\begin{aligned}
\overline{\sigma}_{si} &= \frac{2}{d}\left(\int_0^{R_1} \sigma_{si3}\,dx_i + \int_{R_1}^{R_2} \sigma_{si4}\,dx_i + \int_{R_2}^{d/2} \sigma_{si2}\,dx_i\right) \\
&= \frac{2\sqrt{2}}{d}\left(\int_0^{R_1} \pm\sqrt{\frac{|W_{i3}|}{s_{iip}S_{ip} + s_{iie}S_{ie1} + s_{iim}S_{im3}}}\,dx_i\right. \\
&\left. + \int_{R_1}^{R_2} \pm\sqrt{\frac{|W_{i4}|}{s_{iie}S_{ie2} + s_{iim}S_{im3}}}\,dx_i + \frac{2}{d\times\sqrt{s_{iim}}}\int_{R_2}^{d/2} \pm\sqrt{|W_{i2}|}\,dx_i\right), \\
s_{iiq} &= \frac{1}{E_{iq}}, \quad q = p, e, m,
\end{aligned} \tag{7.70}$$

where the conditions and the analysis in Items 3–6, Section 7.3.1. are considered.

7.4. Methods of Lifetime Prediction

The analytical-computational and analytical-experimental-computational methods of the lifetime prediction presented in Sections 7.4.2. and 7.4.3., respectively, are applicable to the three-component material defined in Item 4, Section 2.1. This three-component material consists of grains with and without a continuous component on a grain surface. The grains with the continuous component and the grains without the continuous component are identical or different regarding their crystal lattices, and thus exhibit identical or different thermal expansion coefficients, respectively.

With regard to analytical modelling of the thermal stresses, this three-component material is replaced by the multi-particle-envelope-matrix system which is analysed in Section 2.1. (see Figure 2.1b). Consequently, as analysed in Section 2.1., we get:

1. The grain with the continuous component on the grain surface and with the thermal expansion coefficient β_p corresponds to the spherical particle of the multi-particle-envelope-matrix system, where $v \in (0, v_{max}\rangle$ is a volume fraction of the spherical particle with the radius R_1. The coefficient v_{max} given by Eq. (2.2) is a function of R_1, t.

2. The grain without the continuous component on the grain surface and with the thermal expansion coefficient β_m corresponds to the infinite matrix of the multi-particle-envelope-matrix system.

3. The continuous component with the thickness t and with the thermal expansion coefficient β_e corresponds to the spherical envelope of the multi-particle-envelope-matrix system.

Finally, the parameters R_1, t, v of the multi-particle-envelope-matrix system represent microstructural parameters of the three-component material defined in Item 4, Section 2.1.. Each of the lifetime prediction methods considers a dependence of $\overline{p_1} = \overline{p_1}(R_1, t, v)$ and $\overline{p_2} = \overline{p_2}(R_1, t, v)$ on these microstructural parameters. The term $\overline{p_i}$ $(i = 1,2)$ represents a mean value of the function $p_i = p_i(\varphi, \nu, R_1, t, v)$ of the variables $\varphi \in \langle 0, \pi/2 \rangle$, $\nu \in \langle 0, \pi/2 \rangle$. The radial stresses $p_1 = p_1(\varphi, \nu, R_1, t, v)$ and $p_2 = p_2(\varphi, \nu, R_1, t, v)$ act at the particle-envelope and matrix-envelope boundaries, respectively. The mean value $\overline{p_i} = \overline{p_i}(R_1, t, v)$ is then defined as [113]

$$\overline{p_i} = \overline{p_i}(R_1, t, v) = \left(\frac{2}{\pi}\right)^2 \int_0^{\pi/2} \int_0^{\pi/2} p_i \, d\varphi \, d\nu, \quad i = 1,2. \tag{7.71}$$

The integration is sufficient to be determined within one eighth of the cubic cell, i.e., for $\varphi \in \langle 0, \pi/2 \rangle$ and $\nu \in \langle 0, \pi/2 \rangle$. This is a consequence of symmetry of the multi-particle-envelope-matrix system due to the matrix infinity and the periodical distribution of the spherical particles and the spherical envelopes (see Figure 2.1b).

7.4.1. Resistive, Quasi-resistive and Contributory Effects of Thermal Stresses

An analysis of the resistive and contributory effects of the thermal stresses considers dependences of $\overline{p_1} = \overline{p_1}(R_1, t, v)$ and $\overline{p_2} = \overline{p_2}(R_1, t, v)$ on the microstructural parameters R_1, t, v of the three-component materials defined in Item 4, Section 2.1., where $v \in (0, v_{max}\rangle$ (see Eq. (2.2)). This analysis is as follows.

In case of the multi-particle-matrix system, the compressive or tensile radial stress $\overline{p_1}(R_1, v) > 0$ or $\overline{p_1}(R_1, v) < 0$ which acts at the particle-matrix boundary corresponds to the condition $\beta_p < \beta_m$ or $\beta_p > \beta_m$ (see Eqs. (3.54)–(3.60)), respectively. The mean value $\overline{p_1} = \overline{p_1}(R_1, v)$ is determined by Eq. (7.71).

In case of the multi-particle-envelope-matrix system, the radial stresses $\overline{p_1}(R_1, t, v) \gtrless 0$ and $\overline{p_2}(R_1, t, v) \gtrless 0$ can or need not correspond to the conditions $\beta_e \gtrless \beta_p$ and $\beta_m \gtrless \beta_e$ (see Eqs. (3.54)–(3.60)), respectively.

The fact that $\overline{p_1}(R_1, t, v) \gtrless 0$ and $\overline{p_2}(R_1, t, v) \gtrless 0$ need not correspond to the conditions $\beta_e \gtrless \beta_p$ and $\beta_m \gtrless \beta_e$, respectively, results from a dependence of $\overline{p_i}$ $(i = 1,2)$ on R_1, t, v. This dependence results from the function $r_c = r_c(R_1, t, v)$ (see Eqs. (3.15)–(3.18)) which is included in Eqs. (4.241), (4.272).

Consequently, the radial stress $\overline{p_i} = \overline{p_i}(R_1, t, v)$ $(i = 1,2)$ can exhibit zero values at R_{1ci}, t_{ci}. Accordingly, R_{1ci}, t_{ci} represent such critical values of the microstructural parameters R_1, t at which $\overline{p_i}$ is transformed from compressive to tensile or vice versa.

If $\overline{p_1}$ is compressive (i.e., $\overline{p_1}(R_1, t, v) > 0$), then radial and tangential thermal stresses in the spherical particle (i.e., in the grain with the envelope on the grain surface) are also compressive. The compressive radial and tangential thermal stresses in the spherical particle thus represent 'resistance' against the compressive radial stress $\sigma_{r1} < 0$. The radial stress σ_{r1} acting at the particle-envelope boundary is explained below.

Similarly, if $\overline{p_1}$ is tensile (i.e., $\overline{p_1}(R_1, t, v) < 0$), then radial and tangential thermal stresses in the spherical particle (i.e., in the grain with the envelope on the grain surface) are also tensile. The tensile radial and tangential thermal stresses in the spherical particle thus represent 'resistance' against the tensile radial stress $\sigma_{r1} > 0$.

If $\overline{p_2}$ is compressive (i.e., $\overline{p_2}(R_1, t, v) > 0$), then radial and tangential thermal stresses in the cell matrix (i.e., in the grain without the envelope on the grain surface) are compressive and tensile, respectively. The compressive radial and tensile tangential thermal stresses in the cell matrix thus represent 'resistance' against the compressive radial stress $\sigma_{r2} < 0$. The radial stress σ_{r2} acting at the matrix-envelope boundary is explained below.

Similarly, if $\overline{p_2}$ is tensile (i.e., $\overline{p_2}(R_1, t, v) < 0$), then radial and tangential thermal stresses in the cell matrix (i.e., in the grain without the envelope on the grain surface) are tensile and compressive, respectively. The tensile radial and compressive tangential thermal stresses in the cell matrix thus represent 'resistance' against the tensile radial stress $\sigma_{r2} > 0$.

The critical microstructural parameters R_{1c1}, t_{c1} for $v \in (0, v_{max}\rangle$ (see Eq. (2.2)) thus represent such critical values of microstructural parameters at which the radial and tangential thermal stresses in the spherical particle change from positive (or negative) to negative (or positive). Accordingly, this resistive effect of the thermal stresses in the spherical particle is transformed to a contributory effect or vice versa.

Similarly, R_{1c2}, t_{c2} for $v \in (0, v_{max}\rangle$ (see Eq. (2.2)) thus represent such critical values of microstructural parameters at which the radial/tangential thermal stresses in the cell matrix change from positive/negative (or negative/positive) to negative/positive (or positive/negative). Accordingly, this resistive effect of the thermal stresses in the cell matrix is transformed to a contributory effect or vice versa.

Due to mechanical loading mentioned below, the following analysis, which is based on the analysis presented above, also considers an quasi-resistive effect.

Let the three-component material be loaded by mechanical loading. Let the mechanical loading be represented by the stresses σ_{1mech}, σ_{2mech}, σ_{3mech} which act along the axes x_1, x_2, x_3 (see Figure 3.1). Let σ_{1mech}, σ_{2mech}, σ_{3mech} induce the radial stresses $\sigma_{r1} = \sigma_{r1}(\varphi, \nu, R_1, t, v)$ and $\sigma_{r2} = \sigma_{r2}(\varphi, \nu, R_1, t, v)$ which act at the particle-envelope and matrix-envelope boundaries, respectively[5]. Let the term $\overline{\sigma_{ri}} = \overline{\sigma_{ri}}(R_1, t, v)$ $(i = 1,2)$ represent a mean value of the function $\overline{\sigma_{ri}} = \overline{\sigma_{ri}}(R_1, t, v)$ of the variables $\varphi \in \langle 0, 2\pi\rangle$, $\nu \in \langle 0, \pi\rangle$.

Consequently, the condition $|\overline{p_i}(R_1, t, v)| = |\overline{\sigma_{ri}}(R_1, t, v)|$ results in the dependence $R_{1Ac} = f_i(t_{Ac}, v, \overline{\sigma_{ri}})$ between the variables R_1, t of the function $\overline{\sigma_{ri}} = \overline{\sigma_{ri}}(R_1, t, v)$ with the parameter $v \in (0, v_{max}\rangle$ (see Eq. (2.2)).

Similarly, the condition $\overline{p_i}(R_1, t, v) = 0$ results in the dependence $R_{1Bc} = f_i(t_{Bc}, v)$ between the variables R_1, t of the function $\overline{p_i} = \overline{p_i}(R_1, t, v)$ with the parameter $v \in (0, v_{max}\rangle$.

---

[5]Such radial stresses (i.e., σ_{r1}, σ_{r2}) can be determined by e.g., the Eshelby's model. The Eshelby's model and its development [1, 2, 91] which are based on the Green's function, ordinary Newtonian potential and biharmonic potential define the disturbance of an applied stress-field in a solid continuum, where the applied stress-field is disturbed due to the presence of inclusions in a solid continuum.

Figure 7.10 shows a schematic illustration of the dependences $R_{1Ac} = f(t_{Ac}, v, \overline{\sigma_r})$ and $R_{1Bc} = f(t_{Bc}, v)$ for $v \in (0, v_{max}\rangle$ (see Eq. (2.2)) which define the areas A, B, C, where $f(t_{Ac}, v, \overline{\sigma_r}) = f_i(t_{Ac}, v, \overline{\sigma_{ri}})$, $f(t_{Bc}, v) = f_i(t_{Bc}, v)$ $(i = 1,2)$.

The area A is characterized by such coordinates (R_1, t) of the microstructural parameters R_1, t for which the thermal stresses exhibit the resistive effect against the mechanical loading, and additionally $|\overline{p_i}(R_1, t, v)| \geq |\overline{\sigma_{ri}}(R_1, t, v)|$, and $\overline{p_i}(R_1, t, v) \times \overline{\sigma_{ri}}(R_1, t, v) < 0$.

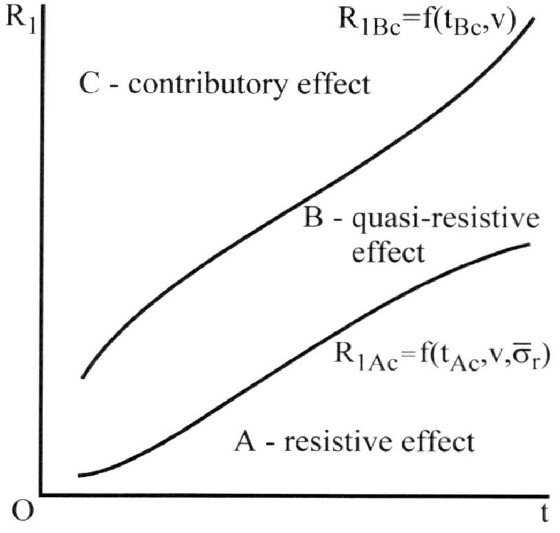

Figure 7.10. A schematic illustration of the dependences $R_{1Ac} = f(t_{Ac}, v, \overline{\sigma_r})$ and $R_{1Bc} = f(t_{Bc}, v)$ which result from the conditions $|\overline{p_i}(R_1, t, v)| = |\overline{\sigma_{ri}}(R_1, t, v)|$ and $\overline{p_i}(R_1, t, v) = 0$, respectively, where $f(t_{Ac}, v, \overline{\sigma_r}) \equiv f_i(t_{Ac}, v, \overline{\sigma_{ri}})$, $f(t_{Bc}, v) \equiv f_i(t_{Bc}, v)$ for $v \in (0, v_{max}\rangle$ (see Eq. (2.2)).

The area B is characterized by such coordinates (R_1, t) of the microstructural parameters R_1, t for which the thermal stresses exhibit the quasi-resistive effect against the mechanical loading, i.e., $|\overline{p_i}(R_1, t, v)| \in \langle 0, |\overline{\sigma_{ri}}(R_1, t, v)|\rangle$, and $\overline{p_i}(R_1, t, v) \times \overline{\sigma_{ri}}(R_1, t, v) \leq 0$.

The area C is characterized by such coordinates (R_1, t) of the microstructural parameters R_1, t for which the thermal stresses exhibit the contributory effect regarding the mechanical loading, and $\overline{p_i}(R_1, t, v) \times \overline{\sigma_{ri}}(R_1, t, v) > 0$.

With regard to Figure 7.10, material scientists are able to design and develop microstructure of the three-component material with such microstructural parameters R_1, t, v which result in this thermal-stress induced resistance against required mechanical loading. The microstructure can be thus tailored regarding the required mechanical loading.

Additionally, let the microstructure be time-dependent. If a time development of the microstructure is determined analytically or experimentally, then the time τ_c when the three-component material exhibits the critical microstructural parameters R_{1c}, t_c for $v \in (0, v_{max}\rangle$ (see Eq. (2.2)) can be determined.

The determination of τ_{c1} and τ_{c2} which are related to $\overline{p_1} = \overline{p_1}(R_1, t, v)$ and $\overline{p_2} = \overline{p_2}(R_1, t, v)$, respectively, is presented in Sections 7.4.2., 7.4.3.

7.4.2. Analytical-Computational Method

The transformation of the resistive effect of the thermal stresses to contributory effect with respect to mechanical loading is considered for the determination of a method of the lifetime prediction. The lifetime prediction method is applicable to e.g., structural steels which are used within high-temperature applications, e.g., superheater and reheater tubing and piping in power plants. With regard to the high-temperature applications, these structural steels are creep-resistant [31, 119]–[129]. Creep-resistant steels represent multi-component materials which exhibit changes of microstructure during years of an application at high temperature. In general, these microstructural changes are a reason of a degradation of strength of creep-resistant steels. The components of a power plant are usually designed with respect to a service life of 20 years (184 080 hours) [122, 123].

The analytical-computational method represents a connection of analytical and computational techniques. The analytical technique is based on the 'resistance-contribution' transformation. The computational technique is represented by a computational simulation of the time development of microstructure at the temperature T. The temperature T is a parameter of this analytical-computational method.

The microstructure of a creep-resistant steel thus exhibits changes during the time-temperature exploitation. An initial state of the microstructure at the beginning of the time-temperature exploitation is characterized by the presence of grains with aperiodically distributed precipitates of more or less defined shape. The grain matrix is ferritic (Fe_{α}-C) or austenitic (Fe_{γ}-C), with the substitutive atoms Mn, Cr, Mo, W. The precipitates are represented by the carbonitride MX (M = Nb,V,Ti; X = C,N), various carbide types of Cr, Mo, V, W, and inter-metallic phases (Laves and sigma phases) [120]–[129]. As presented in [129], the ability of steels to resist creep deformation depends on the presence of such precipitates in a matrix of grains.

Consequently, a final state of the microstructure of the structural steel at the end (regarding the lifetime) of the time-temperature exploitation is characterized by the presence of grains with an envelope on the grain surface as well as by the presence of grains without an envelope on the grain surface.

The matrix of grains without the envelope and the matrix of grains with the envelope are ferritic (Fe_{α}-C) or austenitic (Fe_{γ}-C). The envelope is represented by the carbides $M_{23}C_6$, M_6C, and/or inter-metallic phases (Laves and sigma phases) [120]–[129].

The microstructure of this final state of a creep-resistant steel thus corresponds to the microstructure which is defined in Items 1–3, Section 7.4.

The lifetime prediction method determines the critical time τ_{Ac} and τ_{Bc} when the microstructure exhibits the critical parameters R_{1Ac}, t_{1Ac} and R_{1Bc}, t_{1Bc}, respectively, where $\tau_{Ac} < \tau_{Bc}$. The determination of τ_{Ac}, τ_{Bc} is as follows.

As shown in Figure 7.11, let the Cartesian system $(ORt\tau)$ with the functions $R_{1Ac} = f(t_{Ac}, v, \overline{\sigma_r})$ and $R_{1Bc} = f(t_{Bc}, v)$ in the plane $R-t$ for $v \in (0, v_{max}\rangle$ (see Eq. (2.2)) which are described in Section 7.4.1. (see Figure 7.10) be considered. As analysed in Section 7.4.1., we get $f(t_{Ac}, v, \overline{\sigma_r}) \equiv f_i(t_{Ac}, v, \overline{\sigma_{ri}})$, $f(t_{Bc}, v) \equiv f_i(t_{Bc}, v)$ $(i = 1,2)$.

Let $F(R_1, t, \tau, T) = 0$ in Figure 7.11 represent a function which is determined by a computational simulation of the time development of microstructure of the three-component material defined in Items 1–3, Section 7.4. The grain radius R_1, the thickness t of the component (envelope) on the grain surface, and the time τ represent variables of the function $F(R_1,t,\tau,T) = 0$. The exploitation temperature T ($\equiv T_e$) is a parameter of the function $F(R_1,t,\tau,T) = 0$, where $T < T_r$ and T_r is relaxation temperature (see Section 3.5.1.1). The $F(R_1,t,\tau,T) = 0$ thus defines a time dependence of both R_1 and t at the exploitation temperature T.

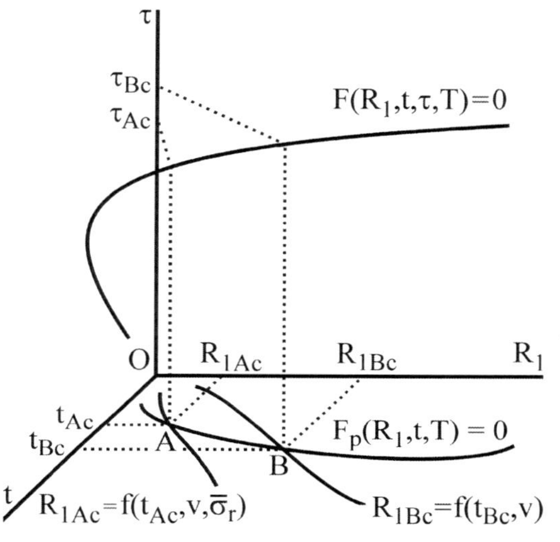

Figure 7.11. The function $F(R_1,t,\tau,T) = 0$ which is determined by a computational simulation represents the time development of microstructure of the three-component material defined in Items 1–3, Section 7.4.. The radius R_1 of the grains which are covered by the envelope on their surfaces, the thickness t of the component (envelope) on the grain surface, and the time τ represent variables of the function $F(R_1,t,\tau,T) = 0$. The exploitation temperature T ($\equiv T_e$) is a parameter of $F(R_1,t,\tau,T) = 0$, where $T < T_r$ and T_r is relaxation temperature (see Section 3.5.1..1). The function $F_p(R_1,t,T) = 0$ represents a projection of $F(R_1,t,\tau,T) = 0$ into the plane $R_1 - t$. The functions $R_{1Ac} = f(t_{Ac},v,\overline{\sigma_r})$, $R_{1Bc} = f(t_{Bc},v)$ for $v \in (0, v_{max}\rangle$ (see Eq. (2.2); Figure 7.10) are described in Section 7.4.1., where $f(t_{Ac},v,\overline{\sigma_r}) \equiv f_i(t_{Ac},v,\overline{\sigma_{ri}})$, $f(t_{Bc},v) \equiv f_i(t_{Bc},v)$ $(i = 1{,}2)$.

Consequently, let the function $F_p(R_1,t,T) = 0$ represent a projection of $F(R_1,t,\tau,T) = 0$ into the plane $R_1 - t$ as shown in Figure 7.11.

The points A and B with the coordinates (R_{1Ac},t_{Ac}) and (R_{1Bc},t_{Bc}) represent intersections of $F_p(R_1,t,T) = 0$ with $R_{1Ac} = f(t_{Ac},v,\overline{\sigma_r})$ and $R_{1Bc} = f(t_{Bc},v)$, respectively, where $f(t_{Ac},v,\overline{\sigma_r}) \equiv f_i(t_{Ac},v,\overline{\sigma_{ri}})$, $f(t_{Bc},v) \equiv f_i(t_{Bc})$ $(i = 1{,}2)$.

The substitution of $R_1 = f(t_{Ac},v,\overline{\sigma_r})$ and $R_1 = f(t_{Bc},v)$ to the function $F_p(R_1,t,T) = 0$ results in the conditions $F_p(f(t_{Ac},v,\overline{\sigma_r}),t_{Ac},T) = 0$ and $F_p(f(t_{Bc},v),t_{Bc},T) = 0$, respectively. The conditions $F_p(f(t_{Ac},\overline{\sigma_r}),t_{Ac},T) = 0$ and $F_p(f(t_{Bc}),t_{Bc},T) = 0$ with the variables t_{Ac} and t_{Bc} are considered for the determination of the t-coordinates, t_{Ac} and t_{Bc}, of the point A and B in the plane $R_1 - t$, respectively.

Consequently, the substitution of the coordinates t_{Ac} and t_{Bc} to $R_{1Ac} = f(t_{Ac},v,\overline{\sigma_r})$ and $R_{1Bc} = f(t_{Bc},v)$ results in the determination of the coordinates R_{1Ac} and R_{1Bc} of the points A and B, respectively.

The substitution of the coordinates (R_{1Ac}, t_{Ac}) and (R_{1Bc}, t_{Bc}) to $F(R_1, t, \tau, T) = 0$ results in the condition $F(R_{1Ac}, t_{Ac}, \tau_{Ac}, T) = 0$ and $F(R_{1Bc}, t_{Bc}, \tau, T) = 0$, respectively. The conditions $F(R_{1Ac}, t_{Ac}, \tau, T) = 0$ and $F(R_{1Bc}, t_{Bc}, \tau_{Bc}, T) = 0$ with the variables τ_{Ac} and τ_{Bc} are considered for the determination of the time τ_{Ac} and τ_{Bc}, respectively, where $\tau_{Ac} = \tau_{Ac}(v, \overline{\sigma_r})$, $\tau_{Bc} = \tau_{Bc}(v)$.

Consequently, the intervals $\tau < \tau_{Ac}(v, \overline{\sigma_r})$, $\tau \in \langle \tau_{Ac}, \tau_{Bc} \rangle$, $\tau > \tau_{Bc}$ defines the time τ which is related to the resistive, quasi-resistive, contributory effects of the thermal stresses against the mechanical loading, respectively. The intervals $\tau < \tau_{Ac}$, $\tau \in \langle \tau_{Ac}, \tau_{Bc} \rangle$, $\tau > \tau_{Bc}$ determine non-critical, quasi-critical, critical time periods of the exploitation, respectively.

The critical time τ_{Ac} and τ_{Bc} is assumed to represent contribution to the total lifetime related to the thermal-stress induced resistive effects of a three-component material.

As analysed in Section 7.4.1., τ_{Ac} is determined for $|\overline{p_1}(R_1, t, v)| = |\overline{\sigma_{r1}}(R_1, t, v)|$ or $|\overline{p_2}(R_1, t, v)| = |\overline{\sigma_{r2}}(R_1, t, v)|$, and then $\tau_{Ac} \to \tau_{Ac1}(v, \overline{\sigma_{r1}})$ or $\tau_{Ac} \to \tau_{Ac2}(v, \overline{\sigma_{r2}})$, respectively, where $\tau_{Aci} = \tau_{Aci}(v, \overline{\sigma_{ri}})$ $(i = 1,2)$.

Similarly, τ_{Bc} is determined for $|\overline{p_1}(R_1, t, v)| = 0$ or $|\overline{p_2}(R_1, t, v)| = 0$, and then $\tau_{Bc} \to \tau_{Bc1}(v)$ or $\tau_{Bc} \to \tau_{Bc2}(v)$, respectively, where $\tau_{Bci} = \tau_{Bci}(v)$ $(i = 1,2)$.

Finally, τ_{Acmin} and τ_{Bcmin} which represent minimal values of the sets $\{\tau_{Ac1}, \tau_{Ac2}\}$ and $\{\tau_{Bc1}, \tau_{Bc2}\}$ are considered to represent the lifetime regarding the 'resistance–(quasi-resistance)' and '(quasi-resistance)–contribution' transformations (see Figure 7.10), respectively.

7.4.3. Analytical-Computational-Experimental Method

The analytical-computational-experimental method represents a connection of an analytical model of the thermal stresses with computational simulation of the time development of microstructure as well as with experimental results. The temperature T is a parameter of this analytical-computational-experimental method. The determination of the analytical-computational-experimental method is as follows.

Let the computational simulation be represented by a time development of the thickness t of the component (envelope) on grain surfaces. Let this time development be represented by the function $t = g(\tau)$, where $\tau = f(t)$ is an inverse function of $t = g(\tau)$. Let $\log\tau = f(\log t)$ be derived by the Taylor series in terms of $\log t$ regarding the time interval $\tau \in \langle 0, 200000 \rangle$ [hour]. Consequently, the function $\log\tau = f(\log t)$ has the form

$$\log\tau = \sum_{i=0}^{n} c_i^{(\tau)} (\log t)^i . \tag{7.72}$$

The experimental results are represented by a time development of the radius R_1 of the grains which are covered by the component (envelope) on their surfaces. Let the dependence $R_1 - \tau$ for $\tau \in \langle 0, 200000 \rangle$ [hour] be derived as

$$\log R_1 = \sum_{j=0}^{m} c_j^{(R_1)} (\log\tau)^j . \tag{7.73}$$

With regard to Eqs. (7.72), (7.73), we get

$$\log R_1 = \sum_{j=0}^{m} c_j^{(R_1)} \left(\sum_{i=0}^{n} c_i^{(\tau)} (\log t)^i \right)^j . \tag{7.74}$$

The analytical model of the thermal stresses is represented by the functions $R_{1Ac} = f(t_{Ac}, v, \overline{\sigma_r})$ and $R_{1Bc} = f(t_{Bc}, v)$ (see Figure 7.10), strictly speaking by the functions $\log R_{1Ac} = \log[f(t_{Ac}, v, \overline{\sigma_r})]$ and $\log R_{1Bc} = \log[f(t_{Bc}, v)]$, where $f(t_{Ac}, v, \overline{\sigma_r}) \equiv f_i(t_{Ac}, v, \overline{\sigma_{ri}})$, $f(t_{Bc}, v) \equiv f_i(t_{Bc}, v)$ $(i = 1,2)$ (see Section 7.4.1.).

With regard to Eq. (7.74), and considering $t \rightarrow t_{Ac}$, we get

$$\log[f(t_{Ac}, v, \overline{\sigma_r})] - \sum_{j=0}^{m} c_j^{(R_1)} \left(\sum_{i=0}^{n} c_i^{(\tau)} (\log t_{Ac})^i \right)^j = 0. \tag{7.75}$$

Similarly, considering $t \rightarrow t_{Bc}$, we get

$$\log[f(t_{Bc}, v)] - \sum_{j=0}^{m} c_j^{(R_1)} \left(\sum_{i=0}^{n} c_i^{(\tau)} (\log t_{Bc})^i \right)^j = 0. \tag{7.76}$$

Consequently, the roots $t_{Ac} = t_{Ac}(v, \overline{\sigma_r})$ and $t_{Bc} = t_{Bc}(v)$ of Eqs. (7.75) and (7.76) are determined by a numerical method.

Substituting $t_{Ac} = t_{Ac}(v, \overline{\sigma_r})$ and $t_{Bc} = t_{Bc}(v)$ to Eq. (7.72), the critical time τ_{Ac} and τ_{Bc} has the form

$$\tau_{Qc} = \exp\left[(\ln 10) \sum_{i=0}^{n} c_i^{(\tau)} (\log t_{Qc})^i \right], \quad Q = A, B. \tag{7.77}$$

Finally, Equations (7.72) and (7.73) can be also determined from experimental results and a computational simulation, respectively.

The analysis concerning τ_{Acmin}, τ_{Bcmin} and the sets $\{\tau_{Ac1}, \tau_{Ac2}\}$ and $\{\tau_{Bc1}, \tau_{Bc2}\}$ (see Section 7.4.2.) is also considered for this analytical-computational-experimental method.

Chapter 8

Selected Applications

Chapter 8 deals with the following illustrative examples of applications:

- an application of the analytical model of the crack formation by the method 2 (see Section 7.1.3..2) to the YBaCuO superconductor with ceramic components (see Section 8.1.) [71]–[74],
- an application of the analytical model of the thermal stresses and the micro- and macro-strengthening (see Section 7.3.2..1) to the SiC-Si_3N_4 two-component material with ceramic components (see Section 8.2.) [27]–[30].

YBaCuO and SiC-Si_3N_4 represent two-component materials of a precipitate-matrix type, which are defined in Item 1, Section 2.1. As presented in Section 7.4.2., the creep-resistant steel at the end (regarding the lifetime) of the time-temperature exploitation represents a three-component material which is defined in Item 4, Section 2.1.

8.1. Crack Formation in YBaCuO Superconductor

As presented in Section 7.1.3..1 (see method 2), the analysis which considers the angle $\nu_{pm} = \nu_{pm}(\varphi)$ can explain the paradoxical behaviour concerning the crack formation which is experimentally observed in real two-component materials [71]–[74]. Strictly speaking, this analysis explains the crack formation in the spherical particle or cell matrix in spite of the condition $\alpha_p < \alpha_m$ or $\alpha_p > \alpha_m$, respectively. This paradoxical behaviour is experimentally observed e.g., in the YBaCuO superconductor [71]–[74].

The papers [71]–[74] present experimental results of the crack formation in the YBaCuO superconductor which represents a two-component material with anisotropic precipitates (Y_2BaCuO_5 particles) and an anisotropic matrix ($YBa_2Cu_3O_7$) [71]–[74]. In spite of the condition $\alpha_p > \alpha_m$, cracks which are observed in the YBaCuO superconductor are formed in the plane x_1x_2 in the $YBa_2Cu_3O_7$ matrix during a cooling process. This paradoxical behaviour of the cracking of the $YBa_2Cu_3O_7$ matrix instead of the cracking of the Y_2BaCuO_5 particle is supported by the following numerical results of this analytical model by the method 2 analysed in Section 7.1.3..1.

Additionally, the Y_2BaCuO_5 particle and $YBa_2Cu_3O_7$ matrix represent ceramic components of the YBaCuO superconductor. YBaCuO then exhibits a high-speed crack propagation during which the modification (change) of $W_{cm}^{(12),2}$ (see Eq. (7.29)) can be assumed to be neglected. Finally, YBaCuO thus represents a two-component material which is suitable for the following illustrative numerical analysis.

As presented in Section 7.1.3..1, the crack formation by the method 2 is required to fulfil the following conditions:

1. $\nu_{pm} \in \langle \nu^*, \pi/2)$ (see Eqs. (3.1)–(3.3), (7.18); Figure 7.7),
2. $\alpha_p(\varphi,\nu) < \alpha_m(\varphi,\nu)$ and $\alpha_p(\varphi,\nu) > \alpha_m(\varphi,\nu)$ for $\nu \in \langle 0, \nu_{pm})$ and $\nu \in (\nu_{pm}, \pi/2\rangle$, respectively,
3. $R_{mc}^{(12),2}(\varphi, v) < R_{mc}^{(23),1}(\varphi, v)$ (see Eq. (7.30)),
4. $R_{mc}^{(12),2}(\varphi, v) < R_{mc}^{(31),1}(\varphi, v)$ (see Eq. (7.30)),
5. $R_{mc}^{(12),2}(\varphi, v) < R_{pc}^{(12),1}(\varphi, v)$ (see Eq. (7.30)).

The following numerical results confirms validity of these conditions.

As presented in Section 4.4., a stress-strain state which is induced by the thermal stresses in the multi-particle-matrix system with anisotropic components is determined by several different solutions. These different solutions result in different values of thermal-stress induced elastic energy of this model system (see Eqs. (3.51), (3.52)). Accordingly, a principle of minimal total potential energy of an elastic solid body [26] is required to be considered (see Section 3.4.).

On the one hand, the determination of such solution which results in minimal thermal-stress induced elastic energy of this model system can be performed numerically. On the other hand, this determination which is time-consuming would require a comprehensive software product. The same is also valid in case of the determination of the function $f_m^{(12)} = f_m^{(12)}(x_{12}, \varphi, R_1, v)$ of the variable $x_{12} \in \langle R_1, x_{0m}\rangle$ (see Eq. (8.2)) which is thus determined for $\varphi = 0, 15, 30, 45$ [deg].

Accordingly, the following numerical determination is performed for the exponents $\lambda_{1q}^{(4.36)(0)}$, $\lambda_{2q}^{(4.36)(0)}$ (see Eq. (4.37)) ($q = p,m$) which represent roots of the characteristic equation (4.36) for $i = 0$.

Y_2BaCuO_5 and $YBa_2Cu_3O_7$ are characterized by a tetragonal crystal lattice. Table 8.1 presents material parameters of a tetragonal crystal lattice along the axis x_{iq} ($i = 1,2,3$) for the Y_2BaCuO_5 particle ($q = p$) and the $YBa_2Cu_3O_7$ matrix ($q = m$), where E_{iq} and μ_{iq} are the Young's modulus and the Poisson's ratio, respectively. In case of the tetragonal crystal lattice, the elastic modulus $s_{ijq} = s_{jiq}$ ($i, j = 1, \ldots, 6$) which is included in Eq. (3.41) has the form [114]

Table 8.1. The Young's modulus E_{iq}, the Poisson's ratio μ_{iq}, the thermal expansion coefficient α_{iq} and the fracture toughness K_{ICiq} of a tetragonal crystal lattice along the axis x_{iq} ($i = 1,2,3$) for the Y_2BaCuO_5 particle ($q = p$) and the $YBa_2Cu_3O_7$ matrix ($q = m$) within the temperature range $\langle T_f, T_r \rangle = \langle 20, 406 \rangle$ [°C] [71]–[74], [130]–[132]

Component Axis x_{iq}	Y_2BaCuO_5 $(q=p)$ $x_{1p}/x_{2p}/x_{3p}$	$YBa_2Cu_3O_7$ $(q=m)$ $x_{1m}/x_{2m}/x_{3m}$
E_{iq} [GPa]	212.8/212.8/213.2	182/182/143
μ_{iq}	0.25/0.25/0.24	0.255/0.255/0.255
α_{iq} $[10^{-5}K^{-1}]$	1.24/1.24/1.22	0.86/0.86/3.2
K_{ICiq} $[MPam^{1/2}]$	-/-/-	0.32/0.32/0.8

$$s_{ijq} = \frac{\delta_{ij} - \mu_{jq}(1-\delta_{ij})}{E_{iq}}, \quad s_{i3+jq} = s_{3+i3+kq} = 0,$$
$$s_{3+i3+iq} = 2\left[(\delta_{1i}+\delta_{2i})(s_{22q}-s_{23q}) + \delta_{3i}(s_{11q}-s_{12q})\right],$$
$$i \neq k; \quad i,j,k = 1,2,3. \tag{8.1}$$

The melting temperature $T_m = 925$°C [71]–[74] represents temperature of the peritectic transformation of the YBaCuO superconductor, and then $T_r = 0.4 \times T_m = 370$°C, $T_f = 20$°C.

Provided that $x_i = x_{iq}$ ($i = 1,2,3$; $q = p,m$) (see Figure 3.1), the angle ν_{pm} is not a function of $\varphi \in \langle 0, \pi/2 \rangle$ due to $\alpha_{1q} = \alpha_{2q}$ ($q = p,m$), and then $\nu_{pm} = 66.34$ [deg], where $\nu_{pm} > \nu^* \in \langle 45, 54.74 \rangle$ [deg] for $\varphi \in \langle 0, \pi/2 \rangle$ (see Eqs. (3.1)–(3.3), (7.18)).

The conditions $\alpha_p(\varphi,\nu) < \alpha_m(\varphi,\nu)$ and $\alpha_p(\varphi,\nu) > \alpha_m(\varphi,\nu)$ are then valid for $\nu \in \langle 0, \nu_{pm})$ and $\nu \in (\nu_{pm}, \pi/2\rangle$, respectively, both for $\varphi \in \langle 0, \pi/2 \rangle$.

With regard to the cracking in the plane x_1x_2, the thermal stresses are equal to zero for $\nu = \nu_{pm}$ due to $\alpha_p(\varphi,\nu) = \alpha_m(\varphi,\nu)$ for $\varphi \in \langle 0, \pi/2 \rangle$, where $\varphi = \angle(x_1, x_{12})$.

With regard to the cracking in the plane x_2x_3, the thermal stresses are equal to zero for $\varphi = 90 - \nu_{pm}$, where $\varphi = \angle(x_2, x_{23})$.

With regard to the cracking in the plane x_3x_1, the thermal stresses are equal to zero for $\varphi = \nu_{pm}$, where $\varphi = \angle(x_3, x_{31})$.

Figure 8.1 shows the critical particle radii $R_{mc}^{(12),2}$, $R_{mc}^{(23),1}$, $R_{mc}^{(31),1}$ as functions of the variable $\varphi \in \langle 0, \pi/2 \rangle$ for $T_r = 20$°C, $T_f = 20$°C, $v = 0.25$. The function $R_{mc}^{(12),2} = R_{mc}^{(12),2}(\varphi, v)$ exhibits mirror symmetry regarding the angle $\varphi = \pi/4$. The functions $R_{mc}^{(23),1} = R_{mc}^{(23),1}(\varphi, v)$ and $R_{mc}^{(31),1} = R_{mc}^{(31)1,}(\varphi, v)$ are mutually mirror-symmetric regarding the angle $\varphi = \pi/4$, Additionally, we get $\lim\limits_{\varphi \to (90-\nu_{pm})^-} R_{mc}^{(23),1}(\varphi) = \infty$ and $\lim\limits_{\varphi \to \nu_{pm}^+} R_{mc}^{(31),1}(\varphi) = \infty$ due to zero thermal stresses for $\varphi = 90 - \nu_{pm}$ and $\varphi = \nu_{pm}$, respectively.

On the one hand, the critical particle radius $R_{pc}^{(12),1} = R_{pc}^{(12),1}(\varphi, v)$ is required to be determined for a verification of the condition $R_{mc}^{(12),2} < R_{pc}^{(12),1}$ (see Eq. (7.30)). On the other hand, the fracture toughness K_{ICip} along the axis x_i ($i = 1,2,3$) of a tetragonal crystal lattice of the Y_2BaCuO_5 particle is not currently known.

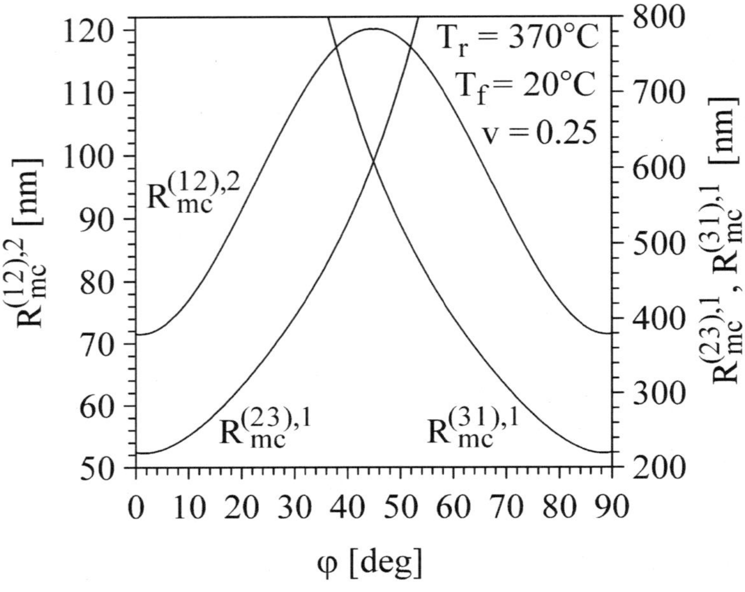

Figure 8.1. The critical particle radii $R_{mc}^{(12),2} = R_{mc}^{(12),2}(\varphi)$, $R_{mc}^{(23),1} = R_{mc}^{(23),1}(\varphi)$, $R_{mc}^{(31),1} = R_{mc}^{(31),1}(\varphi)$ as functions of the variable $\varphi \in \langle 0, \pi/2 \rangle$ for $T_f = 20°C$, $T_r = 370°C$, $v = 0.25$.

However, in addition to an effect of the pinning of an external magnetic field, the Y_2BaCuO_5 particles are also used to increase relatively low fracture toughness of the $YBa_2Cu_3O_7$ matrix. With regard to experimental results in [131, 132], fracture toughness of the YBaCuO superconductor can be thus increased to a value of 2 $MPam^{1/2}$ for $v = 0.4$. Accordingly, fracture toughness of the Y_2BaCuO_5 particle is assumed to be higher than that of the $YBa_2Cu_3O_7$ matrix.

Higher fracture toughness of the Y_2BaCuO_5 particle results in a higher value of $R_{pc}^{(12),1}$, i.e., $R_{pc}^{(12),1} \geq R_{mc}^{(23),1}$, $R_{pc}^{(12),1} \geq R_{mc}^{(31),1}$. Accordingly, the condition defined in Item 5 mentioned above can be assumed to be fulfilled.

As presented in Section 7.1.2., the condition (7.11) ($q = m$) for $W_{cm}^{(ij),2}$ and $R_1 > R_{mc}^{(12),2}$ results in the function $x_{0m} = x_{0m}(\varphi, R_1, v)$ of the variable φ, where R_1, v represent parameters of this function. The dependence $x_{0m} - \varphi$ defines a position of the crack tip point P_0 on the axis x_{12}, where $x_{12} \subset x_1x_2$, $\varphi = \angle(x_1, x_{12}) \in \langle 0, 2\pi \rangle$ (see Figures 7.1b, 7.3 for $i = 1$; $j = 2$; $k = 3$). The point $P_0 \in x_1x_2$ with the coordinate $x_{12} = x_{0m}(\varphi, R_1, v)$ represents the crack tip related to the plane $x_{12}x_3$. The dependence $x_{0m} - \varphi$ thus defines the crack shape in the plane x_1x_2 in the cell matrix.

$$x_{0m} = 1507.6 - 32.6x_1 + 0.26x_1^2 - 0.000714x_1^3, \quad x_1 \in \langle 108, 165 \rangle \ [\text{nm}]; \quad [x_1, x_{0m}] = [\text{nm}]. \tag{8.2}$$

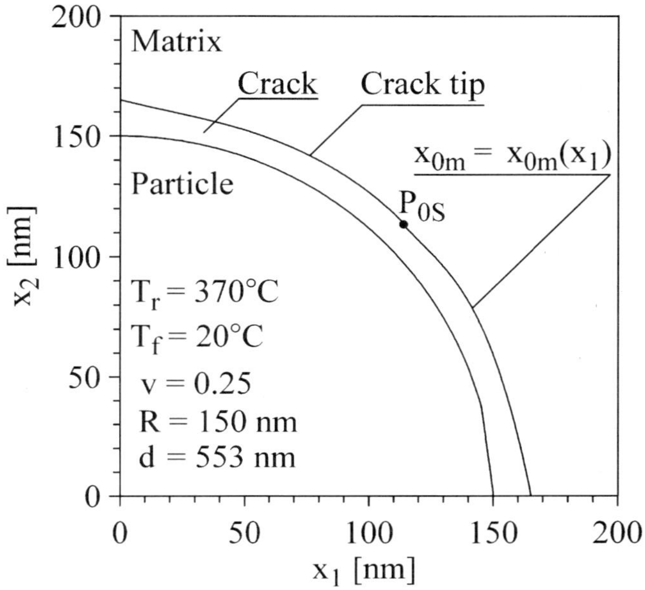

Figure 8.2. The function $x_{0m} = x_{0m}(x_1)$ which thus defines a crack shape in the plane x_1x_2 of the $YBa_2Cu_3O_7$ matrix, where $T_f = 20°C$, $T_r = 370°C$, $v = 0.25$, $d = 553$ nm, R_1,¯ 150 nm and $R_1 > R_{mc}^{(12),2}(\varphi)$ (see Figure 8.1). The function $x_{0m} = x_{0m}(x_1)$ is defined for $x_1 \in \langle 108, 165 \rangle$ [nm], where $x_{1S} = 108$ nm is a coordinate of the point P_{0S}, and $\varphi = \angle\left(x_1, \overline{0P_{0S}}\right) = \pi/4$ (see Figures 7.1b, 7.3 for $i=1$; $j=2$; $k=3$). With regard to the intervals $\varphi \in \langle 0, \pi/4 \rangle$, $\varphi \in \langle \pi/4, \pi/2 \rangle$, this crack shape exhibits mirror symmetry regarding the plane $x_{12}x_k$ for $\varphi = \pi/4$ (see Figure 7.3).

With regard to the intervals $\varphi \in \langle 0, \pi/4 \rangle$, $\varphi \in \langle \pi/4, \pi/2 \rangle$, this crack shape exhibits mirror symmetry regarding the plane $x_{12}x_k$ for $\varphi = \pi/4$ (see Figure 7.3 for $i=1$; $j=2$; $k=3$). Figure 8.3 shows a crack in the plane $x_{12}x_3$ in the $YBa_2Cu_3O_7$ matrix for $\varphi = 0; \pi/12; \pi/6; \pi/4$; where $\varphi = \angle(x_1, x_{12})$, $x_{12} \subset x_1x_2$ (see Figure 7.3). The plane $x_{12}x_3$ is perpendicular to the crack formation plane x_1x_2. A shape of this crack is described by the function $f_m^{(12)} = f_m^{(12)}(x_{12})$ at the parameters $\varphi = 0, \pi/12, \pi/6, \pi/4$; $R_1 = 150$ nm, $v = 0.25$. The function $f_m^{(12)} = f_m^{(12)}(x_{12})$ for $\varphi = 0, \pi/12, \pi/6, \pi/4$ is derived as

$$
\begin{aligned}
f_m^{(12)} &= 3367.686 - 61.26x_{12} + 0.371x_{12}^2 - 0.00075x_{12}^3, \quad \varphi = 0, \\
f_m^{(12)} &= 4143.904 - 76.076x_{12} + 0.466x_{12}^2 - 0.00095x_{12}^3, \quad \varphi = \frac{\pi}{12}, \\
f_m^{(12)} &= 4842.874 - 89.549x_{12} + 0.552x_{12}^2 - 0.00113x_{12}^3, \quad \varphi = \frac{\pi}{6}, \\
f_m^{(12)} &= 6315.58 - 117.88x_{12} + 0.733x_{12}^2 - 0.00152x_{12}^3, \quad \varphi = \frac{\pi}{4}, \\
& x_{12} \geq 150\,[\mathrm{nm}], \quad [x_{12}, f_m] = [\mathrm{nm}].
\end{aligned} \tag{8.3}
$$

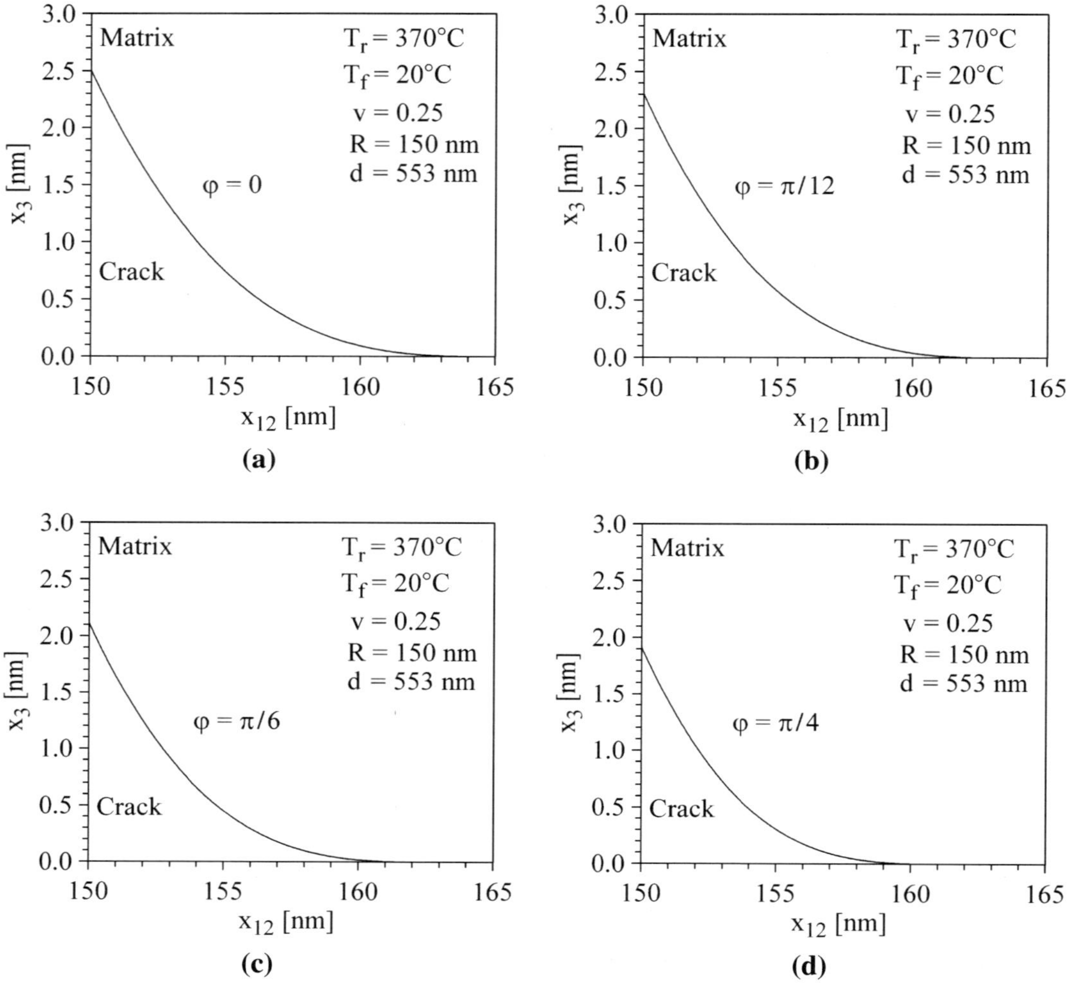

Figure 8.3. The crack in the plane $x_{12}x_3$ in the $YBa_2Cu_3O_7$ matrix for $\varphi = 0, \pi/12, \pi/6, \pi/4$ (see Eq. (8.3)), where $\varphi = \angle(x_1, x_{12})$, $x_{12} \subset x_1x_2$ (see Figure 7.3). The plane $x_{12}x_3$ is perpendicular to the crack formation plane x_1x_2.

8.2. Thermal Stresses and Strengthening in SiC-Si_3N_4 Ceramics

As an illustrative example, this section presents an application of the analytical model of the thermal stresses and thermal-stress induced strengthening (see Section 7.3.) to the SiC-Si_3N_4 ceramics. The SiC-Si_3N_4 ceramics represents a two-component material of the precipitate-matrix type which is defined in Item 1, Section 2.1.. SiC-Si_3N_4 thus consists of the SiC precipitates (particles) and Si_3N_4 grains (matrix), where both components of SiC-Si_3N_4 are ceramic.

In general, ceramic materials are used for such engineering applications, where high strength and wear resistance are required (e.g., engine components and pistons, turbine vanes, components of brakes, cutting tools) [133, 134]. However, ceramics represents a brittle material. Ceramics is then characterized by low fracture toughness and low strength in tension. Accordingly, the high strength is meant at static or quasi-static compressive loading.

Low strength in tension can be increased by the presence of such precipitates so that the condition $\alpha_p < \alpha_m$ is fulfilled, where α_q $(q = p,m)$ is a thermal expansion coefficient of the precipitate (particle) $(q = p)$ and a grain (matrix) $(q = m)$. The condition $\alpha_p < \alpha_m$ then results in the tensile radial stress $\sigma'_{11m} = \sigma_{rm} > 0$ and the compressive tangential stresses $\sigma'_{22m} = \sigma_{\varphi m} < 0$, $\sigma'_{33m} = \sigma_{\nu m} < 0$ (see Figure 3.1) in the matrix (grain). The tensile radial stress and compressive tangential stresses represent 'matrix resistance' against tensile mechanical loading [134]. This resistance is then a reason of an increase of strength in tension as well as an increase of fracture toughness [105]. Finally, a brittle ceramic matrix of grains can be thus strengthened by the presence of the precipitates with $\alpha_p < \alpha_m$.

The SiC-Si_3N_4 two-component material with ceramic components represents such two-component material of the precipitate-matrix type for which the condition $\alpha_p < \alpha_m$ is fulfilled (see Table 8.2). With regard to the following numerical determination, SiC and Si_3N_4 are considered to be both isotropic. This numerical determination is performed for the analytical model of the thermal stresses which is determined by the mathematical procedure 1 in Section 5.3.. The application of the analytical model of the thermal stresses and thermal-stress induced strengthening (see Section 7.3.) to SiC-Si_3N_4 thus includes:

1. the numerical determination of the function $\overline{p_1} = \overline{p_1}(v)$ of the variable $v \in \langle 0, \pi/6 \rangle$ (see Eqs. (2.1), (2.4); Figure 8.4), where v is a particle volume fraction. The radial stress $p_1 = p_1(\varphi, \nu, v)$ acts at the particle-matrix boundary, i.e., at the SiC-Si_3N_4 boundary. The mean value $\overline{p_1} = \overline{p_1}(v)$ regarding the variables $\varphi \in \langle 0, \pi/2 \rangle$ and $\nu \in \langle 0, \pi/2 \rangle$ is given by Eq. (7.71).

2. the numerical determination of courses of the radial, tangential and shear stresses, σ'_{11q}, $\sigma'_{22q} = \sigma'_{33q}$ and σ'_{12q}, σ'_{13q}, respectively, in the SiC particle $(q = p)$ (see Figure 8.5) and the Si_3N_4 matrix $(q = m)$ (see Figure 8.6). These courses are determined along specific directions, i.e., for $(\varphi, \nu) = (0°, 90°)$, $(\varphi, \nu) = (0°, 45°)$, $(\varphi, \nu) = (45°, 90°)$, $(\varphi, \nu) = (45°, \nu^*) = (45°, 54.7°)$, where the angle $\nu^* = \nu^*(\varphi)$ (see Figure 3.2) is given by Eqs. (3.1)–(3.3). The thermal stresses as functions of the spherical coordinates (r, φ, ν) (see Figure 3.1a) are related to the axes $x'_1 = x_r$, $x'_2 = x_\varphi$, $x'_3 = x_\nu$ of the Cartesian system $(Px'_1x'_2x'_3)$ in which the infinitesimal spherical cap (see Figure 3.1b) exhibits only this radial displacement along x'_1 (see Section 3.2.).

3. the numerical determination of the micro-strengthening $\sigma_{s1} = \sigma_{s1}(x_1)$ (see Figure 8.7) and the macro-strengthening $\overline{\sigma}_{s1}$ both along the axis x_1 (see Figure 2.1a) in the SiC-Si_3N_4 cubic cell for $x_1 \in \langle 0, d/2 \rangle$, where $\sigma_{s1} > 0$, $\overline{\sigma}_{s1} > 0$ due to $\alpha_p > \alpha_m$. Due to $\sigma_{s1} > 0$, $\overline{\sigma}_{s1} > 0$, this micro- and macro-strengthening represents a 'resistance' against tensile mechanical loading.

Table 8.2. Material parameters of the SiC-Si_3N_4 ceramics [133, 134]

	Particle ($q = p$) SiC	Matrix ($q = m$) Si_3N_4
E_q [GPa]	360	320
μ_q	0.19	0.28
α_q [$10^{-6}K^{-1}$]	4.15	2.35
T_{mq} [°C]	2730	1900

Table 8.2 presents material parameters of SiC and Si_3N_4. As presented in Section 3.5.1.1, the relaxation temperature $T_r = 665$°C is determined by the formula $T_r = 0.4\,T_m$ [104], where $T_m = T_{mm}$ (see Table 8.2) is melting temperature of the SiC-Si_3N_4 ceramics [133, 134].

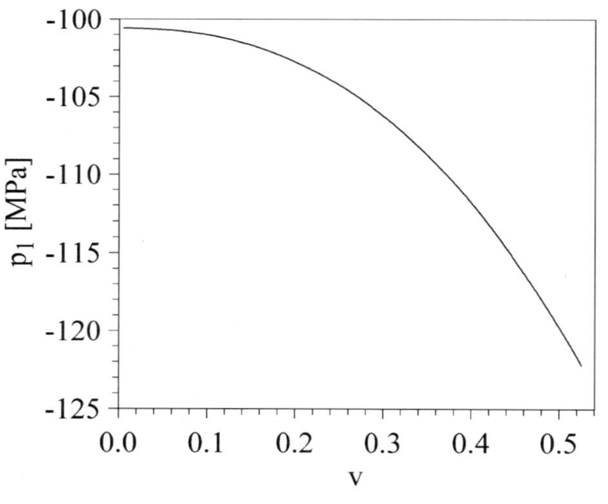

Figure 8.4. The function $\overline{p_1} = \overline{p_1}(v)$ of the volume fraction $v \in (0, \pi/6)$ (see Eqs. (2.1), (2.4)) of the SiC particle, where $R_1 = 200$ nm, $T_f = 20$°C, $T_r = 665$°C. The radial stress $p_1 = p_1(\varphi, \nu, v)$ acts at the SiC-Si_3N_4 boundary. The mean value $\overline{p_1} = \overline{p_1}(v)$ regarding the variables $\varphi \in \langle 0, \pi/2 \rangle$ and $\nu \in \langle 0, \pi/2 \rangle$ is given by Eq. (7.71).

As shown in Figure 8.4, the dependence $\overline{p_1} - v$ for the SiC-Si_3N_4 boundary exhibits a decreasing course for $v \in (0, \pi/6)$, where $R_1 = 200$ nm, $T_f = 20$°C, $T_r = 665$°C.

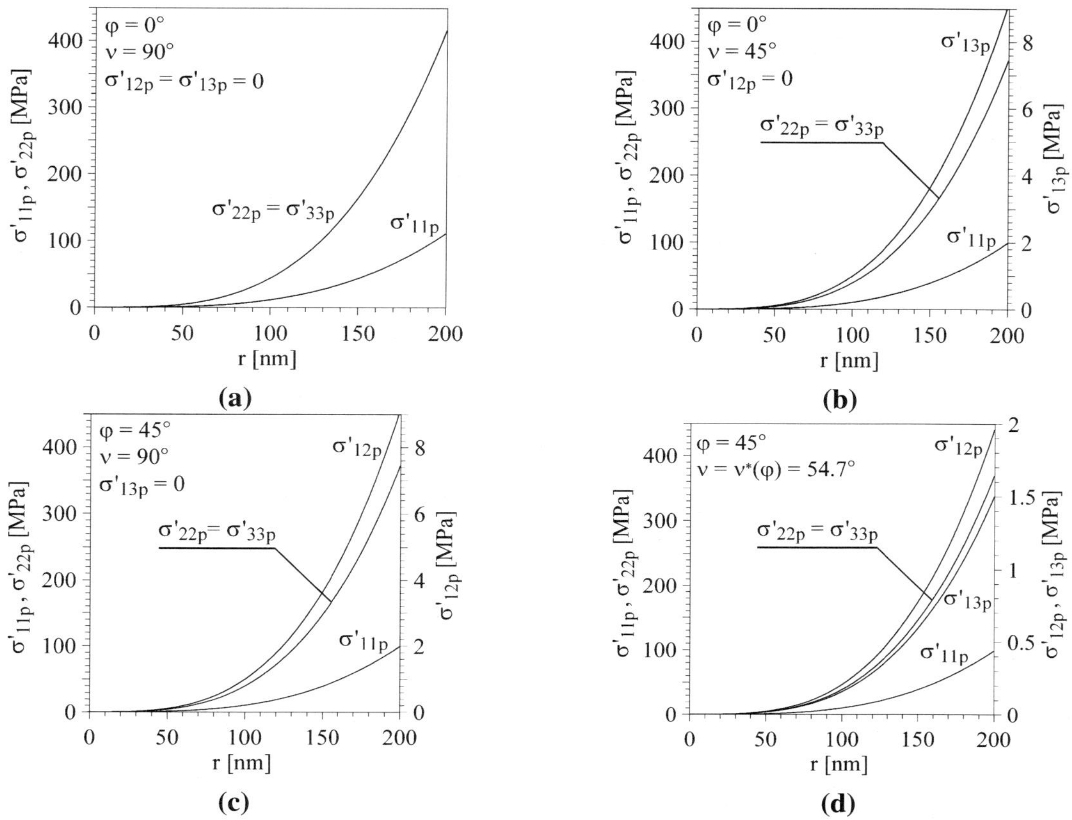

Figure 8.5. Dependences of the radial, tangential and shear stresses, σ'_{11p}, $\sigma'_{22p} = \sigma'_{33p}$ and σ'_{12p}, σ'_{13p}, respectively, on $r \in \langle 0, R_1 \rangle$ in the SiC particle along the following directions: (a) $(\varphi, \nu) = (0°, 90°)$, (b) $(\varphi, \nu) = (0°, 45°)$, (c) $(\varphi, \nu) = (45°, 90°)$, (d) $(\varphi, \nu) = (45°, \nu^*) = (45°, 54.7°)$, where $v = 0.2$, $R_1 = 200$ nm, $T_f = 20$°C, $T_r = 665$°C. The angle $\nu^* = \nu^*(\varphi)$ (see Figure 3.2) is given by Eqs. (3.1)–(3.3).

Additionally, $\overline{p_1} < 0$ is tensile due to $\beta_p > \beta_m$ ($\alpha_p > \alpha_m$) (see Eq. (3.59), Tab. 8.2). Consequently, the dependence $|\overline{p_1}| - v$ exhibits an increasing course for $v \in (0, \pi/6\rangle$.

Figures 8.5 and 8.6 show a courses of σ'_{11q}, $\sigma'_{22q} = \sigma'_{33q}$ (see Eq. (5.2)), σ'_{12q}, σ'_{13q} in the SiC particle ($q = p$) and Si_3N_4 cell matrix ($q = m$) for $r \in \langle 0, R_1 \rangle$ and $r \in \langle R_1, r_c \rangle$ (see Eqs. (3.15)–(3.18)), respectively, where $v = 0.2$, $R_1 = 200$ nm, $d = 551.3$ nm (see Eq. (2.1)). As presented in [134], $v = 0.2$ and $R_1 = 200$ nm represent mean values of a volume fraction and a radius of the SiC particles in the SiC-Si_3N_4 ceramics, respectively.

These dependences, i.e., $\sigma'_{ijp} = \sigma'_{ijp}\left(\varphi, \nu, r^{\lambda_{1p}}\right)$, $\sigma'_{ijm} = \sigma'_{ijm}\left(\varphi, \nu, r^{\lambda_{1m}}, r^{\lambda_{2m}}\right)$ are related to the following directions: $(\varphi, \nu) = (0°, 90°)$ (see Figures 8.5a, 8.6a), $(\varphi, \nu) = (0°, 45°)$ (see Figures 8.5b, 8.6b), $(\varphi, \nu) = (45°, 90°)$ (see Figures 8.5c, 8.6c), $(\varphi, \nu) = (45°, 54.7°)$ (see Figures 8.5d, 8.6d), where $\lambda_{1p} = 4.24$, $\lambda_{1m} = 4.06$, $\lambda_{2m} = -3.06$ (see Eq. (5.12)).

Additionally, the stress σ'_{ijp} ($i, j = 1,2,3$) in the SiC particle is tensile, i.e., $\sigma'_{ijp} > 0$. Similarly, $\sigma'_{11m} > 0$, $\sigma'_{12m} > 0$, $\sigma'_{13m} > 0$ and $\sigma'_{22m} = \sigma'_{33m} < 0$ are tensile and compressive, respectively.

Figure 8.7 shows the micro-strengthening $\sigma_{s1} = \sigma_{s1}(x_1)$ along the axis x_1 (see Eqs. (7.55), (7.56), (7.58), (7.59); Figure 2.1a) in the SiC-Si_3N_4 cubic cell for $x_1 \in \langle 0, d/2 \rangle$, where $v = 0.2$, $R_1 = 200$ nm, $d = 551.3$ nm (see Eq. (2.1)), $T_f = 20$°C, $T_r = 665$°C. Due to $\alpha_p > \alpha_m$, we get $\sigma_{s1} > 0$. Consequently, due to $\sigma_{s1} > 0$, the micro-strengthening $\sigma_{s1} = \sigma_{s1}(x_1)$ represents a 'resistance' against tensile mechanical loading. The macro-strengthening $\overline{\sigma}_{s1} = 45.2$ MPa (see Eq. (7.60)) in the SiC-Si_3N_4 cubic cell also represents a 'resistance' against the tensile mechanical loading.

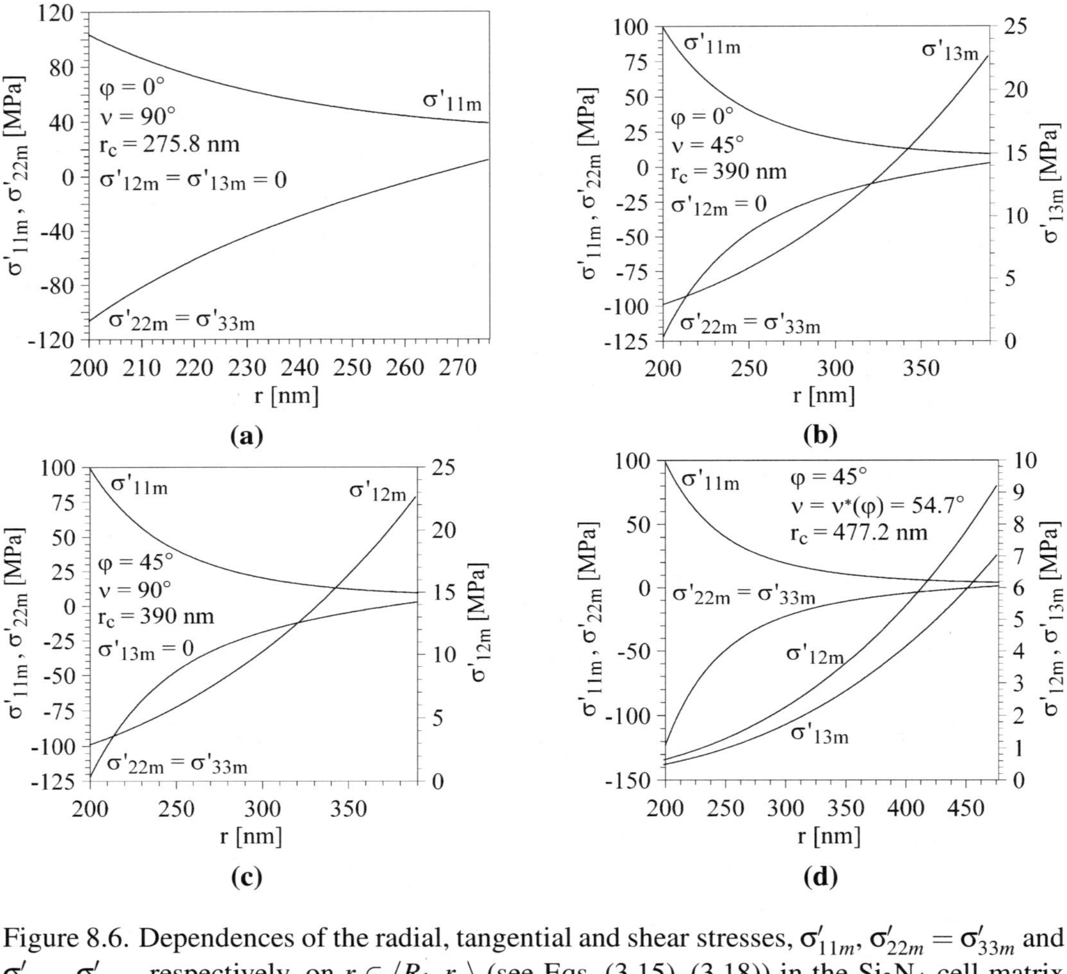

Figure 8.6. Dependences of the radial, tangential and shear stresses, σ'_{11m}, $\sigma'_{22m} = \sigma'_{33m}$ and σ'_{12m}, σ'_{13m}, respectively, on $r \in \langle R_1, r_c \rangle$ (see Eqs. (3.15)–(3.18)) in the Si_3N_4 cell matrix along the following directions: (a) $(\varphi, \nu) = (0°, 90°)$, $r_c = 240.8$ nm; (b) $(\varphi, \nu) = (0°, 45°)$, $r_c = 340.5$ nm; (c) $(\varphi, \nu) = (45°, 90°)$, $r_c = 340.5$ nm; (d) $(\varphi, \nu) = (45°, \nu^*) = (45°, 54.7°)$, $r_c = 244.3$ nm; where $v = 0.2$, $R_1 = 200$ nm, $T_f = 20$°C, $T_r = 665$°C. The angle $\nu^* = \nu^*(\varphi)$ (see Figure 3.2) is given by Eqs. (3.1)–(3.3).

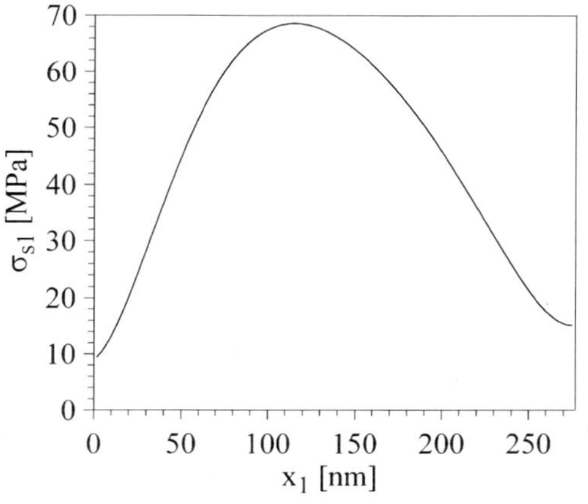

Figure 8.7. The micro-strengthening $\sigma_{s1} = \sigma_{s1}(x_1)$ along the axis x_1 (see Eqs. (7.55), (7.56), (7.58), (7.59); Figure 2.1a) in the SiC-Si_3N_4 cubic cell for $x_1 \in \langle 0, d/2 \rangle$, where $v = 0.2$, $R_1 = 200$ nm, $d = 551.3$ nm (see Eq. (2.1)), $T_f = 20$°C, $T_r = 665$°C.

Chapter 9

Appendix

The elastic moduli $s'_{11}, s'_{12}, \ldots, s'_{56}, s'_{66}$ in the Cartesian system with the axes x'_1, x'_2, x'_3 can be determined by the elastic moduli s_{11}, $s_{12}, \ldots, s_{56}$, s_{66} in the Cartesian system $(Ox_1x_2x_3)$ (see Figure 3.1). With regard to Eqs. (3.41)–(3.43) [114], $s'_{11}, s'_{12}, \ldots, s'_{56}, s'_{66}$ have the forms

$$\begin{aligned}
s'_{11} &= s_{11}(a_{11})^4 + 2s_{12}(a_{11}a_{12})^2 + 2s_{13}(a_{11}a_{13})^2 + 4s_{14}(a_{11})^2 a_{12}a_{13} \\
&+ 4s_{15}(a_{11})^3 a_{13} + 4s_{16}(a_{11})^3 a_{12} + s_{22}(a_{12})^4 + 2s_{23}(a_{12}a_{13})^2 \\
&+ 4s_{24}(a_{12})^3 a_{13} + 4s_{25}a_{11}(a_{12})^2 a_{13} + 4s_{26}a_{11}(a_{12})^3 + s_{33}(a_{13})^4 \\
&+ 4s_{34}a_{12}(a_{13})^3 + 4s_{35}a_{11}(a_{13})^3 + 4s_{36}a_{11}a_{12}(a_{13})^2 + 4s_{44}(a_{12}a_{13})^2 \\
&+ 8s_{45}a_{11}a_{12}(a_{13})^2 + 8s_{46}a_{11}(a_{12})^2 a_{13} + 4s_{55}(a_{11}a_{13})^2 + 8s_{56}(a_{11})^2 a_{12}a_{13} \\
&+ 4s_{66}(a_{11}a_{12})^2, \qquad (9.1)
\end{aligned}$$

$$\begin{aligned}
s'_{12} &= s_{11}(a_{11}a_{21})^2 + s_{12}\left[(a_{11}a_{22})^2 + (a_{12}a_{21})^2\right] + s_{13}\left[(a_{11}a_{23})^2 + (a_{13}a_{21})^2\right] \\
&+ 2s_{14}\left[(a_{11})^2 a_{22}a_{23} + a_{12}a_{13}(a_{21})^2\right] + 2s_{15}\left[(a_{11})^2 a_{21}a_{23} + a_{11}a_{13}(a_{21})^2\right] \\
&+ 2s_{16}\left[(a_{11})^2 a_{21}a_{22} + a_{11}a_{12}(a_{21})^2\right] + s_{22}(a_{12}a_{22})^2 \\
&+ s_{23}\left[(a_{12}a_{23})^2 + (a_{13}a_{22})^2\right] + 2s_{24}\left[(a_{12})^2 a_{22}a_{23} + a_{12}a_{13}(a_{22})^2\right] \\
&+ 2s_{25}\left[(a_{12})^2 a_{21}a_{23} + a_{11}a_{13}(a_{22})^2\right] + 2s_{26}\left[(a_{12})^2 a_{21}a_{22} + a_{11}a_{12}(a_{22})^2\right] \\
&+ s_{33}(a_{13}a_{23})^2 + 2s_{34}\left[(a_{13})^2 a_{22}a_{23} + a_{12}a_{13}(a_{23})^2\right] \\
&+ 2s_{35}\left[(a_{13})^2 a_{21}a_{23} + a_{11}a_{13}(a_{23})^2\right] + 2s_{36}\left[(a_{13})^2 a_{21}a_{22} + a_{11}a_{12}(a_{23})^2\right] \\
&+ 4s_{44}a_{12}a_{13}a_{22}a_{23} + 4s_{45}(a_{12}a_{13}a_{21}a_{23} + a_{11}a_{13}a_{22}a_{23}) \\
&+ 4s_{46}(a_{12}a_{13}a_{21}a_{22} + a_{11}a_{12}a_{22}a_{23}) + 4s_{55}a_{11}a_{13}a_{21}a_{23} \\
&+ 4s_{56}(a_{11}a_{13}a_{21}a_{22} + a_{11}a_{12}a_{21}a_{23}) + 4s_{66}a_{11}a_{12}a_{21}a_{22}, \qquad (9.2)
\end{aligned}$$

$$\begin{aligned}
s'_{13} &= s_{11}\left(a_{11}a_{31}\right)^2 + s_{12}\left[\left(a_{11}a_{32}\right)^2 + \left(a_{12}a_{31}\right)^2\right] + s_{13}\left[\left(a_{11}a_{33}\right)^2 + \left(a_{13}a_{31}\right)^2\right] \\
&+ 2s_{14}\left[\left(a_{11}\right)^2 a_{32}a_{33} + a_{12}a_{13}\left(a_{31}\right)^2\right] + 2s_{15}\left[\left(a_{11}\right)^2 a_{31}a_{33} + a_{11}a_{13}\left(a_{31}\right)^2\right] \\
&+ 2s_{16}\left[\left(a_{11}\right)^2 a_{31}a_{32} + a_{11}a_{12}\left(a_{31}\right)^2\right] + s_{22}\left(a_{12}a_{32}\right)^2 \\
&+ s_{23}\left[\left(a_{12}a_{33}\right)^2 + \left(a_{13}a_{32}\right)^2\right] + 2s_{24}\left[\left(a_{12}\right)^2 a_{32}a_{33} + a_{12}a_{13}\left(a_{32}\right)^2\right] \\
&+ 2s_{25}\left[\left(a_{12}\right)^2 a_{31}a_{33} + a_{11}a_{13}\left(a_{32}\right)^2\right] + 2s_{26}\left[\left(a_{12}\right)^2 a_{31}a_{32} + a_{11}a_{12}\left(a_{32}\right)^2\right] \\
&+ s_{33}\left(a_{13}a_{33}\right)^2 + 2s_{34}\left[\left(a_{13}\right)^2 a_{32}a_{33} + a_{12}a_{13}\left(a_{33}\right)^2\right] \\
&+ 2s_{35}\left[\left(a_{13}\right)^2 a_{31}a_{33} + a_{11}a_{13}\left(a_{33}\right)^2\right] + 2s_{36}\left[\left(a_{13}\right)^2 a_{31}a_{32} + a_{11}a_{12}\left(a_{33}\right)^2\right] \\
&+ 4s_{44}a_{12}a_{13}a_{32}a_{33} + 4s_{45}\left(a_{12}a_{13}a_{31}a_{33} + a_{11}a_{13}a_{32}a_{33}\right) \\
&+ 4s_{46}\left(a_{12}a_{13}a_{31}a_{32} + a_{11}a_{12}a_{32}a_{33}\right) + 4s_{55}a_{11}a_{13}a_{31}a_{33} \\
&+ 4s_{56}\left(a_{11}a_{13}a_{31}a_{32} + a_{11}a_{12}a_{31}a_{33}\right) + 4s_{66}a_{11}a_{12}a_{31}a_{32}, \qquad (9.3)
\end{aligned}$$

$$\begin{aligned}
s'_{14} &= s_{11}\left(a_{11}\right)^2 a_{21}a_{31} + s_{12}\left[\left(a_{11}\right)^2 a_{22}a_{32} + \left(a_{12}\right)^2 a_{21}a_{31}\right] \\
&+ s_{13}\left[\left(a_{11}\right)^2 a_{23}a_{33} + \left(a_{13}\right)^2 a_{21}a_{31}\right] \\
&+ s_{14}\left[\left(a_{11}\right)^2\left(a_{22}a_{33} + a_{23}a_{32}\right) + 2a_{12}a_{13}a_{21}a_{31}\right] \\
&+ s_{15}\left[\left(a_{11}\right)^2\left(a_{21}a_{33} + a_{23}a_{31}\right) + 2a_{11}a_{13}a_{21}a_{31}\right] \\
&+ s_{16}\left[\left(a_{11}\right)^2\left(a_{21}a_{32} + a_{22}a_{31}\right) + 2a_{11}a_{12}a_{21}a_{31}\right] + s_{22}\left(a_{12}\right)^2 a_{22}a_{32} \\
&+ s_{23}\left[\left(a_{12}\right)^2 a_{23}a_{33} + \left(a_{13}\right)^2 a_{22}a_{32}\right] \\
&+ s_{24}\left[\left(a_{12}\right)^2\left(a_{22}a_{33} + a_{23}a_{32}\right) + 2a_{12}a_{13}a_{22}a_{32}\right] \\
&+ s_{25}\left[\left(a_{12}\right)^2\left(a_{21}a_{33} + a_{23}a_{31}\right) + 2a_{11}a_{13}a_{22}a_{32}\right] \\
&+ s_{26}\left[\left(a_{12}\right)^2\left(a_{21}a_{32} + a_{22}a_{31}\right) + 2a_{11}a_{12}a_{22}a_{32}\right] + s_{33}\left(a_{13}\right)^2 a_{23}a_{33} \\
&+ s_{34}\left[\left(a_{13}\right)^2\left(a_{22}a_{33} + a_{23}a_{32}\right) + 2a_{12}a_{13}a_{23}a_{33}\right] \\
&+ s_{35}\left[\left(a_{13}\right)^2\left(a_{21}a_{33} + a_{23}a_{31}\right) + 2a_{11}a_{13}a_{23}a_{33}\right] \\
&+ s_{36}\left[\left(a_{13}\right)^2\left(a_{21}a_{32} + a_{22}a_{31}\right) + 2a_{11}a_{12}a_{23}a_{33}\right] \\
&+ 2s_{44}a_{12}a_{13}\left(a_{22}a_{33} + a_{23}a_{32}\right) \\
&+ 2s_{45}a_{13}\left[a_{12}\left(a_{21}a_{33} + a_{23}a_{31}\right) + a_{11}\left(a_{22}a_{33} + a_{23}a_{32}\right)\right] \\
&+ 2s_{46}a_{12}\left[a_{13}\left(a_{21}a_{32} + a_{22}a_{31}\right) + a_{11}\left(a_{22}a_{33} + a_{23}a_{32}\right)\right] \\
&+ 2s_{55}a_{11}a_{13}\left(a_{21}a_{33} + a_{23}a_{31}\right) \\
&+ 2s_{56}a_{11}\left[a_{12}\left(a_{21}a_{33} + a_{23}a_{31}\right) + a_{13}\left(a_{21}a_{32} + a_{22}a_{31}\right)\right] \\
&+ 2s_{66}a_{11}a_{12}\left(a_{21}a_{32} + a_{22}a_{31}\right), \qquad (9.4)
\end{aligned}$$

$$
\begin{aligned}
s'_{15} &= s_{11}(a_{11})^3 a_{31} + s_{12}a_{11}a_{12}(a_{11}a_{32}+a_{12}a_{31}) + s_{13}a_{11}a_{13}(a_{11}a_{33}+a_{13}a_{31}) \\
&+ s_{14}a_{11}[a_{11}(a_{12}a_{33}+a_{13}a_{32})+2a_{12}a_{13}a_{31}] + s_{15}(a_{11})^2(a_{11}a_{33}+3a_{13}a_{31}) \\
&+ s_{16}(a_{11})^2(a_{11}a_{32}+3a_{12}a_{31}) + s_{22}(a_{12})^3 a_{32} + s_{23}a_{12}a_{13}(a_{12}a_{33}+a_{13}a_{32}) \\
&+ s_{24}(a_{12})^2(a_{12}a_{33}+3a_{13}a_{32}) + s_{25}a_{12}[a_{12}(a_{11}a_{33}+a_{13}a_{31})+2a_{11}a_{13}a_{32}] \\
&+ s_{26}(a_{12})^2(a_{12}a_{31}+3a_{11}a_{32}) + s_{33}a_{13}a_{13}a_{13}a_{33} \\
&+ s_{34}(a_{13})^2(a_{13}a_{32}+3a_{12}a_{33}) + s_{35}(a_{13})^2(a_{13}a_{31}+3a_{11}a_{33}) \\
&+ s_{36}a_{13}[a_{13}(a_{11}a_{32}+a_{12}a_{31})+2a_{11}a_{12}a_{33}] + 2s_{44}a_{12}a_{13}(a_{12}a_{33}+a_{13}a_{32}) \\
&+ 2s_{45}a_{13}[a_{13}(a_{12}a_{31}+a_{11}a_{32})+2a_{11}a_{12}a_{33}] \\
&+ 2s_{46}a_{12}[a_{12}(a_{13}a_{31}+a_{11}a_{33})+2a_{11}a_{13}a_{32}] + 2s_{55}a_{11}a_{13}(a_{11}a_{33}+a_{13}a_{31}) \\
&+ 2s_{56}a_{11}[a_{11}(a_{13}a_{32}+a_{12}a_{33})+2a_{12}a_{13}a_{31}] + 2s_{66}a_{11}a_{12}(a_{11}a_{32}+a_{12}a_{31}),
\end{aligned} \tag{9.5}
$$

$$
\begin{aligned}
s'_{16} &= s_{11}(a_{11})^3 a_{21} + s_{12}a_{11}a_{12}(a_{11}a_{22}+a_{12}a_{21}) + s_{13}a_{11}a_{13}(a_{11}a_{23}+a_{13}a_{21}) \\
&+ s_{14}a_{11}[a_{11}(a_{12}a_{23}+a_{13}a_{22})+2a_{12}a_{13}a_{21}] + s_{15}(a_{11})^2(a_{11}a_{23}+3a_{13}a_{21}) \\
&+ s_{16}(a_{11})^2(a_{11}a_{22}+3a_{12}a_{21}) + s_{22}(a_{12})^3 a_{22} + s_{23}a_{12}a_{13}(a_{12}a_{23}+a_{13}a_{22}) \\
&+ s_{24}(a_{12})^2(a_{12}a_{23}+3a_{13}a_{22}) + s_{25}a_{12}[a_{12}(a_{11}a_{23}+a_{13}a_{21})+2a_{11}a_{13}a_{22}] \\
&+ s_{26}(a_{12})^2(a_{12}a_{21}+3a_{11}a_{22}) + s_{33}(a_{13})^2 a_{23} + s_{34}(a_{13})^2(a_{13}a_{22}+3a_{12}a_{23}) \\
&+ s_{35}(a_{13})^2(a_{13}a_{21}+3a_{11}a_{23}) + s_{36}a_{13}[a_{13}(a_{11}a_{22}+a_{12}a_{21})+2a_{11}a_{12}a_{23}] \\
&+ 2s_{44}a_{12}a_{13}(a_{12}a_{23}+a_{13}a_{22}) + 2s_{45}a_{13}[a_{13}(a_{12}a_{21}+a_{11}a_{22})+2a_{11}a_{12}a_{23}] \\
&+ 2s_{46}a_{12}[a_{12}(a_{11}a_{23}+a_{13}a_{21})+2a_{11}a_{13}a_{22}] + 2s_{55}a_{11}a_{13}(a_{11}a_{23}+a_{13}a_{21}) \\
&+ 2s_{56}a_{11}[a_{11}(a_{13}a_{22}+a_{12}a_{23})+2a_{12}a_{13}a_{21}] + 2s_{66}a_{11}a_{12}(a_{11}a_{22}+a_{12}a_{21}),
\end{aligned} \tag{9.6}
$$

$$
\begin{aligned}
s'_{22} &= s_{11}(a_{21})^4 + 2s_{12}(a_{21}a_{22})^2 + 2s_{13}(a_{21}a_{23})^2 + 4s_{14}(a_{21})^2 a_{22}a_{23} \\
&+ 4s_{15}(a_{21})^3 a_{23} + 4s_{16}(a_{21})^3 a_{22} + s_{22}(a_{22})^4 + 2s_{23}(a_{22}a_{23})^2 \\
&+ 4s_{24}(a_{22})^3 a_{23} + 4s_{25}a_{21}(a_{22})^2 a_{23} + 4s_{26}a_{21}(a_{22})^3 + s_{33}(a_{23})^4 \\
&+ 4s_{34}a_{22}(a_{23})^3 + 4s_{35}a_{21}(a_{23})^3 + 4s_{36}a_{21}a_{22}(a_{23})^2 + 4s_{44}(a_{22}a_{23})^2 \\
&+ 8s_{45}a_{21}a_{22}(a_{23})^2 + 8s_{46}a_{21}(a_{22})^2 a_{23} + 4s_{55}(a_{21}a_{23})^2 + 8s_{56}(a_{21})^2 a_{22}a_{23} \\
&+ 4s_{66}(a_{21}a_{22})^2,
\end{aligned} \tag{9.7}
$$

$$
\begin{aligned}
s'_{23} &= s_{11}(a_{21}a_{31})^2 + s_{12}\left[(a_{21}a_{32})^2+(a_{22}a_{31})^2\right] + s_{13}\left[(a_{21}a_{33})^2+(a_{23}a_{31})^2\right] \\
&+ 2s_{14}\left[(a_{21})^2 a_{32}a_{33} + a_{22}a_{23}(a_{31})^2\right] + 2s_{15}\left[(a_{21})^2 a_{31}a_{33} + a_{21}a_{23}(a_{31})^2\right] \\
&+ 2s_{16}\left[(a_{21})^2 a_{31}a_{32} + a_{21}a_{22}(a_{31})^2\right] + s_{22}(a_{22}a_{32})^2
\end{aligned}
$$

$$+ s_{23}\left[(a_{22}a_{33})^2+(a_{23}a_{32})^2\right]+2s_{24}\left[(a_{22})^2a_{32}a_{33}+a_{22}a_{23}(a_{32})^2\right]$$
$$+ 2s_{25}\left[(a_{22})^2a_{31}a_{33}+a_{21}a_{23}(a_{32})^2\right]+2s_{26}\left[(a_{22})^2a_{31}a_{32}+a_{21}a_{22}(a_{32})^2\right]$$
$$+ s_{33}(a_{23}a_{33})^2+2s_{34}\left[(a_{23})^2a_{32}a_{33}+a_{22}a_{23}(a_{33})^2\right]$$
$$+ 2s_{35}\left[(a_{23})^2a_{31}a_{33}+a_{21}a_{23}(a_{33})^2\right]+2s_{36}\left[(a_{23})^2a_{31}a_{32}+a_{21}a_{22}(a_{33})^2\right]$$
$$+ 4s_{44}a_{22}a_{23}a_{32}a_{33}+4s_{45}a_{23}a_{33}(a_{21}a_{32}+a_{22}a_{31})$$
$$+ 4s_{46}a_{22}a_{32}(a_{21}a_{33}+a_{23}a_{31})+4s_{55}a_{21}a_{23}a_{31}a_{33}$$
$$+4s_{56}a_{21}a_{31}(a_{22}a_{33}+a_{23}a_{32})+4s_{66}a_{21}a_{22}a_{31}a_{32}, \tag{9.8}$$

$$s'_{24}=s_{11}(a_{21})^3a_{31}+s_{12}a_{21}a_{22}(a_{21}a_{32}+a_{22}a_{31})+s_{13}a_{21}a_{23}(a_{21}a_{33}+a_{23}a_{31})$$
$$+ s_{14}a_{21}[a_{21}(a_{22}a_{33}+a_{23}a_{32})+2a_{22}a_{23}a_{31}]+s_{15}(a_{21})^2(a_{21}a_{33}+3a_{23}a_{31})$$
$$+ s_{16}(a_{21}a_{21})^2(a_{21}a_{32}+3a_{22}a_{31})+s_{22}(a_{22})^3a_{32}+s_{23}a_{22}a_{23}(a_{22}a_{33}+a_{23}a_{32})$$
$$+ s_{24}(a_{22})^2(a_{22}a_{33}+3a_{23}a_{32})+s_{25}a_{22}[a_{22}(a_{21}a_{33}+a_{23}a_{31})+2a_{21}a_{23}a_{32}]$$
$$+ s_{26}(a_{22})^2(a_{22}a_{31}+3a_{21}a_{32})+s_{33}(a_{23})^3a_{33}+s_{34}(a_{23})^2(a_{23}a_{32}+3a_{22}a_{33})$$
$$+ s_{35}(a_{23})^2(a_{23}a_{31}+3a_{21}a_{33})+s_{36}a_{23}[a_{23}(a_{21}a_{32}+a_{22}a_{31})+2a_{21}a_{22}a_{33}]$$
$$+ 2s_{44}a_{22}a_{23}(a_{22}a_{33}+a_{23}a_{32})+2s_{45}a_{23}[a_{23}(a_{21}a_{32}+a_{22}a_{31})+2a_{21}a_{22}a_{33}]$$
$$+ 2s_{46}a_{22}[a_{22}(a_{23}a_{31}+a_{21}a_{33})+2a_{21}a_{23}a_{32}]+2s_{55}a_{21}a_{23}(a_{21}a_{33}+a_{23}a_{31})$$
$$+ 2s_{56}a_{21}[a_{21}(a_{23}a_{32}+a_{22}a_{33})+2a_{22}a_{23}a_{31}]+2s_{66}a_{21}a_{22}(a_{21}a_{32}+a_{22}a_{31}), \tag{9.9}$$

$$s'_{25}=s_{11}a_{11}(a_{21})^2a_{31}+s_{12}\left[a_{11}(a_{22})^2a_{31}+a_{12}(a_{21})^2a_{32}\right]$$
$$+ s_{13}\left[a_{13}(a_{21})^2a_{33}+a_{11}(a_{23})^2a_{31}\right]$$
$$+ s_{14}\left[(a_{21})^2(a_{12}a_{33}+a_{13}a_{32})+2a_{11}a_{22}a_{23}a_{31}\right]$$
$$+ s_{15}\left[(a_{21})^2(a_{11}a_{33}+a_{13}a_{31})+2a_{11}a_{21}a_{23}a_{31}\right]$$
$$+ s_{16}\left[(a_{21})^2(a_{11}a_{32}+a_{12}a_{31})+2a_{11}a_{21}a_{22}a_{31}\right]+s_{22}a_{12}(a_{22})^2a_{32}$$
$$+ s_{23}\left[a_{12}(a_{23})^2a_{32}+a_{13}(a_{22})^2a_{33}\right]$$
$$+ s_{24}\left[(a_{22})^2(a_{12}a_{33}+a_{13}a_{32})+2a_{12}a_{22}a_{23}a_{32}\right]$$
$$+ s_{25}\left[(a_{22})^2(a_{11}a_{33}+a_{13}a_{31})+2a_{12}a_{21}a_{23}a_{32}\right]$$
$$+ s_{26}\left[(a_{22})^2(a_{11}a_{32}+a_{12}a_{31})+2a_{12}a_{21}a_{22}a_{32}\right]+s_{33}a_{13}(a_{23})^2a_{33}$$
$$+ s_{34}\left[(a_{23})^2(a_{12}a_{33}+a_{13}a_{32})+2a_{13}a_{22}a_{23}a_{33}\right]$$

$$+ s_{35}\left[(a_{23})^2(a_{11}a_{33}+a_{13}a_{31})+2a_{13}a_{21}a_{23}a_{33}\right]$$
$$+ s_{36}\left[(a_{23})^2(a_{11}a_{32}+a_{12}a_{31})+2a_{13}a_{21}a_{22}a_{33}\right]$$
$$+ 2s_{44}a_{22}a_{23}(a_{12}a_{33}+a_{13}a_{32})$$
$$+ 2s_{45}a_{23}[a_{13}(a_{22}a_{31}+a_{21}a_{32})+a_{33}(a_{11}a_{22}+a_{12}a_{21})]$$
$$+ 2s_{46}[a_{12}a_{22}(a_{23}a_{31}+a_{21}a_{33})+a_{22}a_{32}(a_{11}a_{23}+a_{13}a_{21})]$$
$$+ 2s_{55}a_{21}a_{23}(a_{11}a_{33}+a_{13}a_{31})$$
$$+ 2s_{56}[a_{11}a_{21}(a_{23}a_{32}+a_{22}a_{33})+a_{21}a_{31}(a_{12}a_{23}+a_{13}a_{22})]$$
$$+ 2s_{66}a_{21}a_{22}(a_{11}a_{32}+a_{12}a_{31}), \tag{9.10}$$

$$s'_{26} = s_{11}a_{11}(a_{21})^3+s_{12}a_{21}a_{22}(a_{11}a_{22}+a_{12}a_{21})$$
$$+ s_{13}a_{21}a_{23}(a_{11}a_{23}+a_{13}a_{21})$$
$$+ s_{14}a_{21}[a_{21}(a_{12}a_{23}+a_{13}a_{22})+2a_{11}a_{22}a_{23}]$$
$$+ s_{15}(a_{21})^2(3a_{11}a_{23}+a_{13}a_{21})+s_{16}(a_{21})^2(a_{12}a_{21}+3a_{11}a_{22})$$
$$+ s_{22}a_{22}a_{22}a_{12}a_{22}+s_{23}a_{22}a_{23}(a_{12}a_{23}+a_{13}a_{22})$$
$$+ s_{24}(a_{22})^2(a_{13}a_{22}+3a_{12}a_{23})$$
$$+ s_{25}a_{22}[a_{22}(a_{11}a_{23}+a_{13}a_{21})+2a_{12}a_{21}a_{23}]$$
$$+ s_{26}(a_{22})^2(a_{11}a_{22}+3a_{12}a_{21})+s_{33}(a_{23})^3a_{13}$$
$$+ s_{34}(a_{23})^2(a_{12}a_{23}+3a_{13}a_{22})+s_{35}(a_{23})^2(a_{11}a_{23}+3a_{13}a_{21})$$
$$+ s_{36}a_{23}[a_{23}(a_{11}a_{22}+a_{12}a_{21})+2a_{13}a_{21}a_{22}]$$
$$+ 2s_{44}a_{22}a_{23}(a_{12}a_{23}+a_{13}a_{22})$$
$$+ 2s_{45}a_{23}[a_{23}(a_{11}a_{22}+a_{12}a_{21})+2a_{13}a_{21}a_{22}]$$
$$+ 2s_{46}a_{22}[a_{22}(a_{11}a_{23}+a_{13}a_{21})+2a_{12}a_{21}a_{23}]$$
$$+ 2s_{55}a_{21}a_{23}(a_{11}a_{23}+a_{13}a_{21})$$
$$+ 2s_{56}a_{21}[a_{21}(a_{12}a_{23}+a_{13}a_{22})+2a_{11}a_{22}a_{23}]$$
$$+ 2s_{66}a_{21}a_{22}(a_{11}a_{22}+a_{12}a_{21}), \tag{9.11}$$

$$s'_{33} = s_{11}(a_{31})^4+2s_{12}(a_{31}a_{32})^2+2s_{13}(a_{31}a_{33})^2+4s_{14}(a_{31})^2a_{32}a_{33}$$
$$+ 4s_{15}(a_{31})^3a_{33}+4s_{16}(a_{31})^3a_{32}+s_{22}(a_{32})^4+2s_{23}(a_{32}a_{33})^2$$
$$+ 4s_{24}(a_{32})^3a_{33}+4s_{25}a_{31}(a_{32})^2a_{33}+4s_{26}a_{31}(a_{32})^3+s_{33}(a_{33})^4$$
$$+ 4s_{34}a_{32}(a_{33})^3+4s_{35}a_{31}(a_{33})^3+4s_{36}a_{31}a_{32}(a_{33})^2+4s_{44}(a_{32}a_{33})^2$$
$$+ 8s_{45}a_{31}a_{32}(a_{33})^2+8s_{46}a_{31}(a_{32})^2a_{33}+4s_{55}(a_{31}a_{33})^2$$
$$+ 8s_{56}(a_{31})^2a_{32}a_{33}+4s_{66}(a_{31}a_{32})^2, \tag{9.12}$$

$$
\begin{aligned}
s'_{34} &= s_{11}a_{21}(a_{31})^3 + s_{12}a_{31}a_{32}(a_{21}a_{32} + a_{22}a_{31}) + s_{13}a_{31}a_{33}(a_{21}a_{33} + a_{23}a_{31}) \\
&+ s_{14}a_{31}[a_{31}(a_{22}a_{33} + a_{23}a_{32}) + 2a_{21}a_{32}a_{33}] + s_{15}(a_{31})^2(a_{23}a_{31} + 3a_{21}a_{33}) \\
&+ s_{16}(a_{31})^2(a_{22}a_{31} + 3a_{21}a_{32}) + s_{22}a_{22}(a_{32})^3 + s_{23}a_{32}a_{33}(a_{22}a_{33} + a_{23}a_{32}) \\
&+ s_{24}(a_{32})^2(a_{23}a_{32} + 3a_{22}a_{33}) + s_{25}a_{32}[a_{32}(a_{21}a_{33} + a_{23}a_{31}) + 2a_{22}a_{31}a_{33}] \\
&+ s_{26}(a_{32})^2(a_{21}a_{32} + 3a_{22}a_{31}) + s_{33}a_{23}(a_{33})^3 + s_{34}(a_{33})^2(a_{22}a_{33} + 3a_{23}a_{32}) \\
&+ s_{35}(a_{33})^2(a_{21}a_{33} + 3a_{23}a_{31}) + s_{36}a_{33}[a_{33}(a_{21}a_{32} + a_{22}a_{31}) + 2a_{23}a_{31}a_{32}] \\
&+ 2s_{44}a_{32}a_{33}(a_{22}a_{33} + a_{23}a_{32}) + 2s_{45}a_{33}[a_{33}(a_{21}a_{32} + a_{22}a_{31}) + 2a_{23}a_{31}a_{32}] \\
&+ 2s_{46}a_{32}[a_{32}(a_{21}a_{33} + a_{23}a_{31}) + 2a_{22}a_{31}a_{33}] + 2s_{55}a_{31}a_{33}(a_{21}a_{33} + a_{23}a_{31}) \\
&+ 2s_{56}a_{31}[a_{31}(a_{22}a_{33} + a_{23}a_{32}) + 2a_{21}a_{32}a_{33}] + 2s_{66}a_{31}a_{32}(a_{21}a_{32} + a_{22}a_{31}),
\end{aligned}
\tag{9.13}
$$

$$
\begin{aligned}
s'_{35} &= s_{11}a_{11}(a_{31})^3 + s_{12}a_{31}a_{32}(a_{11}a_{32} + a_{12}a_{31}) + s_{13}a_{31}a_{33}(a_{11}a_{33} + a_{13}a_{31}) \\
&+ s_{14}a_{31}[a_{31}(a_{12}a_{33} + a_{13}a_{32}) + 2a_{11}a_{32}a_{33}] + s_{15}(a_{31})^2(a_{13}a_{31} + 3a_{11}a_{33}) \\
&+ s_{16}(a_{31})^2(a_{12}a_{31} + 3a_{11}a_{32}) + s_{22}a_{12}(a_{32})^3 + s_{23}a_{32}a_{33}(a_{13}a_{32} + a_{12}a_{33}) \\
&+ s_{24}(a_{32})^2(a_{13}a_{32} + 3a_{12}a_{33}) + s_{25}a_{32}[a_{32}(a_{11}a_{33} + a_{13}a_{31}) + 2a_{12}a_{31}a_{33}] \\
&+ s_{26}(a_{32})^2(a_{11}a_{32} + 3a_{12}a_{31}) + s_{33}a_{33}a_{33}a_{13}a_{33} \\
&+ s_{34}(a_{33})^2(a_{12}a_{33} + 3a_{13}a_{32}) + s_{35}(a_{33})^2(a_{11}a_{33} + 3a_{13}a_{31}) \\
&+ s_{36}a_{33}[a_{33}(a_{11}a_{32} + a_{12}a_{31}) + 2a_{13}a_{31}a_{32}] + 2s_{44}a_{32}a_{33}(a_{12}a_{33} + a_{13}a_{32}) \\
&+ 2s_{45}a_{33}[a_{33}(a_{11}a_{32} + a_{12}a_{31}) + 2a_{13}a_{31}a_{32}] \\
&+ 2s_{46}a_{32}[a_{32}(a_{11}a_{33} + a_{13}a_{31}) + 2a_{12}a_{31}a_{33}] + 2s_{55}a_{31}a_{33}(a_{11}a_{33} + a_{13}a_{31}) \\
&+ 2s_{56}a_{31}[a_{31}(a_{12}a_{33} + a_{13}a_{32}) + 2a_{11}a_{32}a_{33}] + 2s_{66}a_{31}a_{32}(a_{11}a_{32} + a_{12}a_{31}),
\end{aligned}
\tag{9.14}
$$

$$
\begin{aligned}
s'_{36} &= s_{11}a_{11}a_{21}(a_{31})^2 + s_{12}\left[a_{11}a_{21}(a_{32})^2 + a_{12}a_{22}(a_{31})^2\right] \\
&+ s_{13}\left[a_{11}a_{21}(a_{33})^2 + a_{13}a_{23}(a_{31})^2\right] \\
&+ s_{14}\left[(a_{31})^2(a_{12}a_{23} + a_{13}a_{22}) + 2a_{11}a_{21}a_{32}a_{33}\right] \\
&+ s_{15}a_{31}[a_{31}(a_{11}a_{23} + a_{13}a_{21}) + 2a_{11}a_{21}a_{33}] \\
&+ s_{16}a_{31}[a_{31}(a_{11}a_{22} + a_{12}a_{21}) + 2a_{11}a_{21}a_{32}] + s_{22}a_{12}a_{22}(a_{32})^2 \\
&+ s_{23}\left[a_{12}a_{22}(a_{33})^2 + a_{13}a_{23}(a_{32})^2\right] \\
&+ s_{24}a_{32}[a_{32}(a_{12}a_{23} + a_{13}a_{22}) + 2a_{12}a_{22}a_{33}] \\
&+ s_{25}\left[(a_{32})^2(a_{11}a_{23} + a_{13}a_{21}) + 2a_{12}a_{22}a_{31}a_{33}\right] \\
&+ s_{26}a_{32}[a_{32}(a_{11}a_{22} + a_{12}a_{21}) + 2a_{12}a_{22}a_{31}] + s_{33}a_{13}a_{23}(a_{33})^2 \\
&+ s_{34}a_{33}[a_{33}(a_{12}a_{23} + a_{13}a_{22}) + 2a_{13}a_{23}a_{32}] \\
&+ s_{35}a_{33}[a_{33}(a_{11}a_{23} + a_{13}a_{21}) + 2a_{13}a_{23}a_{31}] \\
&+ s_{36}\left[(a_{33})^2(a_{11}a_{22} + a_{12}a_{21}) + 2a_{13}a_{23}a_{31}a_{32}\right]
\end{aligned}
$$

$$
\begin{aligned}
&+ s_{44}a_{33}\left[a_{32}\left(a_{12}a_{23}+a_{12}a_{23}\right)+2a_{13}a_{22}a_{32}\right] \\
&+ 2s_{45}a_{33}\left[a_{31}\left(a_{12}a_{23}+a_{13}a_{22}\right)+a_{32}\left(a_{11}a_{23}+a_{13}a_{21}\right)\right] \\
&+ 2s_{46}a_{32}\left[a_{31}\left(a_{12}a_{23}+a_{13}a_{22}\right)+a_{33}\left(a_{11}a_{22}+a_{12}a_{21}\right)\right] \\
&+ 2s_{55}a_{31}a_{33}\left(a_{11}a_{23}+a_{13}a_{21}\right) \\
&+ 2s_{56}a_{31}\left[a_{32}\left(a_{11}a_{23}+a_{13}a_{21}\right)+a_{33}\left(a_{11}a_{22}+a_{12}a_{21}\right)\right] \\
&+ 2s_{66}a_{31}a_{32}\left(a_{11}a_{22}+a_{12}a_{21}\right),
\end{aligned} \tag{9.15}
$$

$$
\begin{aligned}
s'_{44} =\ & s_{11}\left(a_{21}a_{31}\right)^2+2s_{12}a_{21}a_{22}a_{31}a_{32}+2s_{13}a_{21}a_{23}a_{31}a_{33} \\
&+ 2s_{14}a_{21}a_{31}\left(a_{22}a_{33}+a_{23}a_{32}\right)+2s_{15}a_{21}a_{31}\left(a_{21}a_{33}+a_{23}a_{31}\right) \\
&+ 2s_{16}a_{21}a_{31}\left(a_{21}a_{32}+a_{22}a_{31}\right)+s_{22}\left(a_{22}a_{32}\right)^2+2s_{23}a_{22}a_{23}a_{32}a_{33} \\
&+ 2s_{24}a_{22}a_{32}\left(a_{22}a_{33}+a_{23}a_{32}\right)+2s_{25}a_{22}a_{32}\left(a_{21}a_{33}+a_{23}a_{31}\right) \\
&+ 2s_{26}a_{22}a_{32}\left(a_{21}a_{32}+a_{22}a_{31}\right)+s_{33}\left(a_{23}a_{33}\right)^2+2s_{34}a_{23}a_{33}\left(a_{22}a_{33}+a_{23}a_{32}\right) \\
&+ 2s_{35}a_{23}a_{33}\left(a_{21}a_{33}+a_{23}a_{31}\right)+2s_{36}a_{23}a_{33}\left(a_{21}a_{32}+a_{22}a_{31}\right) \\
&+ s_{44}\left(a_{22}a_{33}+a_{23}a_{32}\right)^2 \\
&+ 2s_{45}\left[\left(a_{23}\right)^2a_{31}a_{32}+a_{21}a_{22}\left(a_{33}\right)^2+a_{23}a_{33}\left(a_{21}a_{32}+a_{22}a_{31}\right)\right] \\
&+ 2s_{46}\left[a_{21}a_{23}\left(a_{32}\right)^2+\left(a_{22}\right)^2a_{31}a_{33}+a_{22}a_{32}\left(a_{21}a_{33}+a_{23}a_{31}\right)\right] \\
&+ s_{55}\left(a_{21}a_{33}+a_{23}a_{31}\right)^2 \\
&+ 2s_{56}\left[\left(a_{21}\right)^2a_{32}a_{33}+a_{22}a_{23}\left(a_{31}\right)^2+a_{21}a_{31}\left(a_{23}a_{32}+a_{22}a_{33}\right)\right] \\
&+ s_{66}\left(a_{21}a_{32}+a_{22}a_{31}\right)^2,
\end{aligned} \tag{9.16}
$$

$$
\begin{aligned}
s'_{45} =\ & s_{11}a_{11}a_{21}\left(a_{31}\right)^2+s_{12}a_{31}a_{32}\left(a_{11}a_{22}+a_{12}a_{21}\right) \\
&+ s_{13}a_{31}a_{33}\left(a_{13}a_{21}+a_{11}a_{23}\right) \\
&+ s_{14}a_{31}\left[a_{32}\left(a_{13}a_{21}+a_{11}a_{23}\right)+a_{33}\left(a_{12}a_{21}+a_{11}a_{22}\right)\right] \\
&+ s_{15}a_{31}\left[a_{31}\left(a_{13}a_{21}+a_{11}a_{23}\right)+a_{33}\left(a_{11}a_{21}+a_{11}a_{21}\right)\right] \\
&+ s_{16}a_{31}\left[a_{31}\left(a_{12}a_{21}+a_{11}a_{22}\right)+2a_{11}a_{21}a_{32}\right]+s_{22}a_{12}a_{22}\left(a_{32}\right)^2 \\
&+ s_{23}a_{32}a_{33}\left(a_{13}a_{22}+a_{12}a_{23}\right) \\
&+ s_{24}a_{32}\left[a_{32}\left(a_{12}a_{23}+a_{13}a_{22}\right)+2a_{12}a_{22}a_{33}\right] \\
&+ s_{25}a_{32}\left[a_{12}\left(a_{21}a_{33}+a_{23}a_{31}\right)+a_{22}\left(a_{11}a_{33}+a_{13}a_{31}\right)\right] \\
&+ s_{26}a_{32}\left[a_{32}\left(a_{11}a_{22}+a_{12}a_{21}\right)+2a_{12}a_{22}a_{31}\right]+s_{33}a_{13}a_{23}\left(a_{33}\right)^2 \\
&+ s_{34}a_{33}\left[a_{33}\left(a_{12}a_{23}+a_{13}a_{22}\right)+2a_{13}a_{23}a_{32}\right] \\
&+ s_{35}a_{33}\left[a_{33}\left(a_{11}a_{23}+a_{13}a_{21}\right)+2a_{13}a_{23}a_{31}\right] \\
&+ s_{36}a_{33}\left[a_{13}\left(a_{21}a_{32}+a_{22}a_{31}\right)+a_{23}\left(a_{11}a_{32}+a_{12}a_{31}\right)\right] \\
&+ s_{44}\left[a_{12}a_{22}\left(a_{33}\right)^2+a_{13}a_{23}\left(a_{32}\right)^2+a_{32}a_{33}\left(a_{12}a_{23}+a_{13}a_{22}\right)\right] \\
&+ s_{45}\left[\left(a_{33}\right)^2\left(a_{11}a_{22}+a_{12}a_{21}\right)+a_{31}a_{33}\left(a_{12}a_{23}+a_{13}a_{22}\right)\right.
\end{aligned}
$$

$$+ a_{32}a_{33}(a_{11}a_{23}+a_{13}a_{21})+2a_{13}a_{23}a_{31}a_{32}]$$
$$+ s_{46}\left[(a_{32})^2(a_{11}a_{23}+a_{13}a_{21})+a_{12}a_{32}(a_{23}a_{31}+a_{21}a_{33})\right.$$
$$+ a_{22}a_{32}(a_{11}a_{33}+a_{13}a_{31})+2a_{12}a_{22}a_{31}a_{33}]$$
$$+ s_{55}[a_{11}a_{33}(a_{21}a_{33}+a_{23}a_{31})+a_{13}a_{31}(a_{21}a_{33}+a_{23}a_{31})]$$
$$+ s_{56}\left[(a_{31})^2(a_{12}a_{23}+a_{13}a_{22})+a_{11}a_{31}(a_{22}a_{33}+a_{23}a_{32})\right.$$
$$+ a_{21}a_{31}(a_{12}a_{33}+a_{13}a_{32})+2a_{11}a_{21}a_{32}a_{33}]$$
$$+ s_{66}\left[a_{11}a_{21}(a_{32})^2+a_{12}a_{22}(a_{31})^2+a_{31}a_{32}(a_{12}a_{21}+a_{11}a_{22})\right], \quad (9.17)$$

$$s'_{46} = s_{11}a_{11}(a_{21}a_{21})^2a_{31}+s_{12}a_{21}a_{22}(a_{12}a_{31}+a_{11}a_{32})$$
$$+ s_{13}a_{21}a_{23}(a_{13}a_{31}+a_{11}a_{33})$$
$$+ s_{14}a_{21}[a_{11}(a_{22}a_{33}+a_{23}a_{32})+a_{31}(a_{12}a_{23}+a_{13}a_{22})]$$
$$+ s_{15}a_{21}[a_{21}(a_{13}a_{31}+a_{11}a_{33})+2a_{11}a_{23}a_{31}]$$
$$+ s_{16}a_{21}[a_{21}(a_{11}a_{32}+a_{12}a_{31})+2a_{11}a_{22}a_{31}]+s_{22}a_{12}(a_{22})^2a_{32}$$
$$+ s_{23}a_{22}a_{23}(a_{13}a_{32}+a_{12}a_{33})$$
$$+ s_{24}a_{22}[a_{22}(a_{13}a_{32}+a_{12}a_{33})+2a_{12}a_{23}a_{32}]$$
$$+ s_{25}a_{22}[a_{21}(a_{13}a_{32}+a_{12}a_{33})+a_{23}(a_{11}a_{32}+a_{12}a_{31})]$$
$$+ s_{26}a_{22}[a_{22}(a_{11}a_{32}+a_{12}a_{31})+2a_{12}a_{21}a_{32}]+s_{33}a_{13}(a_{23})^2a_{33}$$
$$+ s_{34}a_{23}[a_{23}(a_{12}a_{33}+a_{13}a_{32})+2a_{13}a_{22}a_{33}]$$
$$+ s_{35}a_{23}[a_{23}(a_{13}a_{31}+a_{11}a_{33})+2a_{13}a_{21}a_{33}]$$
$$+ s_{36}a_{23}[a_{13}(a_{21}a_{32}+a_{22}a_{31})+a_{33}(a_{11}a_{22}+a_{12}a_{21})]$$
$$+ s_{44}[a_{22}a_{33}(a_{12}a_{23}+a_{13}a_{22})+a_{23}a_{32}(a_{12}a_{23}+a_{13}a_{22})]$$
$$+ s_{45}\left[(a_{23})^2(a_{11}a_{32}+a_{12}a_{31})+a_{23}a_{33}(a_{11}a_{22}+a_{12}a_{21})\right.$$
$$+ a_{13}a_{23}(a_{22}a_{31}+a_{21}a_{32})+2a_{13}a_{21}a_{22}a_{33}]$$
$$+ s_{46}\left[(a_{22})^2(a_{11}a_{33}+a_{13}a_{31})+a_{12}a_{22}(a_{21}a_{33}+a_{23}a_{31})\right.$$
$$+ a_{22}a_{32}(a_{11}a_{23}+a_{13}a_{21})+2a_{12}a_{21}a_{23}a_{32}]$$
$$+ s_{55}\left[a_{11}(a_{23})^2a_{31}+a_{13}(a_{21})^2a_{33}+a_{21}a_{23}(a_{11}a_{33}+a_{13}a_{31})\right]$$
$$+ s_{56}\left[(a_{21})^2(a_{12}a_{33}+a_{13}a_{32})+a_{11}a_{21}(a_{22}a_{33}+a_{23}a_{32})\right.$$
$$+ a_{21}a_{31}(a_{12}a_{23}+a_{13}a_{22})+2a_{11}a_{22}a_{23}a_{31}]$$
$$+ s_{66}\left[a_{11}(a_{22})^2a_{31}+a_{12}(a_{21})^2a_{32}+a_{21}a_{22}(a_{11}a_{32}+a_{12}a_{31})\right], \quad (9.18)$$

$$s'_{55} = s_{11}(a_{11}a_{31})^2+2s_{12}a_{11}a_{12}a_{31}a_{32}+2s_{13}a_{11}a_{13}a_{31}a_{33}$$
$$+ 2s_{14}a_{11}a_{31}(a_{12}a_{33}+a_{13}a_{32})+2s_{15}a_{11}a_{31}(a_{11}a_{33}+a_{13}a_{31})$$
$$+ 2s_{16}a_{11}a_{31}(a_{11}a_{32}+a_{12}a_{31})+s_{22}(a_{12}a_{32})^2+2s_{23}a_{12}a_{13}a_{32}a_{33}$$

$$+ 2s_{24}a_{12}a_{32}(a_{12}a_{33}+a_{13}a_{32}) + 2s_{25}a_{12}a_{32}(a_{11}a_{33}+a_{13}a_{31})$$
$$+ 2s_{26}a_{12}a_{32}(a_{11}a_{32}+a_{12}a_{31}) + s_{33}(a_{13}a_{33})^2 + 2s_{34}a_{13}a_{33}(a_{12}a_{33}+a_{13}a_{32})$$
$$+ 2s_{35}a_{13}a_{33}(a_{11}a_{33}+a_{13}a_{31}a_{33}) + 2s_{36}a_{13}a_{33}(a_{11}a_{32}+a_{12}a_{31})$$
$$+ s_{44}(a_{12}a_{33}+a_{13}a_{32})^2$$
$$+ 2s_{45}\left[a_{11}a_{12}(a_{33})^2 + (a_{13})^2 a_{31}a_{32} + a_{13}a_{33}(a_{11}a_{32}+a_{12}a_{31})\right]$$
$$+ 2s_{46}\left[a_{11}a_{13}(a_{32})^2 + (a_{12})^2 a_{31}a_{33} + a_{12}a_{32}(a_{11}a_{33}+a_{13}a_{31})\right]$$
$$+ s_{55}(a_{11}a_{33}+a_{13}a_{31})^2$$
$$+ 2s_{56}\left[(a_{11})^2 a_{32}a_{33} + a_{12}a_{13}(a_{31})^2 + a_{11}a_{31}(a_{13}a_{32}+a_{12}a_{33})\right]$$
$$+ s_{66}(a_{11}a_{32}+a_{12}a_{31})^2, \tag{9.19}$$

$$s'_{56} = s_{11}(a_{11})^2 a_{21}a_{31} + s_{12}a_{11}a_{12}(a_{21}a_{32}+a_{22}a_{31})$$
$$+ s_{13}a_{11}a_{13}(a_{21}a_{33}+a_{23}a_{31})$$
$$+ s_{14}a_{11}[a_{12}(a_{21}a_{33}+a_{23}a_{31}) + a_{13}(a_{21}a_{32}+a_{22}a_{31})]$$
$$+ s_{15}a_{11}[a_{11}(a_{23}a_{31}+a_{21}a_{33}) + 2a_{13}a_{21}a_{31}]$$
$$+ s_{16}a_{11}[a_{11}(a_{21}a_{32}+a_{22}a_{31}) + 2a_{12}a_{21}a_{31}] + s_{22}(a_{12})^2 a_{22}a_{32}$$
$$+ s_{23}a_{12}a_{13}(a_{22}a_{33}+a_{23}a_{32}) + s_{24}a_{12}[a_{12}(a_{22}a_{33}+a_{23}a_{32}) + 2a_{13}a_{22}a_{32}]$$
$$+ s_{25}a_{12}[a_{11}(a_{23}a_{32}+a_{22}a_{33}) + a_{13}(a_{21}a_{32}+a_{22}a_{31})]$$
$$+ s_{26}a_{12}[a_{11}(a_{22}a_{32}+a_{22}a_{32}) + a_{12}(a_{21}a_{32}+a_{22}a_{31})] + s_{33}(a_{13})^2 a_{23}a_{33}$$
$$+ s_{34}a_{13}[a_{12}(a_{23}a_{33}+a_{23}a_{33}) + a_{13}(a_{22}a_{33}+a_{23}a_{32})]$$
$$+ s_{35}a_{13}[a_{13}(a_{21}a_{33}+a_{23}a_{31}) + 2a_{11}a_{23}a_{33}]$$
$$+ s_{36}a_{13}[a_{11}(a_{22}a_{33}+a_{23}a_{32}) + a_{12}(a_{21}a_{33}+a_{23}a_{31})]$$
$$+ s_{44}\left[(a_{12})^2 a_{23}a_{33} + (a_{13})^2 a_{22}a_{32} + a_{12}a_{13}(a_{22}a_{33}+a_{23}a_{32})\right]$$
$$+ s_{45}\left[(a_{13})^2 (a_{21}a_{32}+a_{22}a_{31}) + a_{11}a_{13}(a_{22}a_{33}+a_{23}a_{32})\right.$$
$$\left. + a_{12}a_{13}(a_{21}a_{33}+a_{23}a_{31}) + 2a_{11}a_{12}a_{23}a_{33}\right]$$
$$+ s_{46}\left[(a_{12})^2 (a_{21}a_{33}+a_{23}a_{31}) + a_{11}a_{12}(a_{22}a_{33}+a_{23}a_{32})\right.$$
$$\left. + a_{12}a_{13}(a_{21}a_{32}+a_{22}a_{31}) + 2a_{11}a_{13}a_{22}a_{32}\right]$$
$$+ s_{55}\left[(a_{11})^2 a_{23}a_{33} + (a_{13})^2 a_{21}a_{31} + a_{11}a_{13}(a_{21}a_{33}+a_{23}a_{31})\right]$$
$$+ s_{56}\left[(a_{11})^2 (a_{22}a_{33}+a_{23}a_{32}) + a_{11}a_{12}(a_{21}a_{33}+a_{23}a_{31})\right.$$
$$\left. + a_{11}a_{13}(a_{22}a_{31}+a_{21}a_{32}) + 2a_{12}a_{13}a_{21}a_{31}\right]$$
$$+ s_{66}\left[(a_{11})^2 a_{22}a_{32} + (a_{12})^2 a_{21}a_{31} + a_{11}a_{12}(a_{21}a_{32}+a_{22}a_{31})\right], \tag{9.20}$$

$$s'_{66} = s_{11}(a_{11}a_{21})^2 + 2s_{12}a_{11}a_{12}a_{21}a_{22} + 2s_{13}a_{11}a_{13}a_{21}a_{23}$$
$$+ 2s_{14}a_{11}a_{21}(a_{12}a_{23}+a_{13}a_{22}) + 2s_{15}a_{11}a_{21}(a_{11}a_{23}+a_{13}a_{21})$$
$$+ 2s_{16}a_{11}a_{21}(a_{11}a_{22}+a_{12}a_{21}) + s_{22}(a_{12}a_{22})^2$$

$$
\begin{aligned}
&+2s_{23}a_{12}a_{13}a_{22}a_{23}+2s_{24}a_{12}a_{22}\left(a_{12}a_{23}+a_{13}a_{22}\right)\\
&+2s_{25}a_{12}a_{22}\left(a_{11}a_{23}+a_{13}a_{21}\right)+2s_{26}a_{12}a_{22}\left(a_{11}a_{22}+a_{12}a_{21}\right)\\
&+s_{33}\left(a_{13}a_{23}\right)^2+2s_{34}a_{13}a_{23}\left(a_{12}a_{23}+a_{13}a_{22}\right)\\
&+2s_{35}a_{13}a_{23}\left(a_{11}a_{23}+a_{13}a_{21}\right)+2s_{36}a_{13}a_{23}\left(a_{11}a_{22}+a_{12}a_{21}\right)\\
&+s_{44}\left(a_{12}a_{23}+a_{13}a_{22}\right)^2\\
&+2s_{45}\left[a_{11}a_{12}\left(a_{23}\right)^2+\left(a_{13}\right)^2a_{21}a_{22}+a_{13}a_{23}\left(a_{11}a_{22}+a_{12}a_{21}\right)\right]\\
&+2s_{46}\left[\left(a_{12}\right)^2a_{21}a_{23}+a_{11}a_{13}\left(a_{22}\right)^2+a_{12}a_{22}\left(a_{11}a_{23}+a_{13}a_{21}\right)\right]\\
&+s_{55}\left(a_{11}a_{23}+a_{13}a_{21}\right)^2\\
&+2s_{56}\left[\left(a_{11}\right)^2a_{22}a_{23}+a_{12}a_{13}\left(a_{21}\right)^2+a_{11}a_{21}\left(a_{12}a_{23}+a_{13}a_{22}\right)\right]\\
&+s_{66}\left(a_{11}a_{22}+a_{12}a_{21}\right)^2.
\end{aligned}
\tag{9.21}
$$

The subscript q is related to the spherical particle for $q=p$, the spherical envelope for $q=e$ and the cell matrix for $q=m$. As presented at the end of Section 3.3.5., the transformations $s'_{ij}(\varphi,\nu)\to s'_{ijq}(\varphi,\nu)$, $s_{ij}(\varphi,\nu)\to s_{ijq}(\varphi,\nu)$ (see Eq. (3.41)) are also valid for the coefficient $c_i=c_i(\varphi,\nu)$ $(i=1,\ \ldots,18)$ (see Eqs. (9.22)–(9.27)) as well as for the coefficient $c_j=c_j(\varphi,\nu)$ $(j=19,\ldots,8143)$, i.e., $c_i(\varphi,\nu)\to c_{iq}(\varphi,\nu)$, $c_j(\varphi,\nu)\to c_{jq}(\varphi,\nu)$ (see Eqs. (9.28)–(9.146)).

The coefficient $c_i=c_i(\varphi,\nu)$ $(i=19,\ldots,8143)$, i.e., $c_{iq}=c_{iq}(\varphi,\nu)$ $(q=p,e,m)$ is derived as

$$
c_i=-\frac{s'_{46}\left(s'_{i2}-s'_{i3}\right)-s'_{i4}\left(s'_{26}-s'_{36}\right)}{s'_{45}\left(s'_{36}-s'_{26}\right)+s'_{46}\left(s'_{25}-s'_{35}\right)},\quad i=1-3,
\tag{9.22}
$$

$$
c_{3+i}=\frac{s'_{45}\left(s'_{i2}-s'_{i3}\right)-s'_{i4}\left(s'_{25}-s'_{35}\right)}{s'_{45}\left(s'_{36}-s'_{26}\right)+s'_{46}\left(s'_{25}-s'_{35}\right)},\quad i=1-3,
\tag{9.23}
$$

$$
c_{6+i}=s'_{1i}+c_is'_{15}+c_{3+i}s'_{16},\quad i=1-3,
\tag{9.24}
$$

$$
c_{9+i}=s'_{2i}+c_is'_{25}+c_{3+i}s'_{26},\quad i=1-3,
\tag{9.25}
$$

$$
c_{12+i}=s'_{5i}+c_is'_{55}+c_{3+i}s'_{56},\quad i=1-3,
\tag{9.26}
$$

$$
c_{15+i}=s'_{6i}+c_is'_{56}+c_{3+i}s'_{66},\quad i=1-3,
\tag{9.27}
$$

$$
\begin{aligned}
&c_{18+i}=\frac{1}{c_7c_{12}-c_9c_{10}}\\
&\times\left[2c_{3+i}\left(c_8c_{12}-c_{11}c_9\right)-c_{12}\frac{\partial c_{6+i}}{\partial\varphi}+c_9\frac{\partial c_{9+i}}{\partial\varphi}+c_{15+i}\left(c_{12}-c_9\right)\right],\\
&i=1-3,
\end{aligned}
\tag{9.28}
$$

$$c_{21+i} = \frac{c_{3+i}(c_8c_{12} - c_{11}c_9) + c_{15+i}c_{12}}{c_7c_{12} - c_9c_{10}}, \quad i = 1-3, \tag{9.29}$$

$$\begin{aligned} c_{24+i} &= \frac{1}{c_7c_{12} - c_9c_{10}} \\ &\times \left[2c_{3+i}(c_7c_{11} - c_8c_{10}) + c_{10}\frac{\partial c_{6+i}}{\partial \varphi} - c_7\frac{\partial c_{9+i}}{\partial \varphi} + c_{15+i}(c_7 - c_{10})\right], \\ i &= 1-3, \end{aligned} \tag{9.30}$$

$$c_{27+i} = \frac{c_{3+i}(c_7c_{11} - c_8c_{10}) - c_{15+i}c_{10}}{c_7c_{12} - c_9c_{10}}, \quad i = 1-3, \tag{9.31}$$

$$\begin{aligned} c_{30+i} &= \frac{1}{c_7c_{12} - c_9c_{10}} \\ &\times \left[2c_i(c_9c_{11} - c_8c_{12}) - c_{11}\frac{\partial c_{6+i}}{\partial \nu} + c_8\frac{\partial c_{9+i}}{\partial \nu} + c_{12+i}(c_{11} - c_8)\right], \\ i &= 1-3, \end{aligned} \tag{9.32}$$

$$c_{33+i} = \frac{c_i(c_9c_{11} - c_8c_{12}) + c_{12+i}c_{11}}{c_7c_{12} - c_9c_{10}}, \quad i = 1-3, \tag{9.33}$$

$$\begin{aligned} c_{36+i} &= \frac{1}{c_7c_{12} - c_9c_{10}} \\ &\times \left[2c_i(c_7c_{12} - c_9c_{10}) + c_{10}\frac{\partial c_{6+i}}{\partial \nu} - c_7\frac{\partial c_{9+i}}{\partial \nu} + c_{12+i}(c_7 - c_{10})\right], \\ i &= 1-3, \end{aligned} \tag{9.34}$$

$$c_{39+i} = \frac{c_i(c_7c_{12} - c_9c_{10}) - c_{12+i}c_{10}}{c_7c_{12} - c_9c_{10}}, \quad i = 1-3, \tag{9.35}$$

$$\begin{aligned} c_{42+i+3j} &= c_4c_{18+i+3j} + c_6c_{24+i+3j} + c_1c_{30+i+3j} + c_2c_{36+i+3j} \\ &- (2\delta_{0j} + \delta_{1j})(c_5c_{3+i} + c_3c_i - \delta_{1i}) + \delta_{0j}\left[\frac{\partial c_{3+i}}{\partial \varphi} + \frac{\partial c_i}{\partial \nu} - (\delta_{2i} + \delta_{3i})\right], \\ i &= 1-3; \quad j = 0, 1, \end{aligned} \tag{9.36}$$

$$\begin{aligned} c_{48+i} &= -2c_{18}c_i + \frac{\partial c_{15+i}}{\partial \nu} + c_{16}c_{30+i} + c_{17}c_{36+i} \\ &- \frac{\partial}{\partial \varphi}\left(-2c_9c_i\frac{\partial c_{6+i}}{\partial \nu} + c_1c_{30+i} + c_8c_{36+i}\right) \\ &- \sum_{j=1}^{3}\left(-2c_9c_j + \frac{\partial c_{6+j}}{\partial \nu} + c_7c_{30+j} + c_8c_{36+j}\right)(-2c_{3+i}\delta_{2j} + c_{18+i}\delta_{1j} + c_{24+i}\delta_{3j}), \\ i &= 1-3, \end{aligned} \tag{9.37}$$

$$
\begin{aligned}
c_{51+i} = {} & -4c_{18}c_i\frac{\partial c_{15+i}}{\partial \nu} + c_{16}\left(c_{30+i} + 2c_{33+i}\right) + c_{17}\left(c_{36+i} + 2c_{39+i}\right) \\
& - \frac{\partial}{\partial \varphi}\left(-c_9c_ic_7c_{33+i} + c_8c_{39+i}\right) \\
& - \sum_{j=1}^{3}\left(\frac{\partial c_{6+j}}{\partial \nu} + -2c_9c_j + c_7c_{30+j} + c_8c_{36+j}\right)\left(c_{21+i}\delta_{1j} - c_{3+i}\delta_{2j} + c_{27+i}\delta_{3j}\right) \\
& - \sum_{j=1}^{3}\left[\left(c_{18+i} + c_{21+i}\right)\delta_{1j} - 3c_{3+i}\delta_{2j} + \left(c_{24+i} + c_{27+i}\right)\delta_{3j}\right] \\
& \quad \times\left(-c_9c_j + c_7c_{33+j} + c_8c_{39+j}\right), \quad i = 1-3,
\end{aligned} \tag{9.38}
$$

$$
\begin{aligned}
c_{54+i} = {} & -c_{18}c_i + c_{16}c_{33+i} + c_{17}c_{39+i} \\
& - \sum_{j=1}^{3}\left(-c_9c_j + c_7c_{33+j} + c_8c_{39+j}\right)\left(c_{21+i}\delta_{1j} - c_{3+i}\delta_{2j} + c_{27+i}\delta_{3j}\right), \quad i = 1-3,
\end{aligned} \tag{9.39}
$$

$$
\begin{aligned}
c_{57+i} = {} & -2c_{14}c_{3+i} + \frac{\partial c_{12+i}}{\partial \varphi} + c_{13}c_{18+i} + c_{15}c_{24+i} \\
& - \frac{\partial}{\partial \nu}\left(-2c_8c_{3+i} + \frac{\partial c_{6+i}}{\partial \varphi} + c_7c_{18+i} + c_9c_{24+i}\right) \\
& - \sum_{j=1}^{3}\left(-2c_8c_{3+j} + \frac{\partial c_{6+j}}{\partial \varphi} + c_7c_{18+j} + c_9c_{24+j}\right)\left(c_{30+i}\delta_{1j} + c_{36+i}\delta_{2j} - 2c_i\delta_{3j}\right), \\
& i = 1-3,
\end{aligned} \tag{9.40}
$$

$$
\begin{aligned}
c_{60+i} = {} & -4c_{14}c_{3+i} + \frac{\partial c_{12+i}}{\partial \varphi} + c_{13}\left(c_{18+i} + 2c_{21+i}\right) + c_{15}\left(c_{24+i} + 2c_{27+i}\right) \\
& - \frac{\partial}{\partial \nu}\left(-c_8c_{3+i} + c_7c_{21+i} + c_9c_{27+i}\right) \\
& - \sum_{j=1}^{3}\left(-2c_8c_{3+j} + \frac{\partial c_{6+j}}{\partial \varphi} + c_7c_{18+j} + c_9c_{24+j}\right)\left(c_{33+i}\delta_{1j} + c_{39+i}\delta_{2j} - c_i\delta_{3j}\right) \\
& - \sum_{j=1}^{3}\left[\left(c_{30+i} + c_{33+i}\right)\delta_{1j} + \left(c_{36+i} + c_{39+i}\right)\delta_{2j} - 3c_i\delta_{3j}\right] \\
& \quad \times\left(-c_8c_{3+j} + c_7c_{21+j} + c_9c_{27+j}\right), \quad i = 1-3,
\end{aligned} \tag{9.41}
$$

$$c_{63+i} = -c_{14}c_{3+i} + c_{13}c_{21+i} + c_{15}c_{27+i}$$
$$- \sum_{j=1}^{3} \left(-c_8 c_{3+j} + c_7 c_{21+j} + c_9 c_{27+j}\right)\left(c_{33+i}\delta_{1j} + c_{39+i}\delta_{2j} - c_i\delta_{3j}\right), \quad i = 1-3, \tag{9.42}$$

$$c_{66+i} = -2c_{18}c_i + \frac{\partial c_{15+i}}{\partial \nu} + c_{16}c_{30+i} + c_{17}c_{36+i}$$
$$- \frac{\partial}{\partial \varphi}\left(-2c_{12}c_i + \frac{\partial c_{9+i}}{\partial \nu} + c_{10}c_{30+i} + c_{11}c_{36+i}\right)$$
$$- \sum_{j=1}^{3}\left(-2c_{12}c_j + \frac{\partial c_{9+j}}{\partial \nu} + +c_{10}c_{30+j} + c_{11}c_{36+j}\right)$$
$$\times \left(c_{18+i}\delta_{1j} - 2c_{3+i}\delta_{2j} + c_{24+i}\delta_{3j}\right), \quad i = 1-3, \tag{9.43}$$

$$c_{69+i} = -c_{18}c_i + c_{16}c_{33+i} + c_{17}c_{39+i} - \frac{\partial}{\partial \varphi}\left(-c_{12}c_i + c_{10}c_{33+i} + c_{11}c_{39+i}\right)$$
$$- \sum_{j=1}^{3}\left(-2c_{12}c_j + \frac{\partial c_{9+j}}{\partial \nu} + c_{10}c_{30+j} + c_{11}c_{36+j}\right)\left(c_{21+i}\delta_{1j} - c_{3+i}\delta_{2j} + c_{27+i}\delta_{3j}\right)$$
$$- \sum_{j=1}^{3}\left[\left(c_{18+i} + c_{21+i}\right)\delta_{1j} - 3c_{3+i}\delta_{2j} + \left(c_{24+i} + c_{27+i}\right)\delta_{3j}\right]$$
$$\times \left(-c_{12}c_j + c_{10}c_{33+j} + c_{11}c_{39+j}\right), \quad i = 1-3, \tag{9.44}$$

$$c_{72+i} = -\sum_{j=1}^{3}\left(-c_{12}c_j + c_{10}c_{33+j} + c_{11}c_{39+j}\right)\left(c_{21+i}\delta_{1j} - c_{3+i}\delta_{2j} + c_{27+i}\delta_{3j}\right),$$
$$i = 1-3, \tag{9.45}$$

$$c_{75+i} = -2c_{14}c_{3+i} + \frac{\partial c_{12+i}}{\partial \varphi} + c_{13}c_{18+i} + c_{15}c_{24+i}$$
$$- \frac{\partial}{\partial \nu}\left(-2c_{11}c_{3+i} + \frac{\partial c_{9+i}}{\partial \varphi} + c_{10}c_{18+i} + c_{12}c_{24+i}\right)$$
$$- \sum_{j=1}^{3}\left(-2c_{11}c_{3+j} + \frac{\partial c_{9+j}}{\partial \varphi} + c_{10}c_{18+j} + c_{12}c_{24+j}\right)$$
$$\times \left(c_{30+i}\delta_{1j} + c_{36+i}\delta_{2j} - 2c_i\delta_{3j}\right), \quad i = 1-3, \tag{9.46}$$

$$c_{78+i} = -c_{14}c_{3+i} + c_{13}c_{21+i} + c_{15}c_{27+i} - \frac{\partial}{\partial \nu}\left(-c_{11}c_{3+i} + c_{10}c_{21+i} + c_{12}c_{27+i}\right)$$

$$-\sum_{j=1}^{3}\left(-2c_{11}c_{3+j}+\frac{\partial c_{9+j}}{\partial \varphi}+c_{10}c_{18+j}+c_{12}c_{24+j}\right)(c_{33+i}\delta_{1j}+c_{39+i}\delta_{2j}-c_i\delta_{3j})$$
$$-\sum_{j=1}^{3}\left[(c_{30+i}+c_{33+i})\delta_{1j}+(c_{36+i}+c_{39+i})\delta_{2j}-3c_i\delta_{3j}\right]$$
$$\times(-c_{11}c_{3+j}+c_{10}c_{21+j}+c_{12}c_{27+j}),\quad i=1-3, \tag{9.47}$$

$$c_{81+i}=-\sum_{j=1}^{3}(-c_{11}c_{3+j}+c_{10}c_{21+j}+c_{12}c_{27+j})(c_{33+i}\delta_{1j}+c_{39+i}\delta_{2j}-c_i\delta_{3j}),$$
$$i=1-3, \tag{9.48}$$

$$c_{85+j}=\frac{c_{45-2j}(c_{7+j}-c_{10+j})-c_{43+j}(c_{9-2j}-c_{12-2j})}{c_{44}(c_9-c_{12})-c_{45}(c_8-c_{11})},\quad j=0,1, \tag{9.49}$$

$$c_{86+i+3j}=\frac{c_{45+i}(c_{9-j}-c_{12-j})+c_{9+i}c_{45-j}}{c_{44}(c_9-c_{12})-c_{45}(c_8-c_{11})},\quad i=1-3;\quad j=0,1, \tag{9.50}$$

$$c_{93+i+2j}=-c_{62+9j}c_{88+i}+c_{69+9j}c_{91+i}+c_{65+i+9j},\quad i,j=0,1, \tag{9.51}$$

$$c_{97+i+2j}=c_{61+9j}+c_{62+9j}c_{4+2i}(\delta_{0i}-\delta_{1i})+c_{69+9j}c_{5-2i},\quad i,j=0,1, \tag{9.52}$$

$$c_{101+i+2j}=\frac{c_{93+i}c_{99+j}-c_{95+i}c_{97+j}}{c_{94}c_{95}-c_{93}c_{96}},\quad i,j=0,1, \tag{9.53}$$

$$c_{104+i+3j}=\frac{c_{81+i-9j}c_{93+3j}-c_{72+i+9j}c_{95-j}}{c_{94}c_{95}-c_{93}c_{96}},\quad i=1-3;\quad j=0,1, \tag{9.54}$$

$$c_{111+i}=c_{49+9i}+c_{50+9i}c_{85}+c_{51+9i}c_{86}-c_{47+9i}c_{102}+c_{54+9i}c_{101},\quad i=0,1, \tag{9.55}$$

$$c_{112+i+3j}=\delta_{1i}(c_{46+9j}-c_{47+9j}c_{104}+c_{54+9j}c_{103})-c_{50+9j}c_{2+i}+c_{51+9j}c_{89+i},$$
$$i=1-3;\quad j=0,1, \tag{9.56}$$

$$c_{118+i+3j}=c_{47+9j}c_{107+i}+c_{54+9j}c_{104+i}+c_{54+i+9j},\quad i=1-3;\quad j=0,1, \tag{9.57}$$

$$c_{125+i+2j}=c_{101+2j}c_{115+3i}-c_{102+2j}c_{114+3i}+\delta_{0j}c_{111+i}+\delta_{1j}c_{113+3i},\quad i,j=0,1, \tag{9.58}$$

$$c_{128+i+3j}=c_{104+i}c_{115+3j}+c_{107+i}c_{114+3j}+c_{118+i+3j},\quad i=1-3;\quad j=0,1, \tag{9.59}$$

$$c_{135+i}=c_{4+i}(\delta_{1i}-\delta_{0i})+c_{1+6i}c_{141}+c_{2+6i}c_{142},\quad i=0,1, \tag{9.60}$$

$$c_{137+i}=c_{6-3i}+c_{1+6i}c_{143}+c_{2+6i}c_{144},\quad i=0,1, \tag{9.61}$$

$$c_{139+i}=c_{1+6i}c_{145}+c_{2+i}c_{146},\quad i=0,1, \tag{9.62}$$

$$c_{141+i+2j}=c_{102-i+2j}(\delta_{1i}-\delta_{0i})+c_{109-3i}c_{147+2j}-c_{110-3i}c_{148+2j},\quad i,j=0,1, \tag{9.63}$$

$$c_{145+i}=c_{108-3i}+c_{109-3i}c_{151}-c_{110-3i}c_{152},\quad i=0,1, \tag{9.64}$$

$$c_{147+i}=\frac{c_{126}c_{131-i}-c_{125}c_{134-i}}{c_{130}c_{134}-c_{131}c_{133}},\quad i=0,1, \tag{9.65}$$

$$c_{149+i} = \frac{c_{128}c_{131-i} - c_{127}c_{134-i}}{c_{130}c_{134} - c_{131}c_{133}}, \quad i = 0,1, \tag{9.66}$$

$$c_{151+i} = \frac{c_{132}c_{131-i} - c_{129}c_{134-i}}{c_{130}c_{134} - c_{131}c_{133}}, \quad i = 0,1, \tag{9.67}$$

$$c_{153+i} = c_{1+3i} - c_{2+3i}c_{135} + c_{3+3i}c_{136}, \quad i = 0,1, \tag{9.68}$$

$$c_{155+i+2j} = -c_{2+3i}c_{137+2j} + c_{3+3i}c_{138+2j}, \quad i,j = 0,1, \tag{9.69}$$

$$c_{158+i+3j} = (2c_{9+i} - c_{6+i})\,\delta_{0j} + (c_{42+i} + c_{45+i})\,\delta_{1j}, \quad i = 1-3; \quad j = 0,1, \tag{9.70}$$

$$\begin{aligned} c_{164+i+18j} = {} & \frac{\partial (c_{9+i} - c_{6+i})}{\partial \varphi}\delta_{0j} + \frac{\partial (c_{9+i} - c_{6+i})}{\partial \nu}\delta_{1j} - 3\,(c_{11+j} - c_{8+j})\,c_{i+3j} \\ & + (c_{10} - c_7)\,c_{18+i+12j} + (c_{12-j} - c_{9-j})\,c_{24+i+12j}, \quad i = 1-3; \quad j = 0,1, \end{aligned} \tag{9.71}$$

$$\begin{aligned} c_{167+i+18j} = {} & \frac{\partial c_{9+i}}{\partial \varphi}\delta_{0j} + \frac{\partial c_{9+i}}{\partial \nu}\delta_{1j} - (5c_{11+j} - c_{8+j})\,c_{i+3j} + c_{10}c_{18+i+12j} \\ & + (2c_{10} - c_7)\,c_{21+i+12j} + c_{12-j}c_{24+i+12j} + (2c_{12-j} - c_{9-j})\,c_{27+i+12j}, \\ & i = 1-3; \quad j = 0,1, \end{aligned} \tag{9.72}$$

$$c_{170+i+18j} = -c_{11+j}c_{i+3j} + c_{10}c_{21+i+12j} + c_{12-j}c_{27+i+12j}, \quad i = 1-3; \quad j = 0,1, \tag{9.73}$$

$$\begin{aligned} c_{173+i+18j} = {} & \frac{\partial c_{42+i}}{\partial \varphi}\delta_{0j} + \frac{\partial c_{42+i}}{\partial \nu}\delta_{1j} - 3c_{44+j}c_{i+3j} \\ & + c_{43}c_{18+i+12j} + c_{45-j}c_{24+i+12j}, \quad i = 1-3; \quad j = 0,1, \end{aligned} \tag{9.74}$$

$$\begin{aligned} c_{176+i+18j} = {} & \frac{\partial c_{45+i}}{\partial \varphi}\delta_{0j} + \frac{\partial c_{45+i}}{\partial \nu}\delta_{1j} - (c_{44+j} + 4c_{47+j})\,c_{i+3j} + c_{46}c_{21+i+12j} \\ & + (c_{43} + c_{46})\,c_{21+i+12j} + c_{48-j}c_{24+i+12j} + (c_{45-j} + c_{48-j})\,c_{27+i+12j}, \\ & i = 1-3; \quad j = 0,1, \end{aligned} \tag{9.75}$$

$$c_{179+i+18j} = -c_{47+j}c_{i+3j} + c_{46}c_{21+i+12j} + c_{48-j}c_{27+i+12j}, \quad i = 1-3; \quad j = 0,1, \tag{9.76}$$

$$\begin{aligned} c_{200+i+3j} = {} & c_{159+3j}\delta_{2i} + c_{10+36j}\delta_{3i} + c_{160+3j}c_{139+2i} + c_{161+3j}c_{140+2i} \\ & + c_{11+36j}c_{145+2i} - c_{12+36j}c_{146+2i}, \quad i = 1-3; \quad j = 0,1, \end{aligned} \tag{9.77}$$

$$\begin{aligned} c_{206+i+3j} = {} & c_{165+9j}\delta_{1i} + c_{168+9j}\delta_{2i} + c_{171+9j}\delta_{3i} - c_{166+9j}c_{133+2i} + c_{167-9j}c_{134+2i} \\ & + c_{169+9j}c_{139+2i} + c_{170+9j}c_{140+2i} + c_{172+9j}c_{145+2i} - c_{173+9j}c_{146+2i}, \\ & i = 1-3; \quad j = 0-3, \end{aligned} \tag{9.78}$$

$$c_{218+i+9j} = \sum_{k=0}^{1} c_{158+i+3j}\delta_{jk} + \sum_{k=2}^{5} \left[c_{164+i+9(j-2)} + c_{167+i+9(j-2)}\right]\delta_{jk}$$
$$+ \sum_{k=6}^{9} \left[c_{48+i+9(j-6)} + c_{51+i+9(j-6)}\right]\delta_{jk}, \quad i = 1-3; \quad j = 0-9, \qquad (9.79)$$

$$c_{221+i+9j} = \sum_{k=0}^{1} \left[c_{158+i+3j} + 2c_{9+i+36j}\right]\delta_{jk}$$
$$+ \sum_{k=2}^{5} \left[c_{167+i+9(j-2)} + 2c_{170+i+9(j-2)}\right]\delta_{jk}$$
$$+ \sum_{k=6}^{9} \left[c_{51+i+9(j-6)} + 2c_{54+i+9(j-6)}\right]\delta_{jk}, \quad i = 1-3; \quad j = 0-9, \qquad (9.80)$$

$$c_{224+i+9j} = \sum_{k=0}^{1} c_{9+i+36j}\delta_{jk} + \sum_{k=2}^{5} c_{170+i+9(j-2)}\delta_{jk} + \sum_{k=6}^{9} c_{54+i+9(j-6)}\delta_{jk},$$
$$i = 1-3; \quad j = 0-9, \qquad (9.81)$$

$$c_{308+i+9(j+2k)} = \frac{\partial c_{158+i+3j}}{\partial \varphi}\delta_{0k} + \frac{\partial c_{158+i+3j}}{\partial \nu}\delta_{1k} - 4c_{160+3j+k}c_{i+3k}$$
$$+ c_{159+3j}\left(c_{18+i+12k} + c_{21+i+12k}\right) + c_{161+3j-k}\left(c_{24+i+12k} + c_{27+i+12k}\right),$$
$$i = 1-3; \quad j,k = 0,1, \qquad (9.82)$$

$$c_{311+i+9(j+2k)} = \frac{\partial c_{9+i+36j}}{\partial \varphi}\delta_{0k} + \frac{\partial c_{9+i+36j}}{\partial \nu}\delta_{1k} - \left(c_{160+3j+k} + 5c_{11+36j+k}\right)c_{i+3k}$$
$$+ c_{10+36j}c_{18+i+12k} + \left(c_{159+3j} + 2c_{10+36j}\right)c_{21+i+12k} + c_{12+36j-k}c_{24+i+12k}$$
$$+ \left(c_{161+3j-k} + 2c_{12+36j-k}\right)c_{27+i+12k}, \quad i = 1-3; \quad j,k = 0,1, \qquad (9.83)$$

$$c_{314+i+9(j+2k)} = -c_{11+36j+k}c_{i+3k} + c_{10+36j}c_{21+i+12k} + c_{12+36j-k}c_{27+i+12k},$$
$$i = 1-3; \quad j,k = 0,1, \qquad (9.84)$$

$$c_{344+i+12(j+4k)} = \frac{\partial c_{164+i+9j}}{\partial \varphi}\delta_{0k} + \frac{\partial c_{164+i+9j}}{\partial \nu}\delta_{1k} - 3c_{166+9j+k}c_{i+3k}$$
$$+ c_{165+9j}c_{18+i+12k} + c_{167+9j-k}c_{24+i+12k}, \quad i = 1-3; \quad j = 0-3; \quad k = 0,1,$$
$$(9.85)$$

$$c_{347+i+12(j+4k)} = \frac{\partial c_{167+i+9j}}{\partial \varphi}\delta_{0k} + \frac{\partial c_{167+i+9j}}{\partial \nu}\delta_{1k} - \left(c_{166+9j+k} + 4c_{169+9j+k}\right)c_{i+3k}$$

$$+ c_{168+9j}c_{18+i+12k} + (c_{165+9j} + c_{186+9j})\,c_{21+i+12k} + c_{170+9j-k}c_{24+i+12k}$$
$$+ \left(c_{167+9j-k} + c_{170+9j-k}\right) c_{27+i+12k}, \quad i = 1-3; \quad j = 0-3; \quad k = 0,1, \qquad (9.86)$$

$$c_{350+i+12(j+4k)} = \frac{\partial c_{170+i+9j}}{\partial \varphi}\delta_{0k} + \frac{\partial c_{170+i+9j}}{\partial \nu}\delta_{1k} - \left(c_{169+9j+k} + 5c_{172+9j+k}\right) c_{i+3k}$$
$$+ c_{171+9j}c_{18+i+12k} + (c_{168+9j} + 2c_{171+9j})\,c_{21+i+12k} + c_{173+9j-k}c_{24+i+12k}$$
$$+ \left(c_{170+9j-k} + 2c_{173+9j-k}\right) c_{27+i+12k}, \quad i = 1-3; \quad j = 0-3; \quad k = 0,1, \qquad (9.87)$$

$$c_{353+i+12(j+4k)} = -c_{172+9j+k}c_{i+3k} + c_{171+9j}c_{21+i+12k} + c_{173+9j-k}c_{27+i+12k},$$
$$i = 1-3; \quad j = 0-3; \quad k = 0,1, \qquad (9.88)$$

$$c_{440+i+12(j+4k)} = \frac{\partial c_{48+i+9j}}{\partial \varphi}\delta_{0k} + \frac{\partial c_{48+i+9j}}{\partial \nu}\delta_{1k} - 3c_{50+9j+k}c_{i+3k}$$
$$+ c_{49+9j}c_{18+i+12k} + c_{51+9j-k}c_{24+i+12k}, \quad i = 1-3; \quad j = 0-3; \quad k = 0,1, \qquad (9.89)$$

$$c_{443+i+12(j+4k)} = \frac{\partial c_{51+i+9j}}{\partial \varphi}\delta_{0k} + \frac{\partial c_{51+i+9j}}{\partial \nu}\delta_{1k} - \left(c_{50+9j+k} + 4c_{53+9j+k}\right) c_{i+3k}$$
$$+ c_{52+9j}c_{18+i+12k} + (c_{49+9j} + c_{52+9j})\,c_{21+i+12k} + c_{54+9j-k}c_{24+i+12k}$$
$$+ \left(c_{51+9j-k} + c_{54+9j-k}\right) c_{27+i+12k}, \quad i = 1-3; \quad j = 0-3; \quad k = 0,1, \qquad (9.90)$$

$$c_{446+i+12(j+4k)} = \frac{\partial c_{54+i+9j}}{\partial \varphi}\delta_{0k} + \frac{\partial c_{54+i+9j}}{\partial \nu}\delta_{1k} - \left(c_{53+9j+k} + 5c_{56+9j+k}\right) c_{i+3k}$$
$$+ c_{55+9j}c_{18+i+12k} + (c_{52+9j} + 2c_{55+9j})\,c_{21+i+12k} + c_{57+9j-k}c_{24+i+12k}$$
$$+ \left(c_{54+9j-k} + 2c_{57+9j-k}\right) c_{27+i+12k}, \quad i = 1-3; \quad j = 0-3; \quad k = 0,1, \qquad (9.91)$$

$$c_{449+i+12(j+4k)} = -c_{56+9j+k}c_{i+3k} + c_{55+9j}c_{21+i+12k} + c_{57+9j-k}c_{27+i+12k},$$
$$i = 1-3; \quad j = 0-3; \quad k = 0,1, \qquad (9.92)$$

$$c_{537+4i} = c_{220+9i}c_{141} + c_{221+9i}c_{142} + (c_{223+9i} - 2c_{226+9i})\,c_{147}$$
$$- \ (c_{224+9i} - 2c_{227+9i})\,c_{148}, \quad i = 0-13, \qquad (9.93)$$

$$c_{538+4i} = c_{219+9i} + c_{220+9i}c_{143} + c_{221+9i}c_{144} + c_{226+9i}c_{147} - c_{227+9i}c_{148}$$
$$+ \ (c_{223+9i} - c_{226+9i})\,c_{149} - (c_{224+9i} - c_{227+9i})\,c_{150}, \quad i = 0-13, \qquad (9.94)$$

$$c_{539+4i} = c_{222+9i} + c_{220+9i}c_{145} + c_{221+9i}c_{146} + c_{226+9i}c_{149} - c_{227+9i}c_{150}$$
$$+ c_{223+9i}c_{151} - c_{224+9i}c_{152}, \quad i = 0-13, \qquad (9.95)$$

$$c_{540+4i} = c_{225+9i} + c_{226+9i}c_{151} - c_{227+9i}c_{152}, \quad i = 0-13, \tag{9.96}$$

$$c_{593+4i} = c_{345+12i} - c_{346+12i}c_{135} + c_{347+12i}c_{136} + c_{349+12i}c_{141} + c_{350+12i}c_{142}$$
$$+ (c_{352+12i} - 2c_{355+12i})\,c_{147} - (c_{353+12i} - 2c_{356+12i})\,c_{148}, \quad i = 0-15, \tag{9.97}$$

$$c_{594+4i} = c_{348+12i} - c_{346+12i}c_{137} + c_{347+12i}c_{138} + c_{349+12i}c_{143} + c_{350+12i}c_{144}$$
$$+ c_{355+12i}c_{147} - c_{356+12i}c_{148} + (c_{352+12i} - c_{355+12i})\,c_{149}$$
$$- (c_{353+12i} - c_{356+12i})\,c_{150}, \quad i = 0-15, \tag{9.98}$$

$$c_{595+4i} = c_{351+12i} - c_{346+12i}c_{139} + c_{347+12i}c_{140} + c_{349+12i}c_{145} + c_{350+12i}c_{146}$$
$$+ c_{355+12i}c_{149} - c_{356+12i}c_{150} + c_{352+12i}c_{151} - c_{353+12i}c_{152}, \quad i = 0-15, \tag{9.99}$$

$$c_{596+4i} = c_{354+12i} + c_{355+12i}c_{151} - c_{356+12i}c_{152}, \quad i = 0-15, \tag{9.100}$$

$$c_{656+i+12j} = \sum_{k=0}^{13} c_{218+i+9j}\delta_{jk} + \sum_{k=14}^{29} \left[c_{344+i+12(j-14)} + c_{347+i+12(j-14)}\right]\delta_{jk},$$
$$i = 1-3; \quad j = 0-29, \tag{9.101}$$

$$c_{659+i+12j} = \sum_{k=0}^{13} (c_{218+i+9j} + 2c_{221+i+9j})\,\delta_{jk}$$
$$+ \sum_{k=14}^{29} \left[c_{347+i+12(j-14)} + 2c_{350+i+12(j-14)}\right]\delta_{jk}, \quad i = 1-3; \quad j = 0-29, \tag{9.102}$$

$$c_{662+i+12j} = \sum_{k=0}^{13} (c_{221+i+9j} + 3c_{224+i+9j})\,\delta_{jk}$$
$$+ \sum_{k=14}^{29} \left[c_{350+i+12(j-14)} + 3c_{353+i+12(j-14)}\right]\delta_{jk}, \quad i = 1-3; \quad j = 0-29, \tag{9.103}$$

$$c_{664+i+12j} = \sum_{k=0}^{13} c_{224+i+9j}\delta_{jk} + \sum_{k=14}^{29} c_{353+i+12(j-14)}\delta_{jk}, \quad i = 1-3; \quad j = 0-29, \tag{9.104}$$

$$c_{998+i+12(j+14k)} = \frac{\partial c_{218+i+9j}}{\partial \varphi}\delta_{0k} + \frac{\partial c_{218+i+9j}}{\partial \nu}\delta_{1k} - 4c_{220+9j+k}c_{i+3k}$$
$$+ c_{219+9j}\,(c_{18+i+12k} + c_{21+i+12k}) + c_{221+9j-k}\,(c_{24+i+12k} + c_{27+i+12k}),$$
$$i = 1-3; \quad j = 0-13; \quad k = 0,1, \tag{9.105}$$

$$c_{1001+i+12(j+14k)} = \frac{\partial c_{221+i+9j}}{\partial \varphi}\delta_{0k} + \frac{\partial c_{221+i+9j}}{\partial \nu}\delta_{1k}$$
$$- \left(c_{220+9j+k} + 5c_{223+9j+k}\right) c_{i+3k} + c_{222+9j}c_{18+i+12k}$$
$$+ \left(c_{219+9j} + 2c_{222+9j}\right) c_{21+i+12k} + c_{224+9j-k}c_{24+i+12k}$$
$$+ \left(c_{221+9j-k} + 2c_{224+9j-k}\right) c_{27+i+12k}, \quad i = 1-3; \quad j = 0-13; \quad k = 0,1, \tag{9.106}$$

$$c_{1004+i+12(j+14k)} = \frac{\partial c_{224+i+9j}}{\partial \varphi}\delta_{0k} + \frac{\partial c_{224+i+9j}}{\partial \nu}\delta_{1k}$$
$$- \left(c_{223+9j+k} + 6c_{226+9j+k}\right) c_{i+3k} + c_{225+9j}c_{18+i+12k}$$
$$+ \left(c_{222+9j} + 3c_{225+9j}\right) c_{21+i+12k} + c_{227+9j-k}c_{24+i+12k}$$
$$+ \left(c_{224+9j-k} + 3c_{227+9j-k}\right) c_{27+i+12k}, \quad i = 1-3; \quad j = 0-13; \quad k = 0,1, \tag{9.107}$$

$$c_{1007+i+12(j+14k)} = -c_{226+9j+k}c_{i+3k} + c_{225+9j}c_{21+i+12k} + c_{227+9j-k}c_{27+i+12k},$$
$$i = 1-3; \quad j = 0-13; \quad k = 0,1, \tag{9.108}$$

$$c_{1352+i+15(j+16k)} = \frac{\partial c_{344+i+12j}}{\partial \varphi}\delta_{0k} + \frac{\partial c_{344+i+12j}}{\partial \nu}\delta_{1k} - 3c_{346+12j+k}c_{i+3k}$$
$$+ c_{345+12j}c_{18+i+12k} + c_{347+12j-k}c_{24+i+12k},$$
$$i = 1-3; \quad j = 0-15; \quad k = 0,1, \tag{9.109}$$

$$c_{1355+i+15(j+16k)} = \frac{\partial c_{347+i+12j}}{\partial \varphi}\delta_{0k} + \frac{\partial c_{347+i+12j}}{\partial \nu}\delta_{1k}$$
$$- \left(c_{346+12j+k} + 4c_{349+12j+k}\right) c_{i+3k} + c_{348+12j}c_{18+i+12k}$$
$$+ \left(c_{345+12j} + c_{348+12j}\right) c_{21+i+12k} + c_{350+12j-k}c_{24+i+12k}$$
$$+ \left(c_{347+12j-k} + c_{350+12j-k}\right) c_{27+i+12k}, \quad i = 1-3; \quad j = 0-15; \quad k = 0,1, \tag{9.110}$$

$$c_{1358+i+15(j+16k)} = \frac{\partial c_{350+i+12j}}{\partial \varphi}\delta_{0k} + \frac{\partial c_{350+i+12j}}{\partial \nu}\delta_{1k}$$
$$- \left(c_{349+12j+k} + 5c_{352+12j+k}\right) c_{i+3k} + c_{351+12j}c_{18+i+12k}$$
$$+ \left(c_{348+12j} + 2c_{351+12j}\right) c_{21+i+12k} + c_{353+12j-k}c_{24+i+12k}$$
$$+ \left(c_{350+12j-k} + 2c_{353+12j-k}\right) c_{27+i+12k}, \quad i = 1-3; \quad j = 0-15; \quad k = 0,1, \tag{9.111}$$

$$c_{1361+i+15(j+16k)} = \frac{\partial c_{353+i+12j}}{\partial \varphi}\delta_{0k} + \frac{\partial c_{353+i+12j}}{\partial \nu}\delta_{1k}$$

$$- \left(c_{352+12j+k} + 6c_{355+12j+k}\right) c_{i+3k} + c_{354+12j}c_{18+i+12k}$$
$$+ \left(c_{351+12j} + 3c_{354+12j}\right) c_{21+i+12k} + c_{356+12j-k}c_{24+i+12k}$$
$$+ \left(c_{353+12j-k} + 3c_{356+12j-k}\right) c_{27+i+12k}, \quad i = 1-3; \quad j = 0-15; \quad k = 0,1, \tag{9.112}$$

$$c_{1364+i+15(j+16k)} = -c_{355+12j+k}c_{i+3k} + c_{354+12j}c_{21+i+12k} + c_{356+12j-k}c_{27+i+12k},$$
$$i = 1-3; \quad j = 0-15; \quad k = 0,1, \tag{9.113}$$

$$c_{1833+5i} = c_{658+12i}c_{141} + c_{659+12i}c_{142} + \left[c_{661+12i} - 2\left(c_{664+12i} - 3c_{667+12i}\right)\right] c_{147}$$
$$- \left[c_{662+12i} - 2\left(c_{664+12i} - 3c_{668+12i}\right)\right] c_{148}, \quad i = 0-57, \tag{9.114}$$

$$c_{1834+5i} = c_{657+12i} + c_{658+12i}c_{143} + c_{659+12i}c_{144} + \left(c_{664+12i} - 4c_{667+12i}\right) c_{147}$$
$$- \left(c_{664+12i} - 4c_{668+12i}\right) c_{148} + \left(c_{661+12i} - c_{664+12i} + 2c_{667+12i}\right) c_{149}$$
$$- \left(c_{662+12i} - c_{664+12i} + 2c_{668+12i}\right) c_{150}, \quad i = 0-57, \tag{9.115}$$

$$c_{1835+5i} = c_{660+12i} + c_{658+12i}c_{145} + c_{659+12i}c_{146} + c_{667+12i}c_{147} - c_{668+12i}c_{148}$$
$$+ \left(c_{664+12i} - 2c_{667+12i}\right) c_{149} - \left(c_{664+12i} - 2c_{668+12i}\right) c_{150} + c_{661+12i}c_{151}$$
$$- c_{662+12i}c_{152}, \quad i = 0-57, \tag{9.116}$$

$$c_{1836+5i} = c_{663+12i} + c_{667+12i}c_{149} - c_{668+12i}c_{150} + c_{664+12i}c_{151} - c_{665+12i}c_{152},$$
$$i = 0-57, \tag{9.117}$$

$$c_{1837+5i} = c_{666+12i} + c_{667+12i}c_{151} - c_{668+12i}c_{152}, \quad i = 0-57, \tag{9.118}$$

$$c_{2123+5i} = c_{1353+15i} - c_{1354+15i}c_{135} + c_{1355+15i}c_{136} + c_{1357+15i}c_{141} + c_{1358+15i}c_{142}$$
$$+ \left[c_{1360+15i} - 2\left(c_{1363+15i} - 3c_{1366+15i}\right)\right] c_{147}$$
$$- \left[c_{1361+15i} - 2\left(c_{1364+15i} - 3c_{1367+15i}\right)\right] c_{148}, \quad i = 0-31, \tag{9.119}$$

$$c_{2124+5i} = c_{1356+15i} - c_{1354+15i}c_{137} + c_{1355+15i}c_{138} + c_{1357+15i}c_{143}$$
$$+ c_{1358+15i}c_{144} + \left(c_{1363+15i} - 4c_{1366+15i}\right) c_{147} - \left(c_{1364+15i} - 4c_{1367+15i}\right) c_{148}$$
$$+ \left(c_{1360+15i} - c_{1363+15i} + 2c_{1366+15i}\right) c_{149}$$
$$- \left(c_{1361+15i} - c_{1364+15i} + 2c_{1367+15i}\right) c_{150}, \quad i = 0-31, \tag{9.120}$$

$$c_{2125+5i} = c_{1359+15i} - c_{1354+15i}c_{139} + c_{1355+15i}c_{140} + c_{1357+15i}c_{145} + c_{1358+15i}c_{146}$$

$$+ c_{1366+15i}c_{147} - c_{1367+15i}c_{148} + (c_{1363+15i} - 2c_{1366+15i})\,c_{149}$$
$$- (c_{1364+15i} - 2c_{1367+15i})\,c_{150} + c_{1360+15i}c_{151} - c_{1361+15i}c_{152}, \quad i = 0-31, \tag{9.121}$$

$$c_{2126+5i} = c_{1362+15i} + c_{1366+15i}c_{149} - c_{1367+15i}c_{150} + c_{1363+15i}c_{151} - c_{1364+15i}c_{152},$$
$$i = 0-31, \tag{9.122}$$

$$c_{2127+5i} = c_{1365+15i} + c_{1366+15i}c_{151} - c_{1367+15i}c_{152}, \quad i = 0-31, \tag{9.123}$$

$$c_{2282+i+15j} = \sum_{k=0}^{57} c_{656+i+12j}\delta_{jk} + \sum_{k=58}^{89} (c_{1352+i+15j} + c_{1355+i+15j})\,\delta_{jk}$$
$$+ \sum_{k=90}^{147} \left[\frac{\partial c_{656+i+12j}}{\partial \varphi} + c_{657+12j}(c_{18+i} + c_{21+i})\right.$$
$$\left. - 4c_{658+12j}c_{3+i} + c_{659+12j}(c_{24+i} + c_{27+i})\right]\delta_{jk}$$
$$+ \sum_{k=148}^{205} \left[\frac{\partial c_{656+i+12j}}{\partial \nu} + c_{657+12j}(c_{30+i} + c_{33+i})\right.$$
$$\left. + c_{658+12j}(c_{36+i} + c_{39+i}) - 4c_{659+12j}c_i\right]\delta_{jk},$$
$$i = 1-3; \quad j = 0-205, \tag{9.124}$$

$$c_{2285+i+15j} = \sum_{k=0}^{57} (c_{656+i+12j} + 2c_{659+i+12j})\,\delta_{jk}$$
$$+ \sum_{k=58}^{89} (c_{1355+i+15j} + 2c_{1358+i+15j})\,\delta_{jk}$$
$$+ \sum_{k=90}^{147} \left[\frac{\partial c_{659+i+12j}}{\partial \varphi} + c_{657+12j}c_{21+i} - c_{658+12j}c_{3+i} + c_{659+12j}c_{27+i}\right.$$
$$- 5c_{661+12j}c_{3+i} + c_{660+12j}(c_{18+i} + 2c_{21+i})$$
$$\left. + c_{662+12j}(c_{24+i} + 2c_{27+i})\right]\delta_{jk}$$
$$+ \sum_{k=148}^{205} \left[\frac{\partial c_{659+i+12j}}{\partial \nu} + c_{657+12j}c_{33+i} + c_{658+12j}c_{39+i} - c_{659+12j}c_i\right.$$
$$- 5c_{662+12j}c_i + c_{660+12j}(c_{30+i} + 2c_{33+i})$$
$$\left. + c_{661+12j}(c_{36+i} + 2c_{39+i})\right]\delta_{jk}, \quad i = 1-3; \quad j = 0-205, \tag{9.125}$$

$$c_{2288+i+15j} = \sum_{k=0}^{57} \left(c_{659+i+12j} + 3c_{662+i+12j}\right)\delta_{jk}$$

$$+ \sum_{k=58}^{89} \left(c_{1358+i+15j} + 3c_{1361+i+15j}\right)\delta_{jk}$$

$$+ \sum_{k=90}^{147} \left[\frac{\partial c_{662+i+12j}}{\partial \varphi} + c_{660+12j}c_{21+i} - c_{661+12j}c_{3+i} + c_{662+12j}c_{27+i} - 6c_{664+12j}c_{3+i} + c_{663+12j}\left(c_{18+i} + 3c_{21+i}\right) + c_{665+12j}\left(c_{24+i} + 3c_{27+i}\right)\right]\delta_{jk}$$

$$+ \sum_{k=148}^{205} \left[\frac{\partial c_{662+i+12j}}{\partial \nu} + c_{660+12j}c_{33+i} + c_{661+12j}c_{39+i} - c_{662+12j}c_i - 6c_{665+12j}c_i + c_{663+12j}\left(c_{30+i} + 3c_{33+i}\right) + c_{664+12j}\left(c_{36+i} + 3c_{39+i}\right)\right]\delta_{jk}, \quad i = 1-3; \quad j = 0-205, \tag{9.126}$$

$$c_{2291+i+15j} = \sum_{k=0}^{57} \left(c_{662+i+12j} + 4c_{665+i+12j}\right)\delta_{jk}$$

$$+ \sum_{k=58}^{89} \left(c_{1361+i+15j} + 4c_{1364+i+15j}\right)\delta_{jk}$$

$$+ \sum_{k=90}^{147} \left[\frac{\partial c_{665+i+12j}}{\partial \varphi} + c_{663+12j}c_{21+i} - c_{664+12j}c_{3+i} + c_{665+12j}c_{27+i} - 7c_{667+12j}c_{3+i} + c_{666+12j}\left(c_{18+i} + 4c_{21+i}\right) + c_{668+12j}\left(c_{24+i} + 4c_{27+i}\right)\right]\delta_{jk}$$

$$+ \sum_{k=148}^{205} \left[\frac{\partial c_{665+i+12j}}{\partial \nu} + c_{663+12j}c_{33+i} + c_{664+12j}c_{39+i} - c_{665+12j}c_i - 7c_{668+12j}c_i + c_{666+12j}\left(c_{30+i} + 4c_{33+i}\right) + c_{667+12j}\left(c_{36+i} + 4c_{39+i}\right)\right]\delta_{jk},$$

$$i = 1-3; \quad j = 0-205, \tag{9.127}$$

$$c_{2294+i+15j} = \sum_{k=0}^{57} c_{665+i+12j}\delta_{jk} + \sum_{k=58}^{89} c_{1364+i+15j}\delta_{jk}$$

$$+ \sum_{k=90}^{147} \left(c_{666+12j}c_{21+i} - c_{667+12j}c_{3+i} + c_{668+12j}c_{27+i}\right)\delta_{jk}$$

$$+ \sum_{k=148}^{205} \left(c_{666+12j}c_{33+i} + c_{667+12j}c_{39+i} - c_{668+12j}c_i\right)\delta_{jk},$$

$$i = 1-3; \quad j = 0-205, \tag{9.128}$$

$$c_{5372+i+18j} =$$
$$\sum_{k=0}^{31}\left(\frac{\partial c_{1352+i+15j}}{\partial\varphi} + c_{1353+15j}c_{18+i} - 3c_{1354+15j}c_{3+i} + c_{1355+15j}c_{24+i}\right)\delta_{jk}$$
$$+\sum_{k=32}^{63}\left(\frac{\partial c_{1352+i+15j}}{\partial\nu} + c_{1353+15j}c_{30+i} + c_{1354+15j}c_{36+i} - 3c_{1355+15j}c_i\right)\delta_{jk},$$
$$i = 1-3; \quad j = 0-63, \tag{9.129}$$

$$c_{5375+i+18j} =$$
$$\sum_{k=0}^{31}\left[\frac{\partial c_{1355+i+15j}}{\partial\varphi} + c_{1353+15j}c_{21+i} - c_{1354+15j}c_{3+i} + c_{1355+15j}c_{27+i}\right.$$
$$\left. - 4c_{1357+15j}c_{3+i} + c_{1356+15j}(c_{18+i} + c_{21+i}) + c_{1358+15j}(c_{24+i} + c_{27+i})\right]\delta_{jk}$$
$$+\sum_{k=32}^{63}\left[\frac{\partial c_{1355+i+15j}}{\partial\nu} + c_{1353+15j}c_{33+i} + c_{1354+15j}c_{39+i} - c_{1355+15j}c_i\right.$$
$$\left. - 4c_{1358+15j}c_i + c_{1356+15j}(c_{30+i} + c_{33+i}) + c_{1357+15j}(c_{36+i} + c_{39+i})\right]\delta_{jk},$$
$$i = 1-3; \quad j = 0-63, \tag{9.130}$$

$$c_{5378+i+18j} =$$
$$\sum_{k=0}^{31}\left[\frac{\partial c_{1358+i+15j}}{\partial\varphi} + c_{1356+15j}c_{21+i} - c_{1357+15j}c_{3+i} + c_{1358+15j}c_{27+i}\right.$$
$$- 5c_{1360+15j}c_{3+i} + c_{1359+15j}(c_{18+i} + 2c_{21+i})$$
$$\left. + c_{1361+15j}(c_{24+i} + 2c_{27+i})\right]\delta_{jk}$$
$$+\sum_{k=32}^{63}\left[\frac{\partial c_{1358+i+15j}}{\partial\nu} + c_{1356+15j}c_{33+i} + c_{1357+15j}c_{39+i} - c_{1358+15j}c_i\right.$$
$$- 5c_{1361+15j}c_i + c_{1359+15j}(c_{30+i} + 2c_{33+i})$$
$$\left. + c_{1360+15j}(c_{36+i} + 2c_{39+i})\right]\delta_{jk}, \quad i = 1-3; \quad j = 0-63, \tag{9.131}$$

$$c_{5381+i+18j} =$$
$$\sum_{k=0}^{31}\left[\frac{\partial c_{1361+i+15j}}{\partial\varphi} + c_{1359+15j}c_{21+i} - c_{1360+15j}c_{3+i} + c_{1361+15j}c_{27+i}\right.$$
$$- 6c_{1363+15j}c_{3+i} + c_{1362+15j}(c_{18+i} + 3c_{21+i})$$
$$\left. + c_{1364+15j}(c_{24+i} + 3c_{27+i})\right]\delta_{jk}$$
$$+\sum_{k=32}^{63}\left[\frac{\partial c_{1361+i+15j}}{\partial\nu} + c_{1359+15j}c_{33+i} + c_{1360+15j}c_{39+i} - c_{1361+15j}c_i\right.$$

$$- 6c_{1364+15j}c_i + c_{1362+15j}\left(c_{30+i} + 3c_{33+i}\right)$$
$$\left. + c_{1363+15j}\left(c_{36+i} + 3c_{39+i}\right)\right]\delta_{jk}, \quad i = 1-3; \quad j = 0-63, \quad (9.132)$$

$$c_{5384+i+18j} =$$
$$\sum_{k=0}^{31}\left[\frac{\partial c_{1364+i+15j}}{\partial \varphi} + c_{1362+15j}c_{21+i} - c_{1363+15j}c_{3+i} + c_{1364+15j}c_{27+i}\right.$$
$$- 7c_{1366+15j}c_{3+i} + c_{1365+15j}\left(c_{18+i} + 4c_{21+i}\right)$$
$$\left. + c_{1367+15j}\left(c_{24+i} + 4c_{27+i}\right)\right]\delta_{jk}$$
$$+ \sum_{k=32}^{63}\left[\frac{\partial c_{1364+i+15j}}{\partial \nu} + c_{1362+15j}c_{33+i} + c_{1363+15j}c_{39+i} - c_{1364+15j}c_i\right.$$
$$- 7c_{1367+15j}c_i + c_{1365+15j}\left(c_{30+i} + 4c_{33+i}\right)$$
$$\left. + c_{1366+15j}\left(c_{36+i} + 4c_{39+i}\right)\right]\delta_{jk}, \quad i = 1-3; \quad j = 0-63, \quad (9.133)$$

$$c_{5387+i+18j} = \sum_{k=0}^{31}\left(c_{1365+15j}c_{21+i} - c_{1366+15j}c_{3+i} + c_{1367+15j}c_{27+i}\right)\delta_{jk}$$
$$+ \sum_{k=32}^{63}\left(c_{1365+15j}c_{33+i} + c_{1366+15j}c_{39+i} - c_{1367+15j}c_i\right)\delta_{jk},$$
$$i = 1-3; \quad j = 0-63, \quad (9.134)$$

$$c_{6525+6i} = \sum_{j=0}^{1} c_{2284+15i+j}c_{141+j}$$
$$+ \sum_{j=0}^{1} c_{147+j}\left\{c_{2287+15i+j} - 2\left[c_{2290+15i+j} - 3\left(c_{2293+15i+j} - 4c_{2296+15i+j}\right)\right]\right\}$$
$$\times\left(\delta_{0j} - \delta_{1j}\right), \quad i = 0-205, \quad (9.135)$$

$$c_{6526+6i} = c_{2283+15i} + \sum_{j=0}^{1} c_{2284+15i+j}c_{143+j}$$
$$+ \sum_{j=0}^{1}\left\{2\left[c_{2293+15i+j}\left(c_{149+j} - 2c_{147+j}\right) + 3c_{2296+15i+j}\left(3c_{147+j} - c_{149+j}\right)\right]\right.$$
$$\left. + c_{2287+15i+j}c_{149+j} + c_{2290+15i+j}\left(c_{147+j} - c_{149+j}\right)\right\}\left(\delta_{0j} - \delta_{1j}\right),$$
$$i = 0-205, \quad (9.136)$$

$$c_{6527+6i} = c_{2286+15i} + \sum_{j=0}^{1} c_{2284+15i+j}c_{145+j}$$

$$+\sum_{j=0}^{1}\left[c_{2287+15i+j}c_{151+j}+c_{2290+15i+j}c_{149+j}+c_{2293+15i+j}\left(c_{147+j}-2c_{149+j}\right)\right.$$
$$\left.-\,6c_{2296+15i+j}\left(c_{147+j}-c_{149+j}\right)\right]\left(\delta_{0j}-\delta_{1j}\right),\quad i=0-205, \tag{9.137}$$

$$c_{6528+6i}=c_{2289+15i}$$
$$+\sum_{j=0}^{1}\left[c_{2290+15i+j}c_{151+j}+c_{2293+15i+j}c_{149+j}+c_{2296+15i+j}\left(c_{147+j}-3c_{149+j}\right)\right]$$
$$\times\left(\delta_{0j}-\delta_{1j}\right),\quad i=0-205, \tag{9.138}$$

$$c_{6529+6i}=c_{2292+15i}+\sum_{j=0}^{1}\left(\delta_{0j}-\delta_{1j}\right)\left(c_{2293+15i+j}c_{151+j}+c_{2296+15i+j}c_{149+j}\right),$$
$$i=0-205, \tag{9.139}$$

$$c_{6530+6i}=c_{2295+15i}+\sum_{j=0}^{1}\left(\delta_{0j}-\delta_{1j}\right)c_{2296+15i+j}c_{151+j},\quad i=0-205, \tag{9.140}$$

$$c_{7761+6i}=c_{5373+18i}+\sum_{j=0}^{1}c_{5377+18i+j}c_{141+j}$$
$$+\sum_{j=0}^{1}\left(\left\{c_{5380+18i+j}-2\left[c_{5383+18i+j}-3\left(c_{5386+18i+j}-4c_{5389+18i+j}\right)\right]\right\}c_{147+j}\right.$$
$$\left.-\,c_{5374+18i+j}c_{135+j}\right)\left(\delta_{0j}-\delta_{1j}\right),\quad i=0-63, \tag{9.141}$$

$$c_{7762+6i}=c_{5376+18i}+\sum_{j=0}^{1}c_{5377+18i+j}c_{143+j}$$
$$+\sum_{j=0}^{1}\left[c_{5380+18i+j}c_{149+j}+c_{5383+18i+j}\left(c_{147+j}-c_{149+j}\right)\right.$$
$$+\,2c_{5386+18i+j}\left(c_{149+j}-2c_{147+j}\right)+6c_{5389+18i+j}\left(3c_{147+j}-c_{149+j}\right)$$
$$\left.-\,c_{5374+18i+j}c_{137+j}\right]\left(\delta_{0j}-\delta_{1j}\right),\quad i=0-63, \tag{9.142}$$

$$c_{7763+6i}=c_{5379+18i}+\sum_{j=0}^{1}c_{5377+18i+j}c_{145+j}$$
$$+\sum_{j=0}^{1}\left[c_{5380+18i+j}c_{151+j}+c_{5383+18i+j}c_{149+j}+c_{5386+18i+j}\left(c_{147+j}-2c_{149+j}\right)\right.$$

$$- 6c_{5389+18i+j}\left(c_{147+j} - c_{149+j}\right) - c_{5374+18i+j}c_{139+j}\Big]\left(\delta_{0j} - \delta_{1j}\right),$$
$$i = 0 - 63, \tag{9.143}$$

$$c_{7764+6i} = c_{5382+18i}$$
$$+ \sum_{j=0}^{1}\left[c_{5383+18i+j}c_{151+j} + c_{5386+18i+j}c_{149+j} + c_{5389+18i+j}\left(c_{147+j} - 3c_{149+j}\right)\right]$$
$$\times\left(\delta_{0j} - \delta_{1j}\right), \quad i = 0 - 63, \tag{9.144}$$

$$c_{7765+6i} = c_{5385+18i} + \sum_{j=0}^{1}\left(\delta_{0j} - \delta_{1j}\right)\left(c_{5386+18i+j}c_{151+j} + c_{5389+18i+j}c_{149+j}\right),$$
$$i = 0 - 63, \tag{9.145}$$

$$c_{7766+6i} = c_{5388+18i} + \sum_{j=0}^{1}\left(\delta_{0j} - \delta_{1j}\right)c_{5389+18i+j}c_{151+j}, \quad i = 0 - 63. \tag{9.146}$$

References

[1] Mizutani, T. *J. Mater. Sci.* 1998, 11, 68–72.

[2] Li, S.; Sauer, R.A.; Wang, G. *J. Appl. Mech.* 2007, 74, 770–783.

[3] Li, S.; Sauer, R.A.; Wang, G. *J. Appl. Mech.* 2007, 74, 784–797.

[4] Kushch, V.I. *Int. Appl. Mech.* 2000, 36, 225-233.

[5] Kushch, V.I. *Int. Appl. Mech.* 2000, 36, 623-630.

[6] Kushch, V.I. *Int. Appl. Mech.* 2004, 40, 893-899.

[7] Kushch, V.I. *Int. Appl. Mech.* 2004, 40, 1042-1049.

[8] Kushch, V.I. *Int. J. Sol. Struct.* 2003, 40, 6369-6388.

[9] Kushch, V.I. *Int. J. Sol. Struct.* 2004, 41, 885-906.

[10] Kushch, V.I. *Int. J. Sol. Struct.* 2005, 42, 5491-5512.

[11] Kushch, V.I.; Shmegera, S.V.; Mishnaevsky, L. Jr. *Int. J. Sol. Struct.* 2008, 45, 2758-2784.

[12] Kushch, V.I.; Shmegera, S.V.; Mishnaevsky, L. Jr. *Int. J. Sol. Struct.* 2008, 45, 5103-5117.

[13] Guagliano, M. *J. Mater. Eng. Perform.* 1998, 7, 183–189.

[14] Broberg, K.B. *Comput. Mech.* 1997, 19, 447–452.

[15] Ramakrishnan, N. *Bull. Mater. Sci.* 1997, 20, 885–900.

[16] Babuška, I.; Andersson, B.; Smith, P.J.; Levin, L.; *Computer Methods in Appl. Mech. Eng.* 1999, 172, 27–77.

[17] Byström., J. *Composites: Part B* 2003, 34, 587-592.

[18] Wu, Y.; Dong, Z. *Mater. Sci. Eng. A.* 1995, 203, 314–323.

[19] Sangani, A.S.; Mo, G. *J. Mech. Phys. Sol.* 1997, 45, 2001–2031.

[20] Matouš, K.; Lepš, M.; Zeman, J; Šejnoha, M. *Comp. Meth. Appl. Mech. Eng.* 2000, 190, 1629–1650.

[21] Šejnoha, M.; Zeman, J. *Comp. Meth. Appl. Mech. Eng.* 2002, 191, 5027-5044.

[22] Šejnoha, M.; Zeman, J. *Int. J. Eng. Sci.* 2008, 46, 513-526.

[23] Šejnoha, M.; Zeman, J. *Int. J. Sol. Struct.* 2004, 41, 6549-6571.

[24] Gajdošík, J.; Zeman, J.; Šejnoha, M. *Probal. Eng. Mech.* 2006, 21, 317-329.

[25] Mura, T. *Micromechanics of Defects in Solids*; Martinus Nijhoff Publishers: Dordrecht, NL, 1987, pp 388–392.

[26] Brdička, M.; Samek, L.; Sopko, B. *Mechanics of Continuum*; Academia: Prague, CZ, 2000, pp 205–206.

[27] Dusza, J.; Kašiarová, M.; Hegedűsová, L.; Hnatko, M.; Šajgalík, P. *Mater. Sci. Forum* 2008, 567–568, 177–180.

[28] Hegedűsová, L.; Kašiarová, M.; Dusza, J.; Hnatko, M.; Šajgalík, P. *Int. J. Refract. Met. Hard Mater.* 2009, 27, 438–442.

[29] Hegedűsová, L.; Kašiarová, M.; Csehová, E.; Dusza, J. *J. Eur. Ceram. Soc.* 2010, 30, 1059–1065.

[30] Hald, J. *Int. J. Press. Vess. Pip.* 2008, 85, 30-37.

[31] Výrostková, A.; Homolová, V.; Pecha, J.; Svoboda, M. *Mater. Sci. Eng. A* 2008, 480, 289-298.

[32] Trebuňa, F.; Buršák, M. *Limit States and Fractures*; Technical University: Košice, SK, 2002, pp 116–120.

[33] Trebuňa, F.; Šimčák, F.; Bocko, J.; Trebuňa, P. *Eng. Fail. Anal.* 2010, 17, 787–801.

[34] Trebuňa, F.; Šimčák, F.; Bocko, J.; Trebuňa, P.; Pástor, M.; Šarga, P. *Eng. Fail. Anal.* 2011, 18, 244–255.

[35] Trebuňa, F.; Šimčák, F.; Gmiterko, A. *Metalurgija* 2010, 49, 91–94.

[36] Trebuňa, F.; Šimčák, F.; Bocko, J.; Trebuňa, P. *Eng. Fail. Anal.* 2009, 16, 2054–2065.

[37] Trebuňa, F.; Šimčák, F.; Bocko, J. *Eng. Fail. Anal.* 2009, 16, 26–38.

[38] Mamuzič, T.; Trebuňa, F.; Šimčák, F.; Buršák, M.; Tomčík, J. *Metalurgija* 2007, 46, 267–271.

[39] Trebuňa, F.; Šimčák, F.; Bocko, J.; Trebuňa, P.; Šarga, P. *Proc. 48th Int. Sci. Confer. Exp. Stress Anal.* 2010, 467–474.

[40] Gao, Z.W.; Zhou, Y.H. *Phys. Lett. A* 2008, 372, 5261-5264.

[41] Gudmundson, P. *J. Mech. Phys. Sol.* 1990, 38, 531–551.

[42] Hintermüller, M.; Kovtunenko, V.A.; Kunisch, K. *J. Eng. Math.* 2007, 59, 301-321.

[43] Miller, G.R. *Int. J. Eng. Sci.* 1989, 27, 667–678.

[44] Chang, R.C.; Wang, S.C.; Chao, C.K. *Acta Mech.* 2001, 148, 215–229.

[45] Lekhnitskii, S.G. *Theory of Elasticity of an Anisotropic Elastic Body*; MIR Publishers: Moscow, RU, 1981, pp 36–42.

[46] Ma, L.; Chen, Y. *Acta Mech. Sin.* 2004, 20, 82–88.

[47] Roux, S.; Hild, F. *Int. J. Fract.* 2002, 116, 219-229.

[48] Hutchinson, J.W. *Acta Metall.* 1987, 35, 1605–1619.

[49] Desrumaux, F.; Meraghni, F.; Benzeggagh, M.L. *J. Com. Mat.* 2001, 35, 603–624.

[50] Mori, T.; Tanaka, K. *Acta Metall.* 1973, 21, 571–574.

[51] Qin, Q.H.; Lu, M. *Int. J. Num. Meth. Eng.* 2000, 48, 1071–1088.

[52] Shang, F.; Wang, Z.; Li, Z. *Eng. Fract. Mech.* 1996, 55, 737–750.

[53] Shang, F.; Kuna, M.; Kitamura, T. *Teoret. Appl. Fract. Mech.* 2003, 40, 237-246.

[54] Shang, F.; Kuna, M.; Kitamura, T. *Teoret. Appl. Fract. Mech.* 2003, 40, 247-253.

[55] Budiansky, B.; Hutchinson, J.W.; Lambropoulo, J.C. *Int. J. Sol. Struc.* 1983, 19, 337–355.

[56] Rice, J.R. *Mathematical analysis in the mechanics of fracture*; Fracture: An Advanced Treatise Vol.2 (Mathematical Fundamentals). Academic Press: New York, USA, 1968, pp 191–311.

[57] Koombua, K.; Pidaparti, R.M.; Beatty, M.W. *J. Biomed. Mater. Res. Part A* 2006, 79A, 566-573.

[58] Sládek, J.; Sládek, V.; Solek, P.; Zhang, Ch. *Int. J. Sol. Struc.* 2010, 47, 1381-1391.

[59] Zhang, Ch.; Cui, M.; Wang, J.; Gao, X.W.; Sládek, J.; Sládek, V. *Eng. Frac. Mech.* 2011, 78, 585-604.

[60] Wünsche, M.; Zhang, Ch.; García-Sánchez, F.; Sáez, A.; Sládek, J.; Sládek, V. *Comp. Meth. Appl. Mech. Eng.* 2009, 198, 2812-2824.

[61] Sládek, J.; Sládek, V.; Wen, P.H.; Hon, Y.C. *Eng. Anal. Boundar. Elem.* 2009, 33, 1089-1099.

[62] Sládek, J.; Sládek, V.; Solek, P.; Saez, A. *Int. J. Sol. Struc.* 2008, 45, 4523-4542.

[63] Gao, X.W.; Zhang, Ch.; Sládek, J.; Sládek, V. *Compos. Sci. Technol.* 2008, 68, 1209-1215.

[64] García-Sánchez, F.; Zhang, Ch.; Sládek, J.; Sládek, V. *Comput. Mater. Sci.* 2007, 39, 179-186.

[65] Sládek, J.; Sládek, V; Zhang, Ch. *Comput. Mater. Sci.* 2005, 32, 532-543.

[66] Zhang, Ch.; Sládek, J.; Sládek, V. *Comput. Mater. Sci.* 2005, 32, 611-619.

[67] Sládek, J.; Sládek, V; Zhang, Ch. *Eng. Anal. Boundar. Elem.* 2005, 29, 334-342.

[68] Sládek, J.; Sládek, V; Zhang, Ch. *Eng. Anal. Boundar. Elem.* 2005, 29, 597-609.

[69] Mykhaskiv, V.V.; Zhang, Ch.; Sládek, J.; Sládek, V. *Eng. Anal. Boundar. Elem.* 2006, 30, 167-175.

[70] Soh, A.K.; Liu, J.-X.; Lee, K.L.; Fang, D.-N. *Arch. Appl. Mech.* 2002, 72, 458-469.

[71] Diko, P. *Supercond. Sci. Technol.* 1998, 11, 68–72.

[72] Diko, P. *Supercond. Sci. Technol.* 2004, 17, R45–R58.

[73] Diko, P. *Mater. Sci. Eng. B* 1998, 83, 149–153.

[74] Diko, P.; Krabbes, G. *Supercond. Sci. Technol.* 2003, 16, 90–93.

[75] Lagerpusch, D.; Mohles, V.; Baither, D.; Anczykowski, B.; Nembach, E. *Acta Mater.* 2000, 48, 3647-3656.

[76] Liu, G.; Zhang, G.J.; Ding, X.D.; Sun, J.; Chen, K.H. *Mater. Sci. Eng. A* 2003, 344, 113–124.

[77] Trojanová, Z.; Drozd, Z.; Kúdela, S.; Száraz, Z.; Lukáč, P. *Compos. Sci. Technol.* 2007, 67, 1965-1973.

[78] Nie, J.F.; Muddle, B.C. *Acta Mater.* 2000, 48, 1691–1703.

[79] Farrissey, L.; Schmauder, S.; Dong, M.; Soppa, E.; Poech, M.H.; McHugh, P. *Comput. Mater. Sci.* 1999, 15, 1–10.

[80] Zhang, Z.; Chen, D.L. *Scrip. Mater.* 2006, 54, 1321–1326.

[81] Ramakrishnan, N. *Acta Mater.* 1996, 44, 69–77.

[82] Terentyev, D.A.; Bonny, G.; Malerba, L. *Acta Mater.* 2008, 56, 3229-3235.

[83] Yan-hui, H; Yan-xia, G.; Zhi-yi, L. *Trans. Nonferr. Met. Soc. China* 2010, 20, 863–869.

[84] Suh, Y.S.; Joshi, S.P.; Ramesh, K.T. *Acta Mater.* 2009, 57, 5848-5861.

[85] Lefebvre, S.; Devincre, B.; Hoca, T. *J. Mech. Phys. Sol.* 2007, 55, 788-802.

[86] Queyreau, S.; Monnet, G.; Devincre, B. *Acta Mater.* 2010, 58, 5586-5595.

[87] Blum, W.; Zeng, X.H. *Acta Mater.* 2009, 57, 1966-1974.

[88] Rösler, J.; Bäker, M. *Acta Mater.* 2000, 48, 3553–3567.

[89] Mohles, V. *Mater. Sci. Eng. A.* 2001, 319-321, 201-205.

[90] Zhang, W.X.; Li, L.X.; Wang, T.J. *Comput. Mater. Sci.* 2007, 41, 145-155.

[91] Eshelby, J.D. *Proc. Royal Soc. London A* 1957, 241, 376-396.

[92] Trebuňa, F.; Bigoš, P. *Intensification of Technical Capability of Heavy Supporting Structures*; Technical University: Košice, SK, 1998, pp 56–58.

[93] Trebuňa, F.; Šimčák, F.; Buršák, M.; Bocko, J.; Šarga, P.; Pástor, M.; Trebuňa, P. *Metalurgija* 2007, 46, 41–46.

[94] Trebuňa, F.; Šimčák, F.; Bocko, J.; Šarga, P.; Trebuňa, P.; Pástor, M.; Mihok, J. *Metalurgija* 2008, 47, 133–137.

[95] Trebuňa, F.; Jadlovský, J.; Frankovský, P.; Pástor, M. *Proc. 48th Int. Sci. Confer. Exp. Stress Anal.* 2010, 435–442.

[96] Trebuňa, F.; Hunady, R.; Znamenáková, M. *Proc. 48th Int. Sci. Confer. Exp. Stress Anal.* 2010, 443–449.

[97] Trebuňa, F.; Hunady, R.; Znamenáková, M. *Proc. 48th Int. Sci. Confer. Exp. Stress Anal.* 2010, 451–458.

[98] Shin, H.; Earmme, Y.Y. *Int. J. Fract.* 2004, 126, L35–L40.

[99] Li, J.Y.; Dunn, M.L. *Phil. Magaz. A* 1998, 77, 1341–1350.

[100] Trebuňa, F.; Šimčák, F.; Buršák, M.; Tomčík, J. *Metalurgija* 2007, 46, 191–194.

[101] Trebuňa, F.; Šimčák, F.; Buršák, M.; Bocko, J.; Šarga, P.; Pástor, M.; Trebuňa, P. *Metalurgija* 2007, 46, 201–204.

[102] Trebuňa, F.; Šimčák, F.; Pástor, M. *Proc. 48th Int. Sci. Confer. Exp. Stress Anal.* 2010, 475–482.

[103] Bocko, J.; Trebuňa, F.; Šimčák, F. *Proc. 47th Int. Sci. Confer. Exp. Stress Anal.* 2008, 55–60.

[104] Skočovský, P.; Bokůvka, O.; Palček, P. *Materials Science*; EDIS: Žilina, SK, 1996, pp 75–79.

[105] Awaji, H.; Choi, S.-M.; Yagi, E. *Mech. Mater.* 2002, 34, 411-422.

[106] Timoshenko, S.P.; Goodier, J.N. *Theory of Elasticity*; McGraw-Hill: New York, USA, 1951, pp 213–225.

[107] Davidge, R.W.; Green, T.J. *J. Mater. Sci.* 1968, 3, 629–634.

[108] Kuba, F. *Theory of Elasticity and Selected Applications*; SNTL/Alfa: Prague, CZ, 1982, pp 96–101.

[109] Trebuňa, F.; Šimčák, F.; Jurica, V. *Elasticity and Strength I*; Technical University: Košice, SK, 2005, pp 68–76.

[110] Trebuňa, F.; Šimčák, F.; Jurica, V. *Examples and Problems of Elasticity and Strength I*; Technical University: Košice, SK, 2002, pp 135–138.

[111] Trebuňa, F.; Šimčák, F.; Jurica, V. *Examples and Problems of Elasticity and Strength II*; Technical University: Košice, SK, 2005, pp 73–78.

[112] Hajko, V.; Potocký, L.; Zentko, A. *Magnetization Processes*; Alfa: Bratislava, SK, 1982, pp 37–45.

[113] Rektorys, K. *Review of Applied Mathematics*; SNTL: Prague, CZ, 1973, pp 156–162.

[114] Hearmon, R.F.S. *Introduction to Applied Anisotropic Elasticity*; The Clarendon Press: Oxford, UK, 1961, pp 12–15.

[115] Trebuňa, F.; Jurica, V.; Šimčák, F. *Elasticity and Strength II*; Technical University: Košice, SK, 2000, pp 36–39.

[116] Trebuňa, F.; Šimčák, F.; Bocko, J. *Resolved Examples and Problems of Elasticity and Strength I*; Technical University: Košice, SK, 2009, pp 65–69.

[117] Novikova, S.I. *Measur. Techniq.* 1984, 27, 933–938.

[118] Ceniga, L. *Analytical Models of Thermal Stresses in Composite Materials III*; Nova Science Publishers: New York, USA, pp 98–155.

[119] Abe, F.; Kern, T.-U.; Viswanathan, R. *Creep-Resistant Steels*; Woodhead Publishing Limited: Cambridge, UK, 2008, pp 25–38.

[120] Falat, L. *Microstructure and creep of dissimilar* α/γ *weld-joints*; Institute of Materials Research, Slovak Academy of Sciences: Košice, SK, 2009, pp 45–48.

[121] Ray, A.K.; Tiwari, Y.N.; Roy, P.K.; Chaudhuri, S.; Bose, S.C.; Ghosh, R.N.; Whittenberger, J.D. *Mater. Sci. Eng. A* 2007, 454-455, 679-684.

[122] Masuyama, F. *Int. J. Press. Vess. Pip.* 2006, 83, 819-825.

[123] Dimmler, G.; Weinert, P.; Cerjak, H. *Int. J. Press. Vess. Pip.* 2008, 85, 55-62.

[124] Dimmler, G.; Weinert, P.; Kozeschnik, E.; Cerjak, H. *Mater. Character.* 2003, 51, 341-352.

[125] Falat, L.; Výrostková, A.; Homolová, V.; Svoboda, M. *Eng. Fail. Anal.* 2009, 16, 2114-2120.

[126] Výrostková, A.; Mandziej, S.T. *Weld. World* 2008, 52, 3–26.

[127] Mandziej, S.T.; Výrostková, A.; Solar, M. *Weld. World* 2010, 54, R160–R172.

[128] Robson, J.D.; Bhadeshia, H.K.D.H. *Calphad* 1996, 20, 447–460.

[129] Nakahara, S.; Fisanick, G.J.; Yan, M.F.; Van Dover, R.B.; Boone, T.; Moore, R. *J. Cryst. Growth* 1987, 85, 639–651.

[130] Goyal, A.; Oliver, W.C.; Funkenbusch, P.D.; Kroeger, D.M.; Burns, S.J. *Physica C* 1991, 183, 221–233.

[131] Ledbetter, R.; Lei, M.; Hermann, A.; Sheng, Z. *Physica C* 1994, 225, 397–403.

[132] Pánek, Z.; Figusch, V.; Haviar, M.; Ličko, T.; Šajgalík, P.; Dusza, J. *Structural Ceramics*; R&D Print: Bratislava, SK, 1992, pp 18–24.

[133] Hidvéghy, J.; Dusza, J. *Non-metallic structural materials*; Technical University: Košice, SK, 1998, pp 38–47.

[134] Dusza, J. *Int. J. Mater. Product Technol.* 2000, 15, 292–355.

About the Author

Dr. Ladislav Ceniga was born in 1965 in Košice, Slovak Republic. He graduated from the Faculty of Mechanical Engineering, the Technical University in Košice (1988) (Department of Mechanical Engineering Technology), and from the Faculty of Sciences of the P. J. Šafárik University in Košice (1993) (Department of Physics of Solids), both with distinction. Awarded the prize of the Chancellor of the Technical University in Košice for excellent study results aimed at heat and chemical treatment and magnetic properties of amorphous alloys (1994–2000). He defended a PhD. thesis in Physics of Condensed Matters and Acoustics at the Institute of Experimental Physics of the Slovak Academy of Sciences in Košice (1999). Since 2000 Dr. Ceniga has been employed at the Institute of Materials Research of the Slovak Academy of Sciences in Košice, Dr. Ladislav Ceniga currently works on analytical models of thermal stresses and related thermal-stress induced phenomena in composite materials [1].

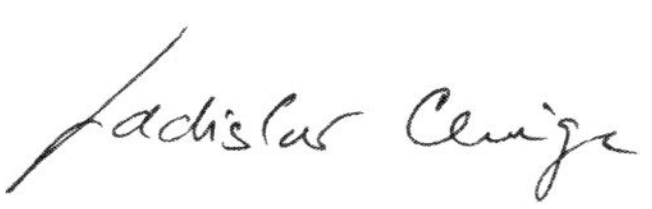

Dr. Ladislav Ceniga, DSc.
Košice, Slovak Republic
November 2014

[1]E-mail addresses: lceniga@yahoo.com, ladislav.ceniga@yahoo.com, ladislavceniga65@gmail.com, lceniga@imr.saske.sk

Index